3rd Euro-Mediterranean Conference on Materials and Renewable Energies 2015 (EMCMRE-3)

Journal of Physics: Conference Series Volume 758

Marrakech, Morocco
2 – 6 November 2015

Editors:

Mustapha Ait Ali
Azzedine Bendounan
Abdelilah Benyoussef
Abdelilah Kaddouri
Abdelkader Kara
Karima Lasri
Jean Lousi Lemaire
Hamid Oughaddou
Abdelilah Slaoui

ISBN: 978-1-5108-3179-7
ISSN: 1742-6588

Printed from e-media with permission by:

Curran Associates, Inc.
57 Morehouse Lane
Red Hook, NY 12571

Printed by Curran Associates, Inc. (2016)

For permission requests, please contact the Institute of Physics
at the address below.

Institute of Physics
Dirac House, Temple Back
Bristol BS1 6BE UK

Phone: 44 1 17 929 7481
Fax: 44 1 17 920 0979

techtracking@iop.org

Additional copies of this publication are available from:

Curran Associates, Inc.
57 Morehouse Lane
Red Hook, NY 12571 USA
Phone: 845-758-0400
Fax: 845-758-2633
Email: curran@proceedings.com
Web: www.proceedings.com

Table of contents

Volume 758

3rd Euro-Mediterranean Conference on Materials and Renewable Energies (EMCMRE-3)

2–6 November 2015, Marrakech, Morocco

Accepted papers received: 21 September 2016
Published online: 19 October 2016

Preface

Papers

Journal of Physics: Conference Series **758** (2016) 011001 doi:10.1088/1742-6596/758/1/011001

Preface

The 3[d] Euro-Med Conference for Materials and Renewable Energies (EMCMRE-3) was held in Marrakech- Morocco November 2-6 2015. This conference is the 3[d] of a series that aims to strengthen relationships between material scientists from all sides of the Mediterranean Sea. The scientific topics of the EMCMRE conference series are important to countries surrounding the Mediterranean Sea with particular attention to novel problems using state of the art experimental techniques as well as advanced theoretical and computational approaches for materials science. The third conference in the series is dedicated to the three main aspects of Renewable Energies, namely harvesting and storing energy as well as novel materials.

In general, the topics of the EMCMRE conference series encompasses the branches of physics, chemistry and computer science that directly impact Renewable Energies. The targeted audience includes Professors, Researchers and Engineers from Universities, Engineering Schools and Research Institutes. The participation of PhD and Masters Students from both sides of the Mediterranean Sea is strongly encouraged and supported with travel grants. The conference includes oral and poster presentations as well as plenary and keynote presentations given by high profile scientists, from all over the world, that will cover the key themes of the conference. The conference also includes a hands-on session on computational tools for materials science, targeted for students and young scientists.

Editorial committee:

The keynote speakers for the EMCMRE-3 are:

Dr. Khalil Amine (Distinguished Fellow and manager, Argonne National Laboratory, USA), for energy storage.
Dr. Daniel Lincot (Director of IRDEP, France), for energy harvesting.
Prof. Thomas Greber (Zurich University, Switzerland), for novel materials.

Journal of Physics: Conference Series **758** (2016) 011001 doi:10.1088/1742-6596/758/1/011001

The invited speakers are:

Energy Storage:
Prof. Emilio Bunel (Division Director, Argonne National Laboratory, USA)
Prof. Falko Schappacher (Münster, Germany)
Prof. Shirley Meng (University of California in San Diago, USA)
Prof. Feng Pan Peking University (China)
Prof. Jim Greenberger (NAATBatt International, USA)
Prof. Zempachi Ogumi (Director, NEDO RIZING HUB , Kyoto University, Japan)
Prof. Tetsuya Osaka (Vice president, Waseda University, Japan)
Prof. Teófilo Rojo (CIC Energigune, Spain)
Dr. Abdelfattah Mahmoud (Jülich, Germany)

Renewable Energy Harvesting
Dr. Mohamed Balli (Sherbrooke University, Canada)
Prof. Sahraoui Chaieb (KAUST, KSA)
Dr. Guido Fratesi (University of Milano, Italy)
Prof. Dario Narducci (University of Milano Bicocca, Italy)
Dr. Abdelilah Slaoui (Director of MaCEPV, Strasbourg, France)
Prof. Laurene Tetard (UCF University, USA)

Novel Materials
Prof. Mike Altman, (HKUST, Hongkong)
Prof. Mohammed Benaissa (Université Mohammed V, Rabat, Morocco)
Dr. Didier Dentel ((Université de Haute-Alsace, France)
Prof. Mohamed Eddaoudi (King Abdullah University of Science and Technology, Saudi Arabia)
Prof. Nasr-Eddine Hamdadou (ENPO, Oran, Algeria)
Dr. Haik Jamgotchian (CINAM, France)
Dr. Andras Kis (EPFL, Switzerland)
Prof. Abdou Lachgar (Wake Forest University, USA)
Prof. Talat S. Rahman (UCF, USA)
Dr. Alistair Rowe (Ecole Polytechnique, France)
Dr. Fausto Sirotti (Synchrotron SOLEIL, France)
Prof. Wlodek Strupinski, (ITME, Poland)
Dr. Mikael Syväjärvi (Linköping University, Sweden)

The conference also hosted a workshop on Computational Methods led by Dr. Ari Seitsonen (CNRS, France) as the Keynote and Prof. Boutheina Kerkeni (IPEST, La Marsa, Tunisia) as an invited speaker. The workshop was attended by thirty students from the Maghreb.

Peer review statement

All papers published in this volume of *Journal of Physics: Conference Series* have been peer reviewed through processes administered by the proceedings Editors. Reviews were conducted by expert referees to the professional and scientific standards expected of a proceedings journal published by IOP Publishing.

Journal of Physics: Conference Series **758** (2016) 012001 doi:10.1088/1742-6596/758/1/012001

Effects of Temperature and Concentration Mono and Polycrystalline Silicon Solar Cells: Extraction Parameters

M Khalis[1], R Masrour[2], G Khrypunov[3], M Kirichenko[3], D Kudiy[3] and M Zazoui[1]

[1]Laboratory of Condensed Matter, F.S.T., University of Hassan II Mohammedia, Casablanca, Avenue Hassan II, BP 146, 28800 Mohammedia, Morocco.
[2]Laboratory of Materials, Processes, Environment and Quality, Cady Ayyed University, National School of Applied Sciences, 63 46000, Safi, Morocco.
[3]National Technical University, 61002 Kharkov, Ukraine.

E-mail: rachidmasrour@hotmail.com

Abstract. The simple and efficient method for the extraction of all the parameters of a solar cell from a single current-voltage curve under one constant illumination level based on the Lambert *W* function. On calculating the lsqcurvefit function with constraints, between the experimental current-voltage characteristic and a theoretical arbitrary characteristic based on Lambert *W*-function. It is significant to understand the effect of the light intensity and temperature on output performance of the crystalline solar cells. The effect of light intensity and temperature on performance parameters of mc-si and pc-si solar cells is discusses. The experiments have been carried out under a solar simulator for various intensity levels in the range 1-2.5 sun and 25–60°C, respectively. The experiment was carried out employing solar cell simulator with varying cell temperature at constant light intensity. The results show that cell temperature has a significant effect on the photovoltaic parameters and it controls the quality and performance of the solar cell. The maximum power and efficiency are found to be decreased with cell temperature and the temperature coefficient of the efficiency and maximum output power is found to be negative.

1. Introduction

Photovoltaic solar paradoxically harm the particularity to withstand the heat. This operation itself produces heat that isn't evacuated if it isn't converted into electrical energy. The electrical performances of a silicon solar cell are very sensitive to temperature [1] Polycrystalline silicon is made through a simpler method and less expensive. Instead of going through the slow and more expensive process of creating a single crystal; mc-si solar cells cost more than pc-si for the same size. Mc-si cells have a higher efficiency 25% in laboratory, standard industrial cells; however, remain limited to 14–18% than pc-si cells 20% in laboratory, commercial limited to 11-15 % [2, 3] These values found in the market are all described in the operating conditions of a solar cell, the spectral distribution of the irradiance and the temperature [4, 5] standard reporting conditions (SRC: illumination =1000 W/m^2, temperature=25°C and AM1. 5 reference spectrum). However, these conditions practically never occur during normal outdoor operation as they do not take into consideration the actual geographical and meteorological conditions at the installation site. The knowledge of solar cell model parameters from measured current-voltage *I-V* characteristics are of vital importance for the quality control and evaluation of the performance of the solar cells. Several authors [6–23] proposed methods to devise ways for extracting the parameters that describe the non-linear electrical model of solar cells. These parameters are usually the saturation current, the series

[2] rachidmasrour@hotmail.com

resistance, the ideality factor, the shunt resistance and the photocurrent. Some of the suggested methods involve both illuminated and dark *I-V* characteristics [24, 25], while others use dynamic measurements [26, 27] or integration procedures [28] based on the computation of the area under the current-voltage curves. Here, we propose a simple fitting method [29] to estimate all the parameters of a solar cell just from a single I-V curve under one constant illumination level based on the Lambert *W* function. In this method, an accurate analytic expression for *I* only depends on five parameters of I_0, I_{ph}, n, R_s and R_{sh}, and is directly used to fit the experimental data. Because this analytic expression is only a function of *V*, the fitting is only one-dimensional and, thus, a much easier method. The proposed method has been used to analyze various solar devices, including Si solar cells, GaAs solar cells [29], 1N4005 diode and Motorola diode [30]. In this paper, we have discussed the temperature and concentration variations effect on the parameters of the solar cells. This will be explained using for two crystalline silicon solar cells as an example, but the concept is also applicable to other types of solar cells. In all the examples given, to design and use solar cells correctly it is evidently necessary to understand how the two parameters, temperature and intensity of radiation, influence their behavior.

2. Theory and Model

The electrical characteristic of a solar cell can be described by the equivalent circuit of the single-diode model, the two-diode model [31], or the three-diode model [32,33]. Among these circuit models, the single-diode model has the simplest form as shown in Figure 1. Although the single-diode mode is simple, it can well describe the characteristics of various solar cells, satisfy most of the applications, and thus becomes the most widely used circuit model [34-51]. In the single-diode model, the relation of the current *I* and the voltage *V* is given as:

$$I = I_{ph} - I_0\left(exp\left(\frac{q(V+R_sI)}{nK_BT}\right) - 1\right) - \frac{V+R_sI}{R_{sh}} \tag{1}$$

Where, *I* is the output current, I_{ph} is the photocurrent, I_0 is the saturation current, R_s is the series resistance, R_{sh} is the shunt resistance, n is the ideality factor, K_B is the Boltzmann constant and *T* is the temperature. Direct parameters extraction from Equation (1) is limited by the nonlinear *I-V* relation and transcendental nature of current equation for a solar cell.

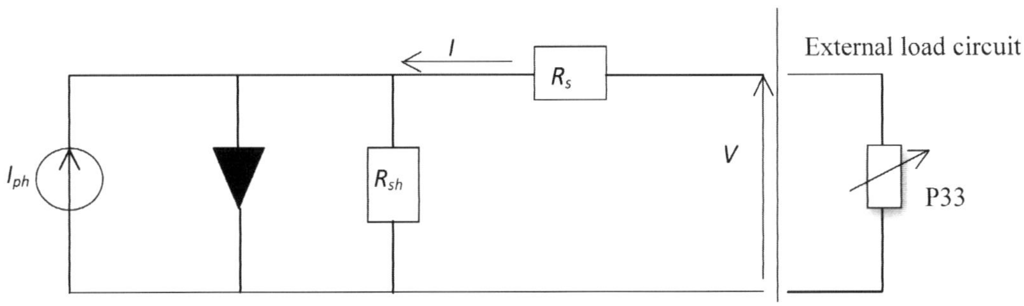

Figure 1. Generic solar cell equivalent circuit (single-diode model).

To find the unknown parameters of I_0, I_{ph}, n, R_s, and R_{sh}, we have used the least squares method (LMS) [52]. The conventional curve fitting methods, the expressions for the target variable are usually implicit functions and include the independent and dependent variables at the same time such as given in Equation (1), the expression for I includes both *V* and *I*. The explicit analytic expressions of Lambert *W* function for *I* or *V* is given in Refs [49-51]:

$$I = \frac{-V}{R_s} + \frac{V_t}{R_s}\left(-LambertW\left(\frac{R_sI_0R_{sh}exp\left(\frac{R_{sh}(R_sI_{ph+})+R_sI_0+V}{nV_t(R_s+R_{sh})}\right)}{nV_t(R_s+R_{sh})}\right)\right) + \frac{R_{sh}(R_sI_{ph+})+R_sI_0+V}{nV_t(R_s+R_{sh})} \tag{2}$$

where $V_t = nK_B/q$.

One such common criterion is the minimization of the sum of the squared the differences between the actual data and the predicted data due to our least squares line.
The ε error thus is defined as:

$$\varepsilon = \sum_{i=1}^{N}\left(I_i - I(V_i, p)\right)^2 \quad (3)$$

where N is the number of measured $I(V)$ pairs denoted by (Vi, Ii); $I(V, p)$ is the theoretical current for voltage V as predicted by a model containing several parameters represented by p that are the variables used to minimize the error. The photocurrent at the level of radiation defined as unity (normally 1 sun AM1.5 =100 (mW/cm^2)) is I_{ph1} the photocurrent at a level of radiation X (concentration factor: X suns) times greater is:

$$I_{ph} = X I_{ph1} \quad (4)$$

If V_{oc1} is the open-circuit voltage at 1 sun, the voltage at X suns is obtained by:

$$V_{oc} = V_{oc1} + \frac{nK_BT}{q}\ln X \quad (5)$$

The energy-conversion efficiency η of a solar cell is defined as the ratio between the maximum electrical power P_m that can be delivered to the load and the power P_{in} of the radiation incident on the cell:

$$\eta = \frac{P_m}{P_{in}} = \frac{I_{mp} V_{mp}}{P_{in}} \quad (6)$$

Where$\eta$is the efficiency, I_{mp} is the current at the maximum power point V_{mp} is the voltage at the maximum power point.
As an illustration of the effect of temperature, let us consider a cell described by an exponential with n= 1, corresponding to behavior dominated by recombination currents in the neutral zones. This is adequate as a first approximation. The characteristic equation for the device is therefore:

$$I = I_{ph} - I_0(T)\left(exp\left(\frac{q(V+R_sI)}{K_BT}\right) - 1\right) \quad (7)$$

The dependence of the reverse bias saturation current on temperature can be written in the following form:

$$I_0 = A\,T^3 exp\left(-\frac{E_{G0}}{K_BT}\right) \quad (8)$$

where A and E_{G0}(the band gap at 0 K) are both approximately constant with respect to temperature.

3. Results and discussion

Monocrystalline silicon test samples was a fragment of full format solar cells made by PSJC «Kvazar» and has area 38,44 cm^2.Test samples made on SHB-2 mark, p-type monocrystalline silicon wafers with thickness about 200 µm, resistance 2 Ω*cm and doping by boron up to10^{16} cm^{-3}. Layers of n+- and p+- type made by thermal diffusion of phosphorous and aluminum up to 10^{19} - 10^{20} cm^{-3} with thickness 0.5 and 3 µm, correspondingly. Antireflextance coating made from Si_3N_4 with thickness 80-100 nm. Front and rear metallization made by standard «screen printed» technology from aluminum and argentum paste with thickness of layers about 40 µm.

Polycrystalline silicon test samples made by «Yingli Green Energy Holding Co» (PRC) with area 19.76 cm^2.Test samples made on p-type polycrystalline silicon wafers with thickness about 230 µm, resistance about 2 Ω*cm and dopant by boron up to10^{16} cm^{-3}. Layers of n± and p± type made by thermal diffusion of phosphorous and aluminum up to 10^{19} - 10^{20} cm^{-3} with thickness 0.5 and 3 µm, correspondingly. Antireflextance coating made from Si_xN_y with thickness 80-100 nm. Front and rear metallization made by standard «screen printed» technology from aluminum and argentum paste with thickness of layers about 40 µm.

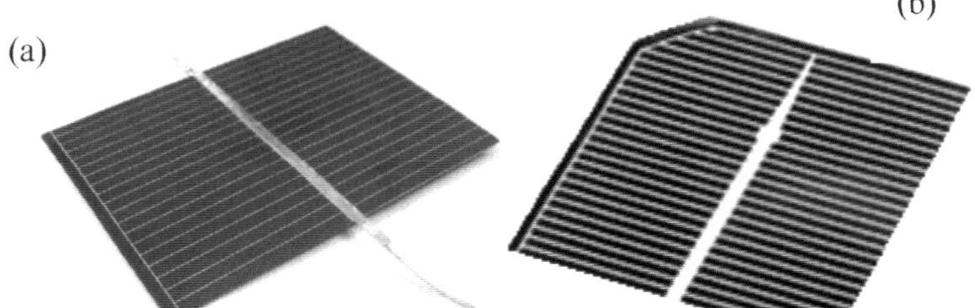

Figure 2. Crystalline Silicon Solar Cells: mono (a) and poly (b)

Journal of Physics: Conference Series **758** (2016) 012001 doi:10.1088/1742-6596/758/1/012001

To illustrate the influence of the concentration of sunlight on R_s, R_{sh}, n, I_{ph},I_0 and efficiency η, a simulation was carried out. The *I-V* characteristics measured under a FALCON EYES SS-110B xenon flash lamp (1000 W/m^2), Digital Oscilloscope RIGOL DS 1064B and variable load resistor P33 served to extract the solar cell model parameters of Equation (1).

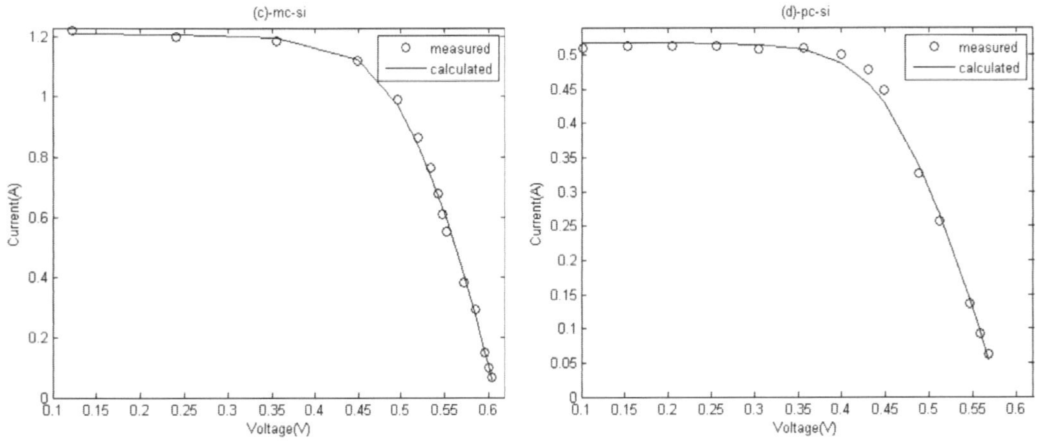

Figure 3. Synthetic and fitted *I-V* characteristic of curves of: (c) mc-si and (d) pc-si for *T*=25 °C cell temperature at 1 sun.

Figure 3(c,d) show the synthetic and fitted *I-V* characteristics for a mc-si solar cell and pc- si solar cell, respectively using least squares method. Table1 show the LMS-extracted parameters for mc-si solar cell and pc-si solar cell under 1sun (1000W/m^2)at 25°C cell temperature, respectively.

Table 1. Parameters extracted using LMS of mc-si solar cell and pc-si solar cell with 1sun at 25°C cell temperature.

Parameters MLS extracted values	mc-si solar cell	pc-si solar cell
$R_{sh}(\Omega)$	20.5	125
$R_s(\Omega)$	0.054	0.164
I_0(A)	4.78x10^{-8}	2.95x10^{-8}
I_{ph}(A)	1.221	0.517
n	1.375	1.339
η %	13.05	10.48

Varying only of the I_{ph} parameter in Equation (1) modified *I-V* characteristics, because I_{ph} is proportional to the power illumination. The experimental results of the two solar cells indicated in Figure 4(e, f) shown the short circuit current I_{sc} increases with increasing the light intensity and decreases with increasing the solar cell voltage. For intensity 2.5sun, the $I_{sc}$is about 3.073 A and 1.299 A for the mc-si and pc-si solar cells respectively. At the same above light intensity, the I_{sc} decreases with increasing the voltage V_{oc} up to 0.62 and 0.59 Volt respectively.

Journal of Physics: Conference Series **758** (2016) 012001 doi:10.1088/1742-6596/758/1/012001

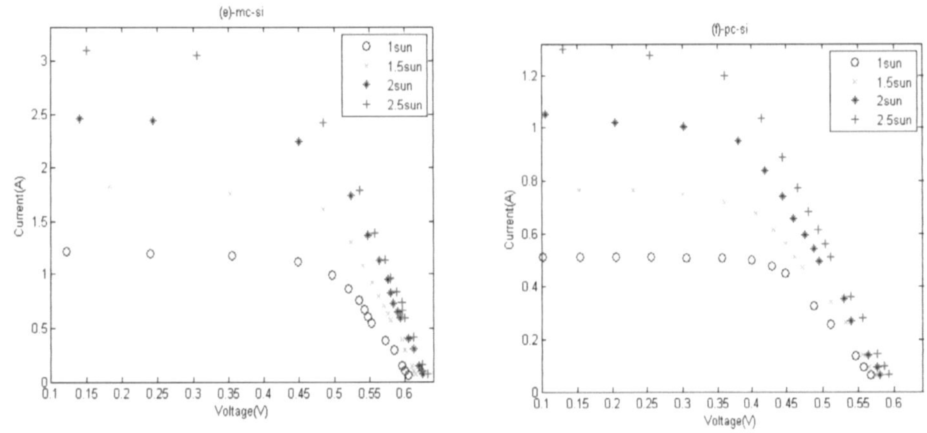

Figure 4. The *I-V* curves of: (e) mc-si and (f) pc-si for *T*=25 °C cell temperature at different light intensities.

We apply our method LMS to determine the five parameters for each light intensity. The results are summarized in Table 2 and in Table 3.

Table 2. Parameters extracted of mc-si solar for *T*=25°Ccell temperature at different light intensities.

X x1000W/m^2	1	1.5	2	2.5
$R_{sh}(\Omega)$	20.44	5000	87.7	769.30
$R_s(\Omega)$	0.054	0.037	0.0349	0.0351
n	1.375	1.563	1.462	1.561
$I_0(x10^{-7}A)$	0.478	4.532	1.677	5.41
$I_{ph}(A)$	1.221	1.823	2.495	3.073
$V_{oc}(V)$	0.604	0.614	0.624	0.632
η%	13.05	13.5	13.17	12.2

Table 3. Parameters extracted of pc-si solar cell for T=25 °C cell temperature at different light intensities.

X x1000W/m^2	1	1.5	2	2.5
$R_{sh}(\Omega)$	1250	135.13	34129	52630
$R_s(\Omega)$	0.164	0.154	0.132	0.133
n	1.339	1.395	1.536	1.320
$I_0(x10^{-8}A)$	2.9	7.3	4.0	3.13
$I_{ph}(A)$	0.517	0.791	1.040	1.299
$V_{oc}(V)$	0.568	0.572	0.580	0.592
η%	10.48	9.344	9.60	8.816

The form of the characteristic efficiency-concentration curve of Figure 4(g) makes evident the great importance of the series resistance in the design of solar cells. The design is influenced considerably in this respect by the levels of concentration at which the device operates.

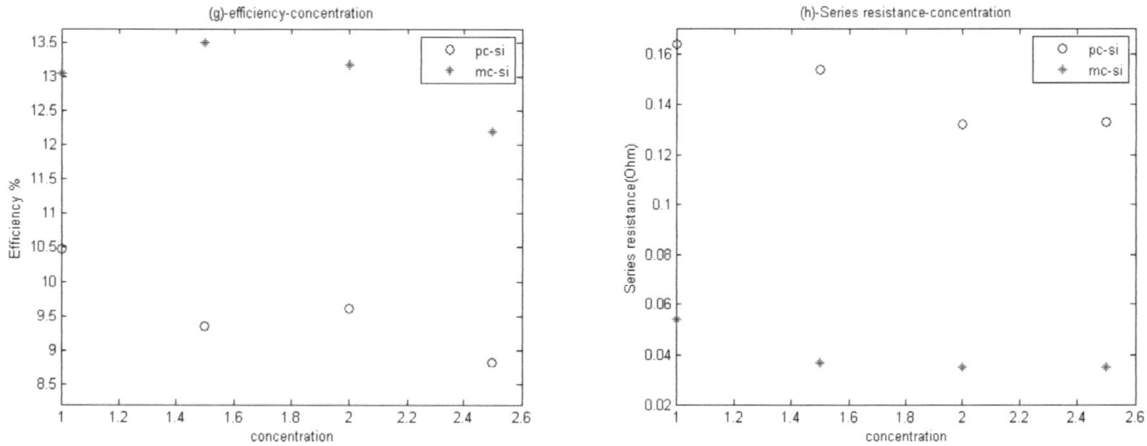

Figure 5.Variation of the efficiency and series resistance with the concentration of the light.

Figure 4(e, f) shows a large concentration, increases the short-circuit current, without affecting the open-circuit voltage. Even when the cell is in short circuit, with zero external voltage, the junction is biased by a voltage of value $I_{sc}R_s$, produced by the passage of current through R_s. This bias produces a current through the diode in the opposite direction to I_{SC}. This can be degraded significantly, resulting in poor efficiency, especially if the current is high as in cells working under optical concentrators. The dependence of efficiency on light intensity, for the two solar cells, is shown in Figure 5(g). It has been found that the efficiency of each cell demonstrated a small decrease with light intensity. So, the efficiency of the mc-is and pc-is solar cells decreased from 10.48 and 13.5 % to 8.816 and 12.2% for light intensity 1 and 2.5 Sun respectively. Figure 5(h) shows the dependence of series resistance with light intensity for the two cells. It has been found that the series resistance, of each solar, decreases with increasing light intensity due to the increase in conductivity of the active layer with the increase in the light intensity [53] So, the series resistance of the mc-is and pc-is solar cells decreased from 0.054 and 0.164 to 0.0351 and 0.133 Ω for light intensity 1 and 2.5 suns respectively. The product $I_{sh}R_s$ of the mc-si and pc-si solar cells increased from0.065934 and 0.084788 (V)to 0.1078623 and 0.172767 Vfor light intensity 1 and 2.5 Sun respectively. In all cases these values are larger than the thermal voltage V_t= 0.025(V). So the effects of series resistance are not negligible and dominate, consequently efficiency decrease as shown in Figure 5(g).

To see the effect of temperature on the behavior of solar cells, we set the value of the load resistance to the value which gives the maximum electric power and we vary the temperature. Table 4 shows the output parameters of solar cells under different temperature.

Table 4. Output parameters of solar cells under different temperature and 1sun.

Sample	Temperature(°C)	V_{mp}(V)	I_{mp}(A)	P_m(W/m^2)	η(%)
	25	0.428	0.475556	1.03E+02	10.30
pc-si	30	0.42	0.466667	9.92E+01	9.92
	35	0.416	0.462222	9.73E+01	9.73
	40	0.412	0.457778	9.54E+01	9.54
	45	0.404	0.448889	9.18E+01	9.18
	50	0.4	0.444444	9.00E+01	9.00
	55	0.39	0.433333	8.55E+01	8.55
	60	0.384	0.426667	8.29E+01	8.29
	25	0.452	1.13	1.33E+02	13.29
mc-si	30	0.448	1.12	1.31E+02	13.05
	35	0.445	1.1125	1.29E+02	12.88
	40	0.44	1.1	1.26E+02	12.59
	45	0.432	1.08	1.21E+02	12.14
	50	0.428	1.07	1.19E+02	11.91
	55	0.424	1.06	1.17E+02	11.69
	60	0.42	1.05	1.15E+02	11.47

Figure 6. shows the variation of the maximum power and efficiency with temperature, in addition to that temperature coefficients are calculated by means of a fit linear and we find $dP_m/dT = -0.55$ W/m^2/°C for both cells, $d\eta/dT = -0.55$(%/°C) for mc-si and $d\eta/dT = -0.558$(%/°C) for pc-si. It has been found that the maximum power of the two solar cells decreases with increasing temperature as shown by the results of the experiment (see Figure 6 (*k*)), where the maximum power solar cells of the mc-si and the pc-si for temperature=25°C was 133 W/m^{2}and 103 W/m^2, respectively. The increasing the temperature to 60°C causes the decrease of the power by 13.53% and 19.51% to reach values 115 W/m^2 and 82.9 W/m^2 respectively.

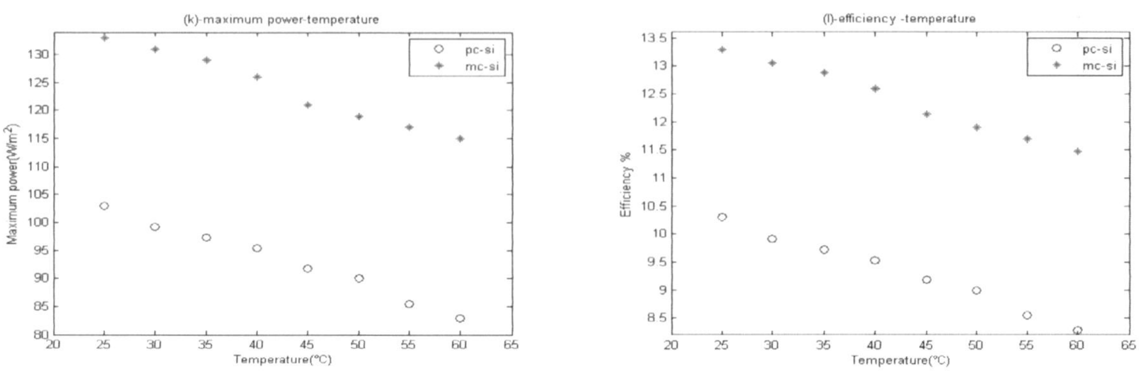

Figure 6. Variation of the maximum power and efficiency with the temperature under 1 sun.

This decrease in maximum power is due to the increase in I_0(see Equation 8) and to the rounding of the elbow of the characteristic *I-V* curve, evident from the effect of increasing *T* in the exponential term of Equation 7, which explains the decrease of the maximum power and consequently the degradation of the cell i.e. decrease efficiency, this is in agreement with the experimental as shows in curve of Figure 6(*l*). The photocurrent I_{ph} increases slightly with temperature, in part because of greater diffusion lengths of the minority carriers and in part because of the narrowing of the band gap that displaces the absorption threshold towards photons of lower energy. This improvement of the photocurrent with temperature is more appreciable in *Si* cells [32]. It is always quite small, however, and can be ignored as a first approximation. Therefore, the change in the characteristic equation of the cell with temperature rises through the exponential term and through $I_0(T)$. We concluded that in the non-concentrated cells the effect of the temperature appears too, but at different values. The results of the two solar cells indicated that light intensity has a dominant effect on current parameters. It is found that photocurrent; short circuit current and maximum current have been increased linearly with increasing light intensity. The conversion efficiency η of solar system decreases. The effect of cell temperature on the photovoltaic parameters of mc-si and pc-si solar cells is undertaken.

4. Conclusion

In this work, we have proposed a simple method to extract all the parameters of a solar cell just from a single *I-V* curve under one constant illumination level. With the help of the Lambert *W* function, the explicit analytic expression for *I* can be obtained. The expression for *I* depends on five parameters ofI_{ph},I_0, n, R_s and R_{sh}. This analytic expression for *I* is directly used in the numerical method to fit the experimental data and then determine the values of the parameters. The Performance of solar cells is dependent on environmental conditions and their output parameters such as output voltage, current, power, and efficiency vary by light intensity and temperature. The experimental results of the two modules indicated that light intensity has a dominant effect on current parameters; the short circuit current of the mc-si and pc-si solar cells increased from 1.221 and 0.517 to 3.073 and 1.299 for light intensity 1 and 2.5 sun respectively, the efficiency of the mc-si and pc-si solar cells decreased from 10.48 and 13.5 % to 8.816 and 12.2% for light intensity 1 and 2.5 sun respectively. The temperature coefficient of the efficiency and maximum output power is found to be negative. The change study of photovoltaic parameters with temperature is also undertaken. The changes are found from -0.55 W/m2/°C, -0.55(%/°C) and -0.558(%/°C) for maximum output power mc-si and pc-si, efficiency mc-si and efficiency pc-si respectively. The results obtained are in good agreement with the available literature.

Acknowledgement:
Characterization current-voltage was performed in the laboratory of Semiconductor Physics and Applications at Kharkov Polytechnic Institute /Ukraine. We extend our gratitude to Vice-rector for Scientific-and-Pedagogical work Gennadiy Khrypunov professor, doctor of science and all staff for their availability.

References

[1] Chandera S, Purohit A, Sharma A, Arvind, Nehra SP and Dhak MS 2015 *Energy. Reports.* **1** 104
[2] Saga T and Asia N P G 2010 *Materials.* **2** 96
[3] Dobrzański LA, Szczęsna M, Szindler M and Drygała A 2013 J. *Achievements in Materials and Manufacturing Engineering.* **59** 67
[4] El-Shaer A, Tadros M T Y and Khalifa M A 2014 *Int. J. Emerg. Tech. Adv. Eng.* **4** 311
[5] Tobnaghi DM, Madatov R and Naderi D, 2013 *Int. J. Adv. Res. Elect. Elect. Inst. Eng. Int.* **2** 6404
[6] Laplaze D and Youm I 1985 *Sol. Cells.* **14** 167
[7] Laplaze D and Youm I 1985 *Sol. Cells.* **14** 179
[8] Kaminski A, Marchand J J and Laugier A 1997 26th IEEE Photovoltaic Specialist Conf, Anaheim, CA, USA
[9] Kaminski A, Marchand J J and Laugier A 1999 *Solid. State. Electron.* **43** 741
[10] Aberle A G, Wenham S R and Green MA 1993 (in 23th IEEE Photovoltaic Specialist Conf, Louisville, KY, USA) p 133
[11] Van Kerschaver E, Einhaus R, Szlufcik J, Nijs J and Merten R 1997 (In Proc of 14th European Community Photovoltaic Solar Energy Conf, Barcelona, Spain) p 2438
[12] Miahle P, Khoury A and Charle J P 1984 *Phys. Stat. Sol.* A. **83** 403
[13] Bashahu M and Habyarimana A 1995 *Renew. Energy.* **6** 129
[14] Chan D S H, Phillps J R and Phang J C H 1986 *Solid. State. Electron.* **29** 329
[15] Flavius-Maxim P and Leonida D T 2010 *CEAI.* **12** 30
[16] Dib S, De la Bardonnie M, Khoury A, Pelanchon F and Mialhe P 1999 *Act. Passive. Electron. Compon.* **22** 157
[17] Bouzidi K, Cheggar M and Nehou N 2007 4th Int Conf on Computer Integrated Manifacturing C I P, Setif, Algeria.
[18] Jain A and Kapoo A 2004 *Sol. Energy. Mater. Sol. Cells.* **81** 269
[19] Haouari M M, Belhamel M, I Tobias and Ruiz JM 2005 *Sol. Energy. Mater. Sol. Cells.* **87** 225
[20] Bashahu M and Habyarimana A 1995 *Renew. Energy.* **6** 129
[21] Malaoui A and El Mansour A 2010 *Revue Energies Renouvelables.* **13** 199
[22] Ortiz-Conde A, Francisco Garcıa Sanchez J and Muc J 2006 *Sol. Energy. Mater. Sol. Cells.* **90** 352
[23] Chegaar M, Ouennoughi Z and Guechi F 2004 *Vacuum.* **75** 367
[24] Chegaar M, Azzouzi G and Mialhe P 2006 *Solid. State. Electron.* **50** 1234
[25] Wolf M and Rauschenbach H 1963 *Adv. Energy. Convers.* **3** 455
[26] Imamura MS and Portschheller J I 1970 (in Proc of 8th IEEE Photovoltaic Specialist Conf, Seattle, WA, USA) 102
[27] Charles JP, Abdelkrim M, Muoy YH, Mialhe P 1981 *Sol. Cells.* **4** 169
[28] Khalis M, Mir Y, Hemine J and Zazoui M 2011 *Eur. Phys. J Appl. Phys.* **54** 10102
[29] Khalis M, Masrour R, Mir Y and Zazoui M 2015 *Int. J. Physical Sciences.* **10** 270
[30] Araujo G L, Sanchez E and Marti M 1982 *Sol. Cells.* **5** 199
[31] Nishioka K, Sakitani N, Uraoka Y and Fuyuki T 2007 *Sol. Energy. Mater. Sol. Cells.* **91** 1222
[32] Mazhari B 2006 *Sol. Energy. Mater. Sol. Cells.* **90** 1021
[33] Wolf M and Rauschenbach H 1963 *Adv. Energy. Convers.* **3** 455
[34] Handy R J 1967 *Solid. State. Electron.* **10** 765
[35] Pysch D, Mette A and Glunz S W 2007 *Sol. Energy. Mater. Sol. Cells.* **91** 1698
[36] Hussein R, Borchert D, Grabosch G and Fahrner W R 2001 *Sol. Energy. Mater. Sol.*

Cells. **69** 123
[37] Radziemska E 2005 *Energy. Convers. Manage.* **46** 1485
[38] Rajkanan K and Shewchun J 1979 *Solid-State. Electron.* **22** 193
[39] Servaites J D, Yeganeh S, Marks T J and Ratner M A 2010 *Adv. Funct. Mater.* **20** 97
[40] Mialhe P, Khoury A and Charles J P 1984 *Phys. Status. Solidi.* (a). **83** 403
[41] Schilinsky P, Waldauf C, Hauch J and Brabec C J 2004 *J. Appl. Phys.* 95 16
[42] Ishibashi K, Kimura Y and Niwano M 2008 *J. Appl. Phys.* **103** 094507
[43] Chegaar M, Ouennoughi Z and Hoffmann A 2001 *Solid-State. Electron.* **45** 293
[44] Kaminski A, Marchand J J and Laugier A 1999 *Solid-State. Electron.* **43** 741
[45] Murayama M and Mori T 2006 *Jpn. J. Appl. Phys.* **45** 542
[46] Jain A and Kapoor A 2005 *Sol. Energy. Mater. Sol. Cells.* **86** 197
[47] Ortiz C A and Garcıa S F J 2005 *Solid-State. Electron.* **49** 465
[48] Ortiz-Conde A, Garcıa Sanchez F J and Muci J 2006 *Sol. Energy. Mater. Sol. Cells.* **90** 352
[49] Kim J Y, Lee K, Coates N E, Moses D, Nguyen T, Dante M and Heeger A 2007 *Science* **317**, , 222
[50] Yakimov A and Forrest S R 2002 *Appl. Phys. Lett.* **80** 1667
[51] Zhang C, Zhang J, Hao Y, Lin Z and Zhu C 2011 *J. Appl. Phys.* **110** 064504
[52] PLus S M 1998 (Mathsoft Inc., Seattle, "*Getting Started with Splus* 5.0" Washington).
[53] Cuce E, Cuce P and Bali T 2013 *Applied. Energy.* **111** 374

First-principles study of Cl-terminated silicon nanoribbons electronic properties

I Djabri[1] , NE Derradji[1,2] , and R Belghit[1]
[1] Lesims,University Badji Mokhtar BP12 Annaba(Algeria)
[2] Preparatory school of technical science BP218 Annaba (Algeria)

E-mail: djabri_issam@yahoo.fr n.derradji@epst-annaba.dz

Abstract. Based on density functional pseudopotential calculations, the electronic properties of Cl- -terminated silicene nanoribbons (SiNRs) -with either zigzag edge (ZSiNRs) or armchair edge (ASiNRs)- are studied. The results show that all the Cl-terminated ZSiNRs present a zero gaps except for the 2-, 3- and 4-ZSiNRs, which have narrow band gaps. We also found degenerate flat edge-state bands at Fermi level (E_F),which give rise to a very large density of states (DOS). The Cl-terminated ASiNRs are semiconductors regardless of their widths and the band gaps exhibit oscillatory behavior allowing the classification into three distinct families as 3k, (3k + 1) and (3k + 2), where k is a positive integer. Furthermore, the band gaps are the largest for the 3k group, moderate for (3k+1) and the smallest for members of the (3k + 2). Because of the σ–π mixing effect, the total density of states (DOS) and projected density of states (PDOS) of Cl-terminated ASiNR show that the Cl atoms affect the width between the top of the valance band or the bottom of the conduction band. The charge density distributions show that there is covalent bonding between all sorts of Si–Si bonds, with ionic bonding between the edge Cl atom and the nearest Si atom. This results from the electro-negativity of Si atoms being smaller than that of Cl atoms.

1. Introduction

Graphene, a two-dimensional (2D) hexagonal lattice of single layer of sp^2 hybridized carbon atoms [1-2]. Graphene has attracted much attention and research interest. It has many potential applications in many fields [3-4] and unique physical properties, such as its high-speed carrier mobilities [5], its massless relativistic Dirac fermion behavior [6]. Quasi-one-dimensional (1D) derivatives of graphene, called graphene nanoribbons (GNRs) have been of interest to scientists for many years both theoretically [7-8] and experimentally [9-10]. GNRs are semiconductors having interesting electronic and magnetic properties dependent on their geometry [11-12], size [11-13], edge-passivation pattern [8-14], as well as external electronic field [15-16].

Scientists have been paying their attentions to another two-dimensional material, silicene, has been fabricated [17-18]. Silicene is a two-dimensional hexagonal lattice of silicon (Si) atoms, but contrary to graphene, silicene has a buckled atomic structure (0.44 Å) [19].Electronic structure of the two-dimensional silicene is similar to that of graphene. However, silicene also possesses some unique

[1]djabri_issam@yahoo.fr

properties which carbon has not [7]. A new material, quasi one-dimensional (1D) derivatives of silicene, called Silicene nanoribbons (SiNRs), (namely, graphene-like Si [20]) has been a subject of interest. The Silicene nanoribbons was successfully epitaxially grown on Ag surface in both theoretically [21-22] and experimentally [22-23-24-25]. Recently, The first-principles calculations investigated the effect of the H-termination on the stabilities and electronic properties of GNRs [26], SiNRs [27], also F-terminated silicon nanoribbons [27] with zigzag shaped edges and armchair shaped edges on both sides. The results show that different terminations can affect the electronic properties and give rise to novel physical phenomena.

2. Calculation methods and models

The results have been performed using the VASP [28-29] code based on the density functional theory (DFT) [30-31]. Ultrasoft Vanderbilt-type pseudopotentials [32] were used to describe the interactions between ions and electrons. The generalized gradient approximation (GGAPW91) according to Perdew and Wang [33] was applied to evaluate the exchange-correlation energies of all examined structures. The plane-wave cutoff energy is set to be 350eV. Supercells are used to simulate the isolated nanostructures, and the distance between images is chosen to be 15 Å in order to avoid interactions between the ribbon and its periodic images. Geometric structures of the SiNRs are fully relaxed to minimize the total energy of the system until a precision of 10^{-4} is reached. The k-point is sampled due to the Monkhorst–Pack automatic generation scheme with their origin at Γ point, together with a Gaussian smearing broadening of 0.2 eV. We investigate for infinitely long single-layer SiNRs and set the periodicity of the SiNRs along the ribbon axis. Our calculations are set by spin-polarized calculations. We define N_z-ZSiNR and N_a-ASiNR as the zigzag and armchair SiNRs, where N_z and N_a represent zigzag chains and dimer lines across the ribbon width for the zigzag and armchair edges. As examples, the geometry structures of the 5-ZSiNR and 7-ASiNR terminated with Cl atoms are shown in Fig. 1(a) and (b), respectively. The area between two red lines represents the primitive periodic cell.

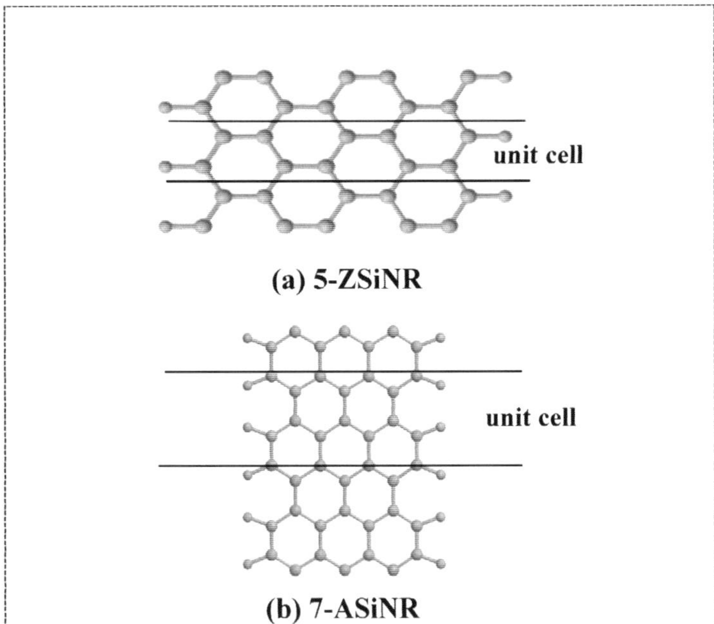

Figure 1. The geometry structures of (a) 5-ZSiNR and (b) 7-ASiNR terminated with Cl atoms. The mandarin and green balls represent Si and Cl atoms balls, respectively. The area between two red lines represents the primitive unit cell. (a) 5-ZSiNR and (b) 7-ASiNR.

Journal of Physics: Conference Series **758** (2016) 012002 doi:10.1088/1742-6596/758/1/012002

3. Results and discussions

We studied the effect of Cl-termination on the electronic properties of SiNRs with the width N_z = 2–14 for the ZSiNRs and N_a = 2–14 for the ASiNRs. As examples, the band structures of Cl-terminated N_z -ZSiNR with the width N_z = 2–7 and N_a -ASiNR with the width N a = 3–8 are shown in Fig. 2 (a) and (b), respectively.

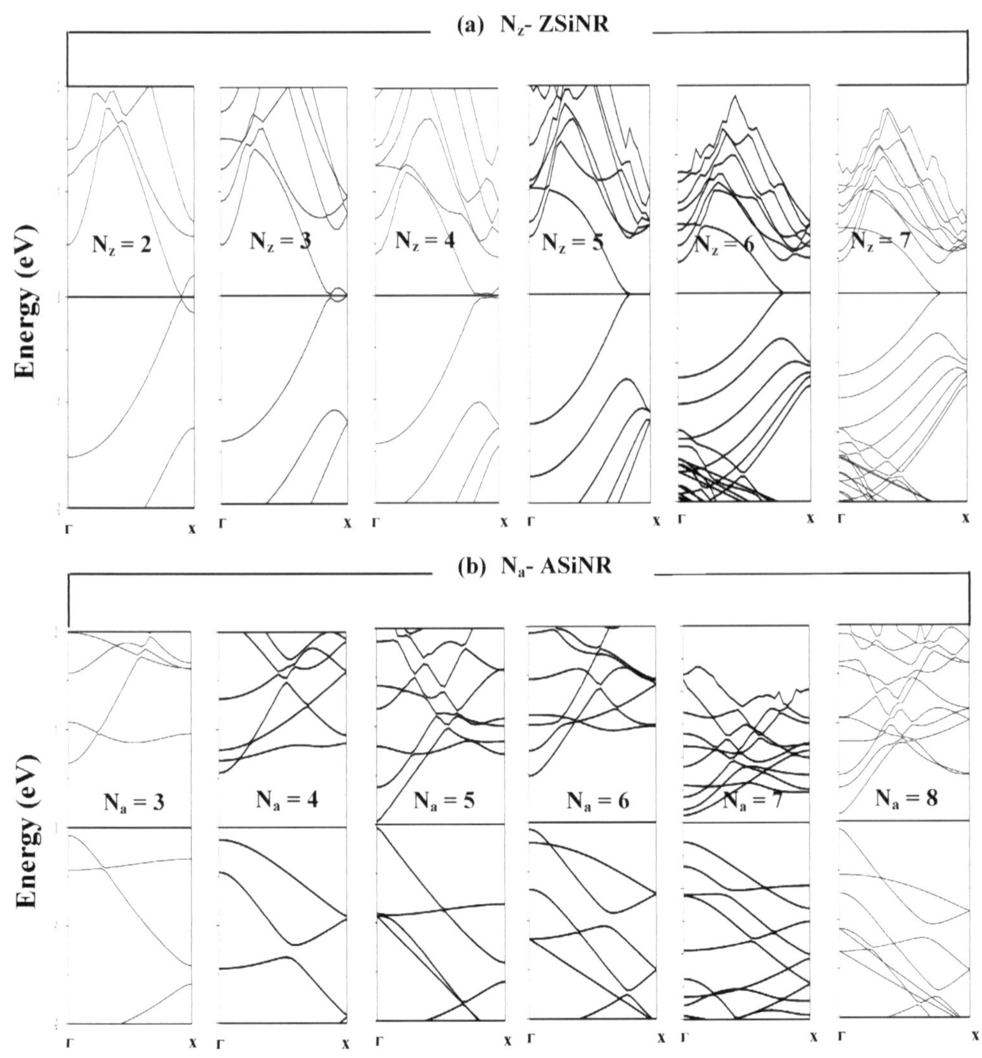

Figure 2. The band structures of Cl-terminated (a) ZSiNRs with Nz = 2–7 and (b) ASiNRs with Na = 3–8. The Fermi level E_F is set to zero and is indicated by the horizontal lines. (a) N_z-ZSiNR (b) N_a-ASiNR.

It can be seen that the 2-, 3- and 4-ZSiNR have narrow band gaps because these ZSiNRs are too narrow to separate the edge states breakthrough from both edges. After N_z = 5, The band gaps of the

ZSiNRs are zero , a bonding (the lowest conduction band and the highest valence band) coincide near the X point and degenerate flat edge-state bands at Fermi level E_F ,which give rise to a very large spin polarized total density (DOS up and DOS down) of states for 5-ZSiNR terminated with Cl atoms at Fermi level E_F (see Fig.3).This phenomenon demonstrating magnetic properties of the Cl-terminated 5-ZSiNR . The value of the total magnetic moment per super cell is 0.16 μ_B.

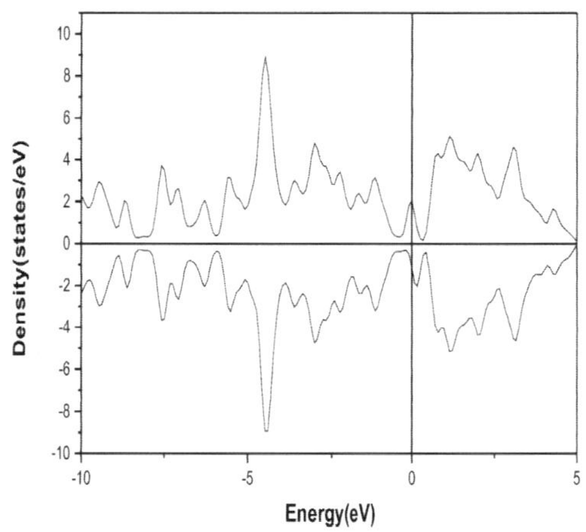

Figure 3. The spin polarized total density states of 5-ZSiNR terminated with Cl atoms, The Fermi level E_F is set to zero and is indicated by the vertical lines.

The variations of the band gap with ribbon width with N_z = 2–14 of Cl-terminated ZSiNR are summarized in Fig.4 (a) , It can be clearly seen that all the Cl-terminated ZGNRs having zero gaps except for the 2-, 3- and 4-ZSiNRs with very small direct gaps of 0.0081, 0.005 and 0.0057 eV, respectively. For the band structures of Cl-terminated ASiNRs with N_a= 3-8 it can be seen that the bottom of the conduction band and the top of the valance band appearing at Γ point are always separated, representing a direct semiconductor character .the variation of the band gap as a function of the ribbon width exhibits oscillatory behavior, allowing the classifcation of Cl-terminated ASiNRs with N_a = 3–14 into three distinct families, i.e., 3k, 3k + 1 and 3k + 2(where k is a positive integer), as shown in Fig. 4(b). The band gaps are the largest for the N a = 3k group, moderate for N_a = 3k+1 and the smallest for members of the N_a =3k + 2 group. For example, the band gap of 6-ASiNR is 0.477 eV, those of 7-ASiNR and 8-ASiNR are 0.11 eV and 0.07 eV, respectively. So that, The Cl-terminated ASiNRs are semiconductors regardless of their widths.

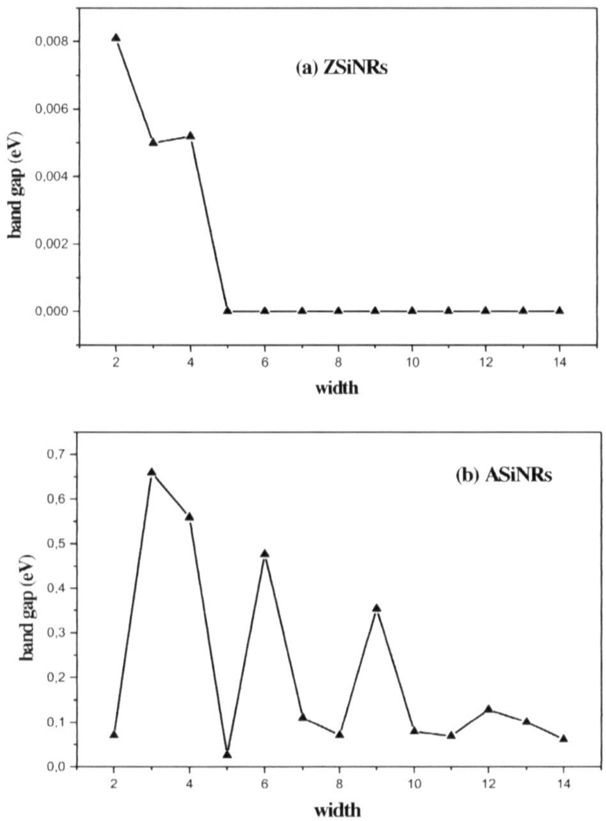

Figure 4. Variation of energy band gap as a function of the width for (a) ZSiNRs and (b) ASiNRs.

The total density of states (DOS) (a) and projected density of states (PDOS) on Si 3p atom (b) and Cl 3p atom (c) of Cl-terminated 7-ASiNR.given in Fig. 5 . It can be seen that, the p states of Cl atom contributes and affects the width between the top of the valance band or the bottom of the conduction band. This is because the interaction between non-bonding 3p electrons of Cl atoms and 3p state of Si atoms, produce an important orbital mixing with the σ valence electrons What's called his phenomenon is called the σ-π mixing effect[40]. Such a phenomena was also observed by S.Y.You et al. in the Si halides nanowires[41].

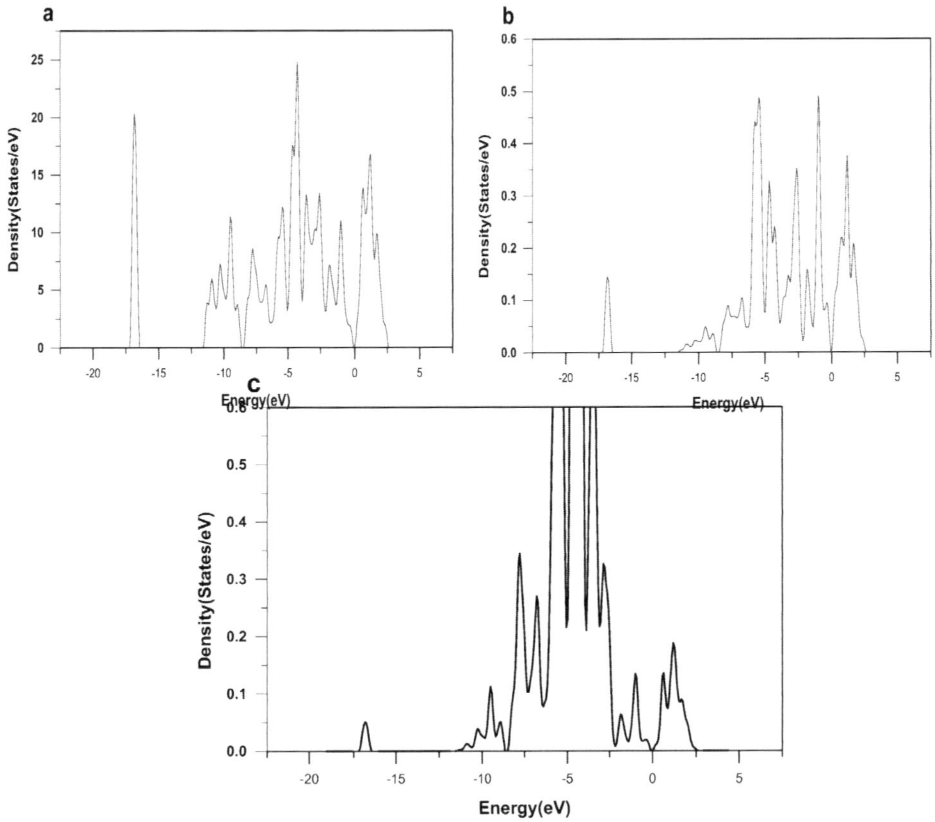

Figure 5. The total density of states (DOS)(a),projected density of states(PDOS) on Si 3p atom (b) and Cl 3p atom (c)of Cl-terminated 7-ASiNR.The Fermi level E_F is set as energy zero.

The charge-densities contours of 5-ZSiNR and 7-ASiNR terminated with the Cl atoms in the (0 0 1) crystallographic plane respectively, are presented in Figure 6. The values of the charge density are given in (e / Å ^{3}). Around Si atoms the density values from about 0 to 0.5 (e / Å ^{3}) are smaller than that from 0.75 to 2 (e / Å ^{3}) around the terminated Cl atoms. The electro-negativity of Si atoms is smaller (1.8) than that of Cl atoms (3.16), where the Si atoms lose the electrons while the Cl atom gain the electrons in both SiNRs. So that the charge density distribution indicates that there is covalent bonding between all sorts of the Si–Si bonds, ionic bonding resulted between the edge Cl atom and the nearest Si atom. The values of the charge density distribution of the edge zigzag Si–Si bond (indicated by the red arrow) for 5-ZSiNR is smaller than that of the edge dimer Si–Si bond (indicated by the black arrow) for 7-ASiNR. This is because the edge zigzag Si–Si bond with only one adjacent Cl atom but the edge dimer Si–Si bond with two adjacent Cl atoms, the bond direction of the nanoribbon axis, the bond position from the nanoribbon edge and the flatband that appeared in 5-ZSiNR but does not appear in 7-ASiNR terminated with the Cl atoms at the Fermi level E_F.DFT calculations have shown that the two types SiNRs with H-terminated [27].and also with F-terminated, the result show that different terminations can affect the electronic properties of SiNRs [39].The Cl- terminated SiNRs has a advantage because of its narrower band gap compared to others one.

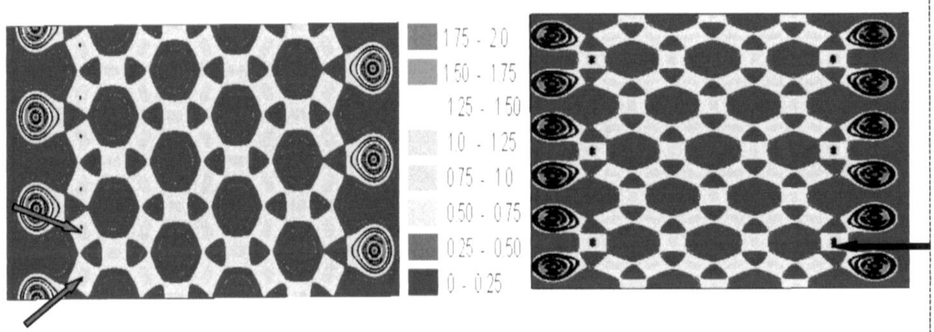

Figure 6. The charge density contours of (a) 5-ZSiNR and (b) 7-ASiNR terminated with Cl atoms.

.

4. Conclusions

To conclude, we have carried out density functional calculations on bare 5-ZSiNR or 7-ASiNR and the effect of Cl-termination on the electronic and properties of the two types SiNRs. The following conclusions are obtained, firstly, the cl-terminated ZSiNR having zero gaps except for the narrow 2-, 3- and 4-ZSiNRs with small direct gaps and degenerate flat edge-state bands at Fermi level E_F ,which give rise to a very large spin polarized total density (DOS up and DOS down) of states demonstrating magnetic properties of Cl-ZSiNR. Secondly, The Cl-terminated ASiNRs are semiconductors regardless of their widths and the band gaps present an oscillation behavior. Furthermore, the band gaps exhibit periodic features with the band gaps are the largest for the N a = 3k group, moderate for N_a = 3k+1 and the smallest for members of the N_a =3k + 2 group. The DOS and PDOS analyses show that the p states of Cl atom contributes and affects the width between the top of the valance band or the bottom of the conduction band. because of the σ-π mixing effect the interaction between non-bonding 3p electrons of Cl atoms and 3p state of Si atoms, produce an important orbital mixing with the σ valence electrons.The bonding features are analyzed by using the charge-densities contours in the (0 0 1) crystallographic plane. We found that covalent bonding between all sorts of the Si–Si bonds, ionic bonding resulted between the edge Cl atom and the nearest Si atom.

Acknowledgments

The authors would like to acknowledge the University Badji Mokhtar and preparatory school for providing financial support for this research, and also the University of Central Florida.

References

[1] K.S. Novoselov, A.K. Geim, S.V. Morozov, D. Jiang, Y. Zhang, S.V. Dubonos, I.V. Grigorieva,and A.A. Firsov, Science 306, 666 (2004).

[2] C. Berger, Z. Song, X. Li, X. Wu, N. Brown, C. Naud, D. Mayou, T. Li, J. Hass, A. N. Marchenkov, E. H. Conrad, P. N. First, and W. A. de Heer, Science 312, 1191 (2006).

[3] Y. Kong, D.L. Zhao, L. Z.Bai, Z. M.Shen, Material Letters 65, 2739–2741 (2011).

[4] V. Singh, D. Joung, L. Zhai, S. Das, S.I. Khondaker, S. Seal, Progress in Material Science 56, 1178–1271 (2011).

[5] S.V. Morozov, K.S. Novoselov, M.I. Katsnelson, F. Schedin,D. Elias, J.A. Jaszczak, A.K. Geim, Phys. Rev. Lett. 100, 016602 (2008).

[6] K.S. Novoselov, A. K.Geim, S.V. Morozov, D. Jiang, M.I. Katsnelson, I.V. Grigorieva, S.V. Dubonos, A.A. Firsov, Nature 438,197(2005).

[7] H. Lee, Y. W. Son, N. Park, S. Han, and J. Yu, Phys. Rev. B 72,174431 (2005).

[8] O. Hod, V. Barone, J. E. Peralta, and G. E. Scuseria, Nano Lett. 7, 2295 (2007).

[9] M. Y. Han, B. Özyilmaz, Y. Zhang, and P. Kim, Phys. Rev. Lett. 98, 206805 (2007).

[10] X. Li, X. Wang, L. Zhang, S. Lee, and H. Dai, Science 319, 1229 (2008).

[11] K. Nakada, M. Fujita, G. Dresselhaus, M.S. Dresselhaus, Phys. Rev. B: Condens. Matter 54, 17954 (1996).

[12] V. Barone, O. Hod, G.E. Scuseria, Nano Lett. 6, 2748 (2006).

[13] Y.W. Son, M.L. Cohen, S.G. Louie, Phys. Rev. Lett.97,216803 (2006).

[14] M. Wang, C.M. Li, Nanoscale 3, 2324. (2011).

[15] Y.W.Son, M.L. Cohen, S.G. Louie, Nature 444 ,347 (2006).

[16] H. Raza, E.C. Kan, Phys. Rev. B: Condens. Matter Mater. Phys.77, 245434 (2008).

[17] B. Aufray, A. Kara, S. Vizzini, H. Oughaddou, C. Landri, B. Ealet, G. Le Lay, Appl.Phys. Lett. 96, 183102 (2010).

[18] A. Kara, H. Enriquez, A. P. Seitsonen, L.C. Lew Yan Voon,, S. Vizzini, B. Aufray, H. Oughaddou, Surf. Sci. Rep. 67, 1-18, (2012).

[19] S. Cahangirov, M .Topsakal, E. AktÜrk, H. Sahin and S. Ciraci Phys. Rev. Lett. 102, 236804, (2009).

[20] G.G. Guzman- Verri, L.C. Lew Yan Voon, Phys. Rev. B 76 , 075131(2007).

[21] A. Kara, S. Vizzini, C. Lèandri, B. Ealet, H. Oughaddou, B.Aufray, and G. Le Lay, J. Phys. Condens. Matter 22 , 045004 (2010).

[22] Y. Ding, J. Ni, Appl. Phys. Lett. 95, 083115, (2009).

[23] C. Léandri, H. Oughaddou, B. Aufray, J. M. Gay, G. Le Lay, A. Ranguis, and Y.

Garreau, Surf. Sci. 601, 262 (2007).

[24] A. Kara, C. Lèandri, M. E. Dàvila, P. De Padova, B. Ealet, H. Oughaddou, B. Aufray, and G. Le Lay, J. Supercond. Novel Magn. 22, 259 (2009).

[25] P. De Padova, C. Léandri, S. Vizzini, C. Quaresima, P. Perfetti, B. Olivieri, H. Oughaddou, B. Aufray, and G. Le Lay, Nano lett. 8, 2299 (2008).

[26] Y. Miyamoto, K. Nakada, M. Fujita, Phys. Rev. B 59, 159858 (1999).

[27] Yu-Ling Song, Yan Zhang, Jian-Min Zhang, Dao-Bang Lu, Applied Surface Science 256, 6313 -6317, (2010).

[28] G. Kresse, J. Furthmüller, Physical Review B 54, 11169, (1996).

[29] G. Kresse, D. Joubert, Physical Review B 59, 1758, (1999).

[30] P. Hohenberg, W. Kohn, Physical Review B 136 864, (1964).

[31] W. Kohn, L.J. Sham, Physical Review A140, 1133, (1965).

[32] D. Vanderbilt, Physical Review B 41, 7892, (1990).

[33] J.P. Perdew, J.A. Chevary, S.H. Vosko, K.A. Jackson, M.A. Pederson, D.J. Singh, C. Fiolhais, Physical Review B 46, 6671,(1992).

[34] C. Lian, J.Ni, Physica B407, 4695–4699, (2012).

[35] N. Gao, W. T. Zheng , Q. Jiang, *Phys. Chem. Chem. Phys.*14, 257–261. (2012).

[36] P. De Padova, C.Quaresima, P.Perfetti, B.Olivieri, C.Leandri, B.Aufray, S.Vizzini, G.Le Lay, Nano Lett.8, 271, (2008).

[37] K. Nakada, M. Fujita, G. Dresselhaus, M.S. Dresselhaus, Phys. Rev. B 54 ,17954,(1996).

[38] S.Cahangirov, M.Topsakal,E.Akturk,H.sahin,and S.Ciraci PRL 102 236804 (2009).

[39] Yan-Ni Yang ,Jian-min Zhang , Ke-Wei Xu ,Physica E,57 21-27,(2014).

[40] H.Motoyama,K.Takeda,K.Shiraishi,Proc.Mat.Res.Soc.Symp.486(1997)385.

[41] S.Y.You,Y.Wang,Chin.J.semicond.27(2006)1927

$NaTaO_3$/MCM-48 composites for photocatalytic conversion of organic molecules

K Senevirathne,[1,2,6] C-K Xia,[1,7] S P Adhikari,[1,5] L Zhang,[3] R T Williams,[1,4] and A Lachgar[1,5,8]

[1]Center for Energy, Environment, and Sustainability, Wake Forest University, Winston-Salem, NC 27109 USA
[2]Department of Chemistry, Florida A & M University, Tallahassee, FL 32307
[3]Joint School of Nanoscience and Nanoengineering, North Carolina Agricultural and Technical State University and the University of North Carolina at Greensboro, Greensboro, NC 27401
[4]Department of Physics, Wake Forest University, Winston-Salem, NC 27109
[5]Department of Chemistry, Wake Forest University, Winston-Salem, NC 27109

Abstract
The synthesis, characterization, and photocatalytic activity of a series of $NaTaO_3$/MCM-48 composite photocatalysts prepared by sol-gel method is reported. The composite catalysts were investigated by powder X-ray diffraction (XRD), scanning electron microscopy (SEM), and N_2 adsorption-desorption porosimetry. Their photocatalytic activity for conversion of organic molecules was tested using *p*-nitroaniline as a model pollutant. The surface area was found to be in the order of 235-964 m^2/g and 35-407 m^2/g for photocatalysts calcined at 500 °C and 850 °C, respectively. The surface area was found to depend on the MCM silica content. The pore size for samples calcined at 500 °C ranges from 2.6-3.2 nm, and becomes significantly larger at 850 °C (12.6-30.0 nm) for most composites. All composites have been found to be photocatalytically active towards *p*-nitroaniline conversion into *p*-phenylenediamine under uv-visible irradiation. The conversion of *p*-nitroaniline per weight of $NaTaO_3$ in the composite photocatalysts is found to be higher than that of pristine $NaTaO_3$. The composites lose 12% of their original activity after three consecutive catalytic cycles.

Keywords: Composites, photocatalysts, sodium tantalate, *p*-nitroaniline conversion

[6] Corresponding authors:
Keerthi Senevirathne
E-mail:keerthi.senevirathne@famu.edu
Department of Chemistry, 1530, S. martin Luther King Jr. Blvd, Tallahassee, FL 32307

[7] Current affiliation: School of Chemistry and Chemical Engineering, Jiangsu University, Zhenjiang, 212013, China

[8] Abdou Lachgar
Email: lachgar@wfu.edu
Department of Chemistry, Salem Hall, P.O. Box 7486, Winston-Salem, NC 27109-7486

1. Introduction

Many efforts have been devoted to developing heterogeneous photocatalysts to harness solar energy for hydrogen generation, pollution remediation, and organic molecule conversion[1] . Among the large number of photocatalysts, $NaTaO_3$ with perovskite-type structure was prepared using a wide range of synthetic methods such as solid state reactions[2], hydrothermal synthesis[3], and template-assisted sol-gel synthesis[4], and was extensively studied in photochemical[5, 6] or photoelectrochemical[7, 8] hydrogen generation via water splitting. In addition, there are a few reports in the literature that discuss its application in pollution remediation[9-11]. Yokoi and co-workers have applied a confined space synthesis method to make nanoparticulate $NaTaO_3$ and explore the effect of photocatalysts' morphology on its activity, and found nanostructured $NaTaO_3$ was three times as active as $NaTaO_3$ prepared by high temperature solid state reactions[4]. Perovskite type $NaTaO_3$ is a catalyst with significant water splitting ability under uv irradiation because of its bandgap energy and band structure. Bandgap energy calculated for $NaTaO_3$ by applying Kubelka-Munk method to diffuse reflectance data is 4.0 eV[12]. The band structure of $NaTaO_3$ is positioned in such a way that it can catalyze both oxidation and reduction reactions in water splitting[13]. Kato and co-workers studied Lanthanum (La^{3+}) doped $NaTaO_3$ and observed the highest quantum yield ever reported– (56%)- under UV irradiation at 270 nm[14]. Li and co-workers used the same material in order to evaluate the structural and optical effects of La^{3+} doping and evaluated its photocatalytic activity using decomposition of a model dye, Safranine T[10]. They attributed the enhanced photocatalytic activity to easy excitation of electrons from the valence band to the conduction band due to La^{3+} doping, in contrast to Kudo's conclusion that smaller particle size and ordered surface nanostep structure is responsible for the observed enhanced activity. Yi and co-workers utilized 1-D nanofibers of $NaTaO_3$ prepared by electrospining to evaluate photodegradation of methylene blue and concluded that the 1-D nature of the photocatalyst effectively enhances the photodecomposition of dye molecules[11]. Mohamed and co-workers reported the synthesis of Pd decorated $NaTaO_3$ photocatalyst and its use in environmental remediation of aqueous nitrate solutions, and observed that Pd/$NaTaO_3$ was photocatalytically active under visible light with $\lambda > 420$ nm even though its band gap data suggest a band edge at about 360 nm[15].

It is commonly acknowledged that high surface area benefits photocatalytic activity due to the availability of a larger number of active sites and enhanced substrate adsorption on the catalyst surface[16]. High crystallinity of the photocatalyst is also desired, since it minimizes photoexcited electron-hole pair recombination by decreasing the defects concentration. However, it is difficult to achieve both high surface area and crystallinity by regular synthesis methods because latter is achieved at higher temperatures, which generally leads to larger particle size, hence lower surface area. There are several advantages of using a mesoporous host material to support photocatalysts. High surface area, long range order, and uniform pore size allow uniform dispersion and prevent agglomeration of photocatalysts during high temperature synthesis and photocatalytic testing. In many instances, confinement of photocatalysts within the host pore structure tends to longer life time of the exciton, leading to enhanced catalytic activity[17]. MCM-48 (Mobile crystalline mesoporous) silica is highly ordered, high surface area silica template, which is a good candidate for catalytic applications due to its favorable mass transfer kinetics compared to other silica templates such as MCM-41[18]. Peng and co-workers studied the encapsulation of CdS visible light photocatalyst in MCM-48 and its photocatalytic activity towards water splitting, and found the composite to be highly active with 16.6% quantum yield[19].

Since surface area and catalyst distribution seem to be important factors that affect catalytic properties, we utilized a high surface area silica template (MCM-48) to homogenously disperse $NaTaO_3$ catalyst and form a composite. Specifically, $NaTaO_3$/MCM-48 composite photocatalysts were prepared by applying a sol-gel synthesis method and subsequent calcination in air. The effect of synthesis conditions on structural integrity, pore size distribution, and surface area of MCM-48 and $NaTaO_3$/MCM-48 composite catalysts were investigated by applying a wide range of analytical techniques including Powder X-ray diffraction (PXRD), BET surface area, thermogravimetric analysis

(TGA), and electron microscopy (EM). The photocatalytic activity of $NaTaO_3$/MCM-48 composite catalysts was subsequently evaluated using *p*-nitroaniline (*p*-NA) conversion into *p*-phenylenediamine (*p*-PDA) as a model reaction. The *p*-PDA is an important component in many industrial chemicals such as rubber antioxidants, thermoplastics, and textile fibers[20]. Developing a facile and economic route to synthesize *p*-PDA would be of interest due to rigorous synthetic conditions required in most commonly used commercial hydrogenation methods.

2. Experimental

2.1. Materials

Sodium acetate (NaAc, Aldrich), tantalum chloride ($TaCl_5$, Aldrich), citric acid (Fisher Scientific), absolute ethanol (EtOH, 200 proof, Pharmco-AAPER), dihydrogenhexachloroplatinate hexahydrate (H_2PtCl_6-$6H_2O$, Aldrich), ethylene glycol, *p*-nitroaniline (4-$C_6N_2O_2H_6$, Aldrich), cetyltrimethylammonium bromide (CTAB, 98%, Alfa Aesar), and tetraethylorthosilicate (TEOS, 98%) were used without further purification. De-ionized water was used in all the preparations.

2.2 Materials Synthesis

2.2.1. MCM-48. Mesoporous MCM-48 silica was synthesized following a procedure outlined in the literature[21]. Briefly, 1.2 g of surfactant CTAB was mixed with 50 mL of de-ionized water by vigorous stirring. Then, 25 mL of EtOH and 6 mL of 36% aq. ammonia were added to the surfactant solution and stirred for 10 minutes. Next, 1.8 mL (8 mmol) of TEOS was added and the mixture was stirred for 4 hours at room temperature to form a colorless precipitates, which was collected by centrifugation, washed 3 times with de-ionized water, and dried at 80 °C overnight. The product was ground, and then calcined at 550 °C for 6 hours in air with heating and cooling rates of 3 °C /min.

2.2.2. $NaTaO_3$. A literature reported sol-gel method was used for the synthesis of $NaTaO_3$ [22]. Sodium acetate (NaAc) and tantalum chloride were used as precursors, and citric acid (CA) was used as complexing agent. In a typical experiment, NaAc (0.23 g, 2.79 mmol) and CA (2.93 g, 14 mmol) were dissolved in a minimum amount of de-ionized water (~5 mL). The required amount of $TaCl_5$ (1.0 g, 2.79 mmol) was dissolved in 10 mL of EtOH. Then, NaAc-CA mixture was added dropwise to $TaCl_5$ solution and stirred for one hour. The mixture was heated in an oil bath at 80 °C until gelation occurred and a colorless precipitate was formed., The product was dried, then calcined at 350 °C for one hour and 500 °C for 3 hours in air at a heating and cooling rate of 3 °C / min to remove all organic components and to form $NaTaO_3$.

2.2.3. $NaTaO_3$/MCM-48 composite synthesis. Required amounts of MCM-48 silica,NaAc, $TaCl_5$, and citric acid for the preparation of $NaTaO_3$ / MCM-48 composite were calculated using the MCM-48/$NaTaO_3$ mass ratio. Precursor solutions for $NaTaO_3$ were prepared as described in the $NaTaO_3$ synthesis section. The required amount of MCM-48 was placed in a beaker and the $NaTaO_3$ precursor solution was added dropwise. The mixture was stirred for 1 hour prior to heating at 80 °C in an oil bath until gelation and dry material was formed. The gel was calcined at 350 °C for 1 hour and then at 500 °C or 850 °C for 3 hours at a heating and cooling rate of 3 °C / min.

2.2.4. Pt deposition. 1 wt% of Pt used as co-catalyst was deposited on each catalyst by microwave-assisted polyol reduction method. 0.25 g of photocatalyst was mixed with 10 mL of ethylene glycol and sonicated for 1-2 hours in order to uniformly disperse the catalyst. Next, 0.007 g of Pt precursor ($H_2PtCl_6.6H_2O$) was dissolved in ~2 mL of ethylene glycol and mixed with dispersed photocatalyst by stirring overnight. Reduction of Pt precursor was carried out under microwave irradiation for 3 minutes at 180 °C. The sample was recovered by centrifugation, and several washings with de-ionized water were performed to remove ethylene glycol. The sample was dried in an oven at 70 °C overnight.

2.3 Physical Characterization

2.3.1. Powder X-ray diffraction (PXRD). A Bruker-AXS D2 Phaser powder X-ray diffractometer (PXRD) equipped with a Ni-filtered Cu K sealed X-ray tube (=1.54184 Å) and a Lynxeye position-sensitive detector was used for X-ray analysis. The diffractometer was operated at 30 kV and 10 mA with a step width of 0.015° and a scan rate of 1.5°/min. PXRD data were collected at room temperature and analyzed using the Bruker-AXS EVA software package.

2.3.2. Scanning Electron Microscopy (SEM). A Carl Zeiss Auriga-BU FIB FESEM scanning electron microscope was employed to examine morphologies of all $NaTaO_3$/MCM-48 composite samples. A small amount of sample was spread on a carbon tape attached to an aluminum stub. The stubs were then mounted on a sample holder and examined.

2.3.3. Thermogravimetric analysis (TGA). A Perkin Elmer Pyris 1 thermogravimetric analyzer was used to determine thermal stability and crystallization temperature of $NaTaO_3$/MCM-48 composite. The composite was heated up to 1000 °C with a temperature ramp rate of 5°C min^{-1} in a continuous air flow.

2.3.4. Surface area and porosity analysis. A Quantachrome ASiQC 0100-2 Surface Area Analyzer was used in surface area and porosity analysis at 77 K. Microcrystalline samples were outgassed at 120 °C for 2 hours prior to analysis. The samples' specific surface area and pore size distribution were evaluated using the Brunauer-Emmet-Teller (BET) method. The pore size distribution was analyzed using a function, which applies the Density Functional Theory (DFT) built into ASiQwin software.

2.3.5. Elemental analysis. Elemental analysis of synthesized catalysts was carried out using inductive coupled plasma (ICP) and atomic absorption spectrometric (AAS) analysis. A Prodigy High Dispersion ICP OES (Teledyne Leeman Labs, USA) instrument and a calibration curve method were used to quantify the amounts of Na, Ta, and Si in solid catalysts. The solids were dissolved in Conc. nitric acid in an acid digestion autoclave by heating at 200 °C for 24 hours.

2.4 Photocatalytic experiments

Photocatalytic activity of the prepared compounds was evaluated using the following procedures. The photocatalytic reaction was carried out using an outer-irradiation reaction photo-cell connected to an inert gas line (argon). Photocatalyst powder (0.1 g) was dispersed in 30 mL of 20 ppm aqueous *p*-nitroaniline solution and stirred for ~30 min in an argon atmosphere, prior to irradiation. The suspension was then irradiated with a 200 W super-pressure mercury lamp (calculated power density, 10.2 W/cm^2) to initiate the conversion reaction. At a given irradiation time interval, a small aliquot (1.0 mL) was drawn from the system by a syringe, centrifuged to remove any advertently trapped catalyst, and analyzed with an Agilent 8453 UV-visible spectrophotometer. The conversion of *p*-nitroaniline was monitored by measuring the absorbance at λ_{max} = 238 and 380 nm, which corresponds to product and precursor peaks, respectively. Photocatalytic reproducibility tests were carried out as follows. After first catalytic testing, the solid catalyst was recovered by centrifugation and washed with de-ionized water in order to remove residual *p*-nitroaniline and/or *p*-phenylenediamine followed by drying in vacuum.

3. Results and Discussion

3.1. Structural analysis

The structure of mesoporous MCM-48 silica was confirmed by low angle PXRD. Figure 1 shows PXRD patterns of MCM-48 silica and $NaTaO_3$/MCM-48 composites calcined at 500 and 850 °C in air. The PXRD pattern of MCM-48 silica shows two significant peaks at 2θ = 2.7° and 3.2°, corresponding to (211) and (220) crystallographic planes, respectively. It is evident from Figure 1a that high calcination temperature of MCM-48 silica has an effect on intensity of the PXRD peaks; peak intensity decreases with increasing calcination temperature due to a decrease in scattering contrast between pores and pore walls[21]. High temperature treatment by itself does not destroy the MCM-48 silica structure. However, the presence of the complexing agent (i.e. citric acid) coupled with high temperature calcination has a detrimental effect on the MCM-48 silica structure and causes the cubic structure to collapse. As shown in Figure 1b, the signature PXRD reflection that appears at $2\theta = 2.7°$ for MCM-48 silica is absent from MCM-48 silica treated with citric acid and high temperature. Similarly, incorporation of $NaTaO_3$ also causes changes in the silica MCM-48 structure.

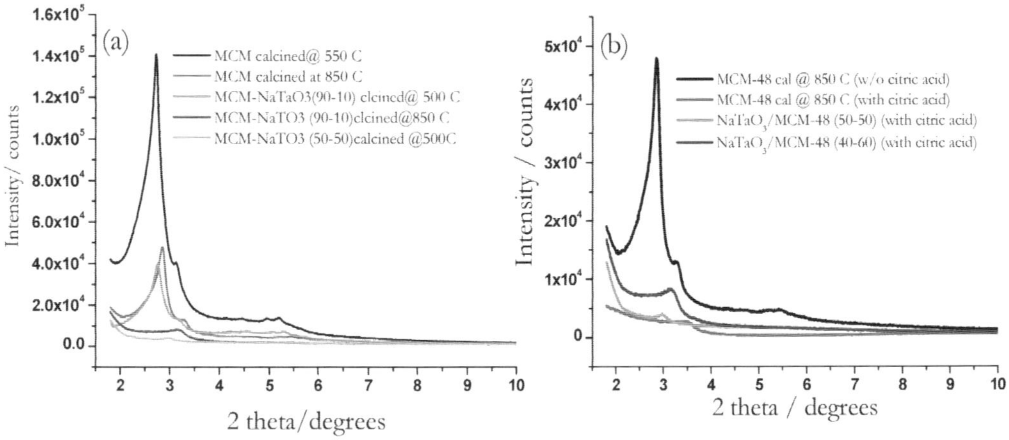

Figure 1. (a) Low angle X-ray diffraction patterns of silica MCM-48 and $NaTaO_3$-MCM composite catalysts calcined at different temperatures. (b) Figure shows XRD patterns of silica MCM-48 prepared in the presence and absence of citric acid along with composites calcined at 850 °C.

According to the literature and our experiments, $NaTaO_3$ prepared by sol-gel method and calcined at 500 °C is crystalline. Therefore, initially we used this temperature as the calcination temperature of composite catalysts. However, the composite catalysts prepared at this temperature were amorphous and no detectable XRD peaks were observed. Even though crystallinity is a highly important factor in photocatalysis, Tüysüz and co-workers reported photocatalytic water splitting using porous but amorphous $NaTaO_3$ and reported that better catalytic activity was observed for the amorphous over the crystalline phase[23]. In order to discern the calcination temperature that makes crystalline composite catalysts, thermogravimetric analysis (TGA) data was acquired (Figure 2). When $NaTaO_3$ was heated in air three prominent weight losses at 250-500 °C, 500 - 550°C, and 800 - 850 °C are observed (Figure 2a). The major weight loss between 250-500 °C may be attributed to loss of carbon, which is coming from the citric acid complexing agent. The second weight loss observed at 500 °C corresponds to calcination of $NaTaO_3$, as reported in the literature. This weight loss is not visible in the $NaTaO_3$/MCM composite catalyst (Figure 2b) which is consistent with PXRD data, which shows amorphous structure. After the third weight loss approximately at 850 °C, a stable phase was observed. Therefore, we selected 850 °C as the calcination temperature of composite catalysts based on the second calcination temperature, i.e. third weight loss, of $NaTaO_3$ (Figure 2a). It should be noted that the structure of MCM-48 slightly changes, according to XRD data (Figure 1a) and this might be the reason why we see a weight loss at 850 C for composites, but not pure $NaTaO_3$.

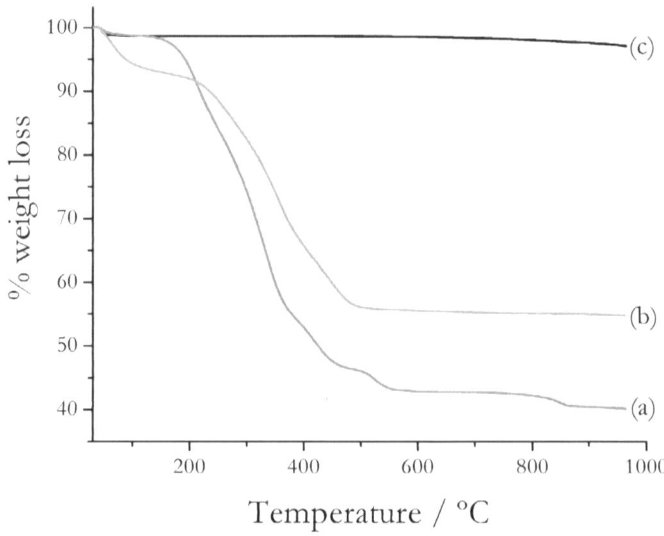

Figure 2. Thermogravimetric plots of (a) $NaTaO_3$, (b) $NaTaO_3$/MCM-48, (c) silica MCM-48 heated in air.

Powder X-ray diffraction was employed to evaluate the structure and crystallinity of $NaTaO_3$/MCM-48 composite catalysts. Figure 3 represents the PXRD plots of $NaTaO_3$/MCM-48 composites calcined at 850 °C for 3 hours in air along with $NaTaO_3$ calcined at 500 °C. The PXRD pattern of $NaTaO_3$/MCM-48 composites matches well, in terms of peak positions and intensity; with orthorhombic $NaTaO_3$ (PDF # 00-025-0863). Interestingly, composite catalysts calcined at 500 °C in air are amorphous (not shown) even though $NaTaO_3$ prepared without any silica MCM-48 is crystalline at that temperature. Since crystallinity is an important parameter in photocatalysts, higher temperature calcination is essential in order to form a crystalline phase of $NaTaO_3$. The crystallite size of $NaTaO_3$ in the composites calcined at 850 °C was calculated by applying the Debye-Scherer equation (Eqn. 1)[24].

$$B = \frac{0.9\,\lambda}{\text{FWHM x Cos}\,\theta} \qquad \text{Eqn 1}$$

Where B is crystallite size, λ is the wavelength, FWHM is full width at half maximum of the peak at 2θ = 46.59° chosen because background interference originating from silica is minimal above 2θ = 40°. The calculated crystallite size of pure $NaTaO_3$ calcined at 500 °C is approximately 23 nm while crystallite size of $NaTaO_3$ in $NaTaO_3$/MCM-48 composites ranges from 11 to 22 nm depending on $NaTaO_3$ content. The $NaTaO_3$/MCM-48 composites with 10% and 50% $NaTaO_3$ exhibited the smallest and largest $NaTaO_3$ crystallite size, respectively. Apparently, crystallite size of $NaTaO_3$ decreases with the increase of silica MCM-48 content.

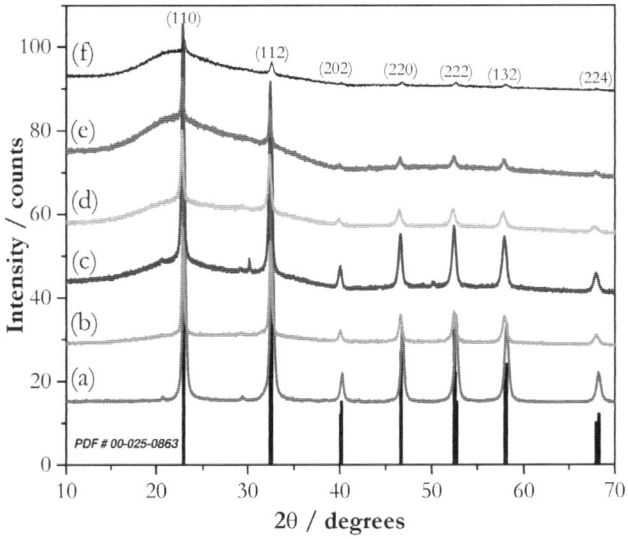

Figure 3. Powder X-ray diffraction patterns of $NaTaO_3$/MCM-48 composites calcined at 850 °C for 3 hours in air. The $NaTaO_3$/MCM-48 ratio changes as (a) 100/0 calcined at 500 °C, (b) 50/50, (c) 40/60, (d) 30/70, (e) 20/80, and (f) 10/90.

3.2. BET Surface area and pore size distribution analysis

Surface area and porosity of silica MCM-48, $NaTaO_3$, and $NaTaO_3$/MCM-48 composite materials were studied by nitrogen adsorption-desorption isotherms data. Figure 4 represents physisorption isotherms of silica MCM-48, $NaTaO_3$, and $NaTaO_3$/MCM-48 composites samples. Silica MCM-48 shows typical reversible type IV isotherm even though the hysteresis loop is not prominent and higher P/P_0 region is a spike instead of a plateau. However, the inflection point, which clearly distinguishes the transition from monolayer adsorption to multilayer adsorption, is prominent. Additionally, this marked inflection, which is caused by capillary condensation, is an indication of the mesoporous nature of silica MCM-48[21]. The sharpness of the inflection remained unchanged when MCM-48 silica was calcined at 850 °C (Figure 4e) indicating that highly ordered mesoporous structure of MCM-48 silica is not significantly altered even after calcination, which supports PXRD data. The surface area of MCM-48 synthesized in this work is in the range of literature reported values (1270 vs 1281 m^2/g)[18]. However, compared to MCM-48 calcined at 500 °C, the adsorption-desorption isotherms for MCM-48 calcined at 850 °C (Figure 4e and Table 1) showed lower N_2 adsorption and lower BET surface area,1270 vs. 1011 $m^2.g^{-1}$ (Figure 4f and Table 1)]. However, isotherms for $NaTaO_3$ and $NaTaO_3$/MCM-48 composites (Figure 4c-d) calcined at 850 °C show a plateau at lower P/P_0 region. This suggests that high temperature treatment diminishes the mesoporosity and attractive interactions between N_2 adsorbate and the $NaTaO_3$ surface are weaker than interactions among N_2 molecules.

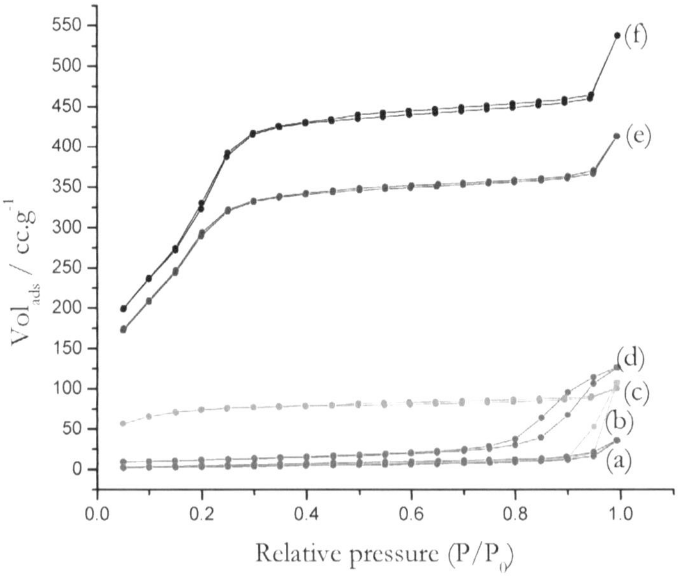

Figure 4. Nitrogen adsorption-desorption isotherms of $NaTaO_3$/MCM-48 (50/50 ratio) (a), $NaTaO_3$ (b), and silica MCM-48 (e) heated at 850 °C for 3 hours in air. (c), (d), and (f) represent $NaTaO_3$/MCM-48 (50/50 ratio), $NaTaO_3$, and silica MCM-48, respectively, calcined at 500 °C for 3 hours in air.

The BET surface area, pore size, and pore volume of composite photocatalysts are tabulated in Table 1. The average pore diameter and volume were calculated by applying a density functional theory (DFT) model, in which silica with cylindrical pores is considered to be the adsorbent, into adsorption branch. DFT model provides a more realistic pore size distribution, In this model, the local fluid structure near curved solid walls is clearly defined and pore distribution is determined by intermolecular potentials of the fluid-fluid and solid fluid interactions[25]. In contrast classical macroscopic theories such as Barrett- Joyner- Halenda (BJH) do not provide a realistic description of filling of micro and narrow mesopores and underestimate pore size distribution. In this method, pore size was found to be largely uniform for all samples calcined at 500 °C, in the range of 26-32 Å (Figure 5a). The average pore size of MCM-48 silica calcined at 500 °C is 32 Å and decreases to ~29 Å upon calcination at 850 °C, but it is still in the meso regime. In contrast, pure $NaTaO_3$ possesses relatively larger inter-grain mesopores (126-294 Å) after processing in the 500-850 °C temperature range. As shown in Table 1, the pore sizes of $NaTaO_3$/MCM-48 composite catalysts calcined at 500 °C are similar to that of silica MCM-48 except at higher $NaTaO_3$ loading (40-50 % loading). However, pore volume shows a systematic decrease upon increase in $NaTaO_3$ content (Figure 5a). Nevertheless the pore structure of all composite catalysts calcined at 850 °C, except those with $NaTaO_3$ loading 10 and 20%, have collapsed resulting in a relatively small pore volume.

Table 1. Surface area, pore diameter, and pore volume of $NaTaO_3$/MCM-48 composites.

Nominal $NaTaO_3$ /MCM-48 ratio	BET surface area / $m^2 . g^{-1}$		Pore diameter (Å) (adsorption)		Pore volume / $cm^3 . g^-$ (adsorption)	
	500 °C	850 °C	500 °C	850 °C	500 °C	850 °C
MCM	1270	1011	32	29	0.767	0.590
$NaTaO_3$	40	15	126	294	0.185	0.135
10/90	964	407	32	26	0.549	0.246
20/80	866	139	32	26	0.512	0.133
30/70	538	96	32	300	0.318	0.041
40/60	278	68	26	300	0.185	0.037
50/50	235	35	26	300	0.145	0.048

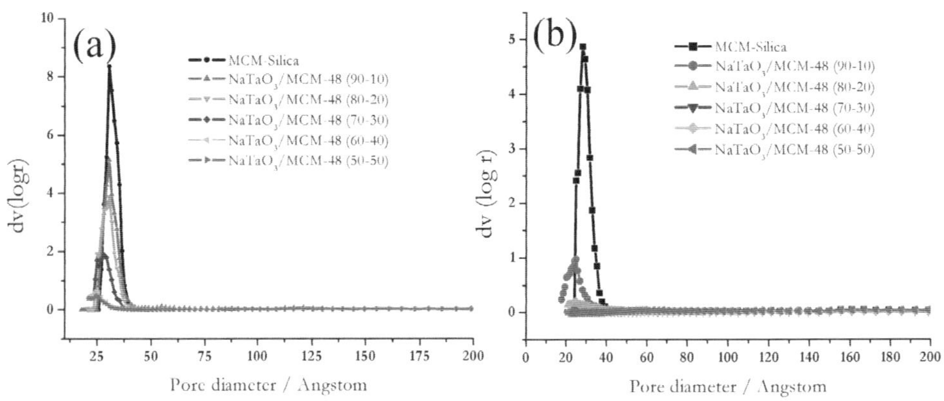

Figure 5. Pore size distribution of MCM-48 and composites calcined at (a) 500 °C and (b) 850 °C

3.3. Morphological characterization

SEM images of MCM-48 and $NaTaO_3$/MCM-48 composites are shown in Figure 6. MCM-48 calcined at 500 °C in air is spherical in shape and polydispersed. The particle size ranges from about 0.2 – 1 micron. Calcination of MCM-48 at 850 °C does not significantly change the morphology (Figure 6(b)). Similarly, the composite with 10% $NaTaO_3$ retains its spherical morphology. However, a significant change in morphology is evident in the composite with higher $NaTaO_3$ content (50%) (Figure 6(c)), which consists of a mixture of large particles and some spheres indicating 50% loading is too much. It is possible that excess $NaTaO_3$ catalyst crystallizes outside the MCM-48 pores.

Figure 6. SEM images of MCM-48 calcined at 500 °C (a) and 850 °C (b). Figure (c) and (d) show 50-50 and 10-90 $NaTaO_3$/MCM-48 composites calcined at 850 °C.

3.4. Photocatalytic activity

Photocatalytic activity of the Pt/$NaTaO_3$/MCM-48 composite catalysts was evaluated by investigating the conversion of *p*-nitroaniline (*p*-NA) to *p*-phenylenediamine (*p*-PDA) under uv irradiation using a 20 ppm aqueous solution of *p*-nitroaniline. The experiments were carried out at room temperature with continuous argon purge at atmospheric pressure, in the absence of hole-scavengers. The conversion analysis of *p*-NA solution was made spectrophotometrically by assessing its decrease in absorbance at 380 nm. The Figure 7 shows the evolution of absorbance change of *p*-NA as a function of time during photocatalytic reduction. The photocatalytic activity, in terms of *p*-NA conversion was calculated using the decrease in absorbance at 380 nm. As shown in Figure 7a, relative change in the absorbance of *p*-NA in the absence of photocatalyst is negligible, which indicates that light irradiation alone has no effect on *p*-NA conversion. In the presence of photocatalysts, a significant decrease of the absorbance at 380 nm along with simultaneous appearance of an absorption band at 238 nm, the characteristic absorbance peak of *p*-PDA, was observed (Figure 7b-d).

Journal of Physics: Conference Series **758** (2016) 012003 doi:10.1088/1742-6596/758/1/012003

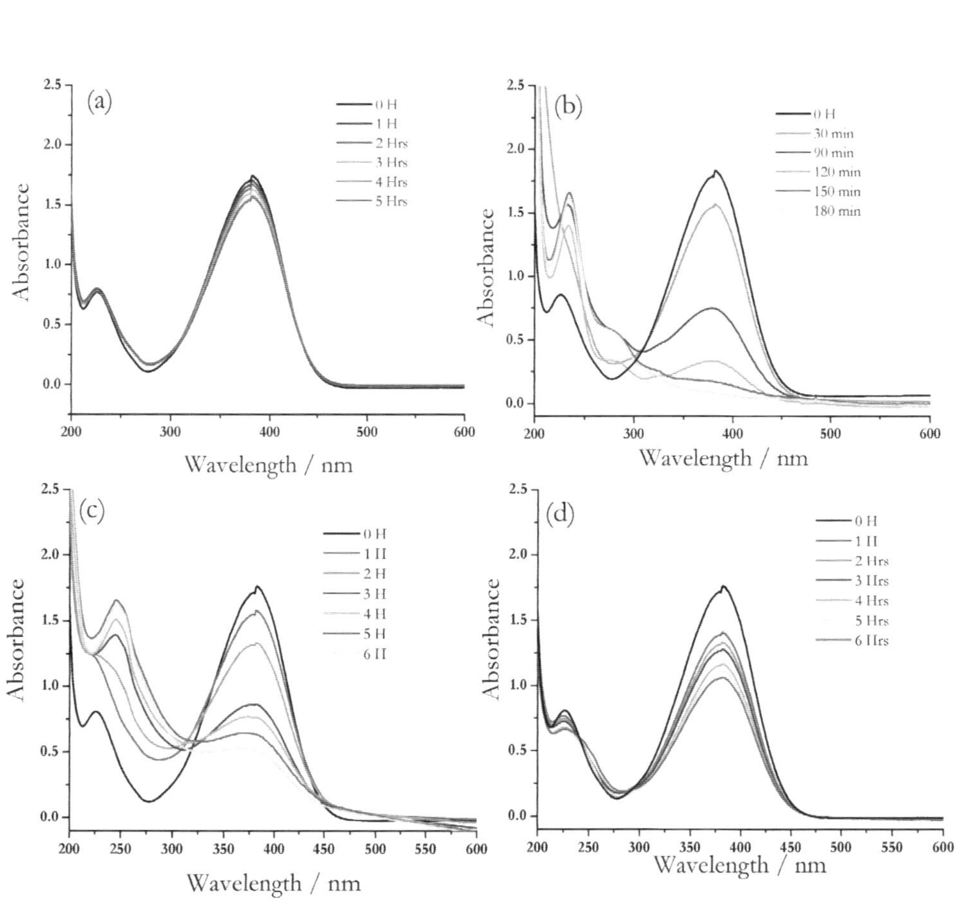

Figure 7. Relative concentration change of *p*-NA in the presence of Pt /($NaTaO_3$/MCM-48) composite catalysts. (a)p-NA without catalyst, (b) $NaTaO_3$ only, (c) $NaTaO_3$/MCM-48 (50-50), and (d) $NaTaO_3$/MCM-48 (10-90).

The plot of C/C_0 as a function of irradiation time, where C is *p*-NA concentration in reaction mixture and C_0 is the initial concentration of *p*-NA, is shown in Figure 8a. The relative concentration change of *p*-NA in the absence of a catalyst is approximately 10 % after 5 hours of irradiation. The concentration change in the presence of a composite catalyst ranges from 30 to 70%. The $NaTaO_3$/MCM-48 composites with 10-90 and 50-50 compositions gave the lowest and highest *p*-NA concentration change, respectively, after 5 hours of irradiation. As expected, relative concentration change increases with higher loading of $NaTaO_3$. If we just compare the data shown in Figure 8a as a function of total weight of catalyst (0.10 g) but not as a function of amount of $NaTaO_3$, pure $NaTaO_3$ causes the highest percent conversion in a given time. In order to correlate the weight of $NaTaO_3$ and the percent conversion and to evaluate photocatalytic efficiency, we calculated and compared the conversion by each composite catalyst as a function of composite weight and $NaTaO_3$ weight. This calculation is similar to turnover number (TON) and turnover frequency (TOF), but we used number of grams instead of number of active sites in the calculation. The turnover number is defined as the number of *p*-NA molecules converted into *p*-PDA per number of active sites of catalyst while TOF is defined as turnover number per second. The change in the concentration of *p*-NA after 3 hours of irradiation was used in the calculation. The amount of $NaTaO_3$ catalyst present in the SiO_2 matrix was analyzed by ICP spectrometer. Two samples with highest (50%) and lowest (10%) $NaTaO_3$ content were chosen for elemental analysis. The $NaTaO_3$:MCM-48 ratio obtained for the sample with 50%

nominal $NaTaO_3$ content was 46:54 while it was 8:92 for the sample with 10% $NaTaO_3$. A summary of conversions calculated as a function of composite weight and simple $NaTaO_3$ weight for different $NaTaO_3$/MCM-48 composite catalysts is tabulated in Table 2. The simple $NaTaO_3$ shows the highest conversion if the overall weight ($NaTaO_3$ + MCM) is used in the calculation. However, conversion numbers show a significant increase in composite catalysts when the calculations are based on just the $NaTaO_3$ content in the composite, not overall weight. In this case, it is clearly evident from the data that the conversion as a function of $NaTaO_3$ content increases with respect to decrease in $NaTaO_3$ content.

The reusability is a vital issue in applications of a photocatalyst. Therefore, reusability of composite photocatalysts was evaluated using the catalyst with 50% $NaTaO_3$ in $NaTaO_3$/MCM-48 composite under the same reaction conditions applied for photocatalytic activity testing. Based on relative concentration data acquired in 3 attempts for the same catalyst (Figure 8b), it is seen that the catalyst loses approximately 12 % of its initial activity after 3 consecutive attempts, calculated using data acquired after 5 hours of irradiation. SEM image of 50/50 MCM/NaTaO3 spent catalyst was acquired (Figure 9) in order to discern whether there is a significant change in morphology due to uv irradiation. It is evident that irradiation is not detrimental even though there is slight coagulation, which may cause drop in surface area. In fact this might be a reason why we see 12% activity reduction upon cycling.

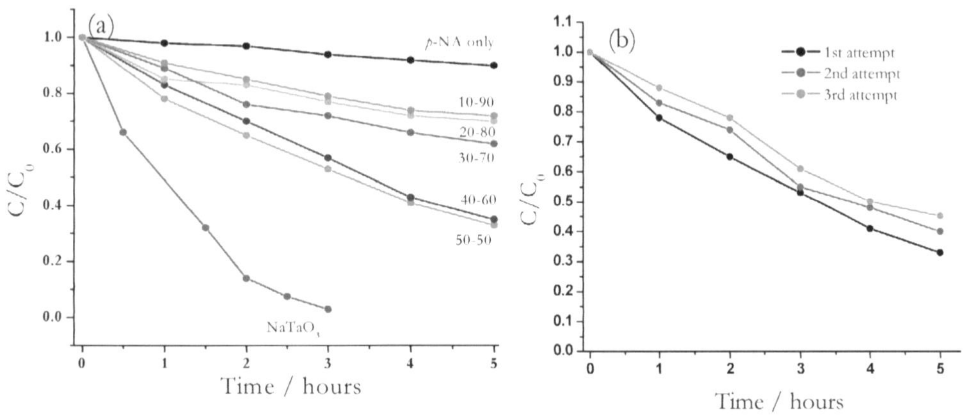

Figure 8. Plot of C/C_0 of composite photocatalysts as a function of time (a) and reproducibility of photocatalytic data for Pt /($NaTaO_3$/MCM-48) (b).

Table 2. Summary of conversion numbers of $NaTaO_3$/MCM-48 composite catalysts calculated from absorbance data at 3 hours of irradiation.

	Conversion per composite weight		Conversion per $NaTaO_3$ weight	
$NaTaO_3$:MCM	(molecules /g)	(molecules /g.s)	(molecules /g_{NaTaO3})	(molecules/g_{NaTaO3} .s
100 : 0	2.49×10^{19}	2.30×10^{15}	2.49×10^{19}	2.30×10^{15}
50 : 50	1.33×10^{19}	1.24×10^{15}	2.66×10^{19}	2.46×10^{15}
40 : 60	1.07×10^{19}	9.87×10^{14}	2.67×10^{19}	2.48×10^{15}
30 : 70	7.30×10^{18}	6.78×10^{14}	2.53×10^{19}	2.34×10^{15}
20 : 80	5.98×10^{18}	5.54×10^{14}	2.99×10^{19}	2.77×10^{15}
10 : 90	5.47×10^{18}	5.06×10^{14}	5.47×10^{19}	5.06×10^{15}

Figure 9. SEM image of 50/50 MCM/$NaTaO_3$ catalysts acquired after photocatalytic testing

Conclusions

Sol-gel method was used to synthesize a series of $NaTaO_3$/MCM-48 silica composite photocatalysts. Their physicochemical properties and photocatalytic activities in *p*-nitroaniline conversion into *p*-phenylenediamine were measured. The calculated conversion of *p*-nitroaniline by the composite photocatalysts is higher than that for pristine $NaTaO_3$ catalyst, which indicates a higher activity of conversion per gram of $NaTaO_3$ in composites. Nevertheless, it is difficult to find a clear correlation between the surface area and the activity even though the overall activity of composite catalysts is higher. Photocatalytic activity towards p-NA conversion is reproducible and the catalyst loses only 12 % of its activity after 3 consecutive attempts.

Acknowledgement

This work has been financially supported by the Wake Forest University Center for Energy, Environment, and Sustainability (CEES), and by the Triad Interuniversity Project (TIP) of Wake Forest University, North Carolina A & T University, University of North Carolina Greensboro, and Winston Salem State University.

References

[1] Kudo A and Miseki Y 2009 *Chem. Soc. Rev.* **38** 253-8
[2] Kudo A and Kato H 2000 *Phys. Lett.* **331** 373-7
[3] He Y, Zhu Y and Wu N 2004 *Solid State Chem.* **177** 3868-72
[4] T. Yokoi, T Sakuma K, Domen K, Tatsumi T and Kondo, K 2011 *Phys. Chem. Chem. Phys.* **13** 2563-70
[5] Osterloh F E 2008 *Chem. Mater* 20 35-54
[6] Parachino A, Laporte V, Sivula K, Grätzel M and Thimsen E, 2011 *Nat. Mater.* **10** 456-461
[7] Esswein A J, McMurdo M J, Ross P N, Bell A T and Tilley T D, 2009 *J. Phys. Chem. C* **113** 15068-72
[8] Rossmeisl J, Qu Z W, Zhu H, Kroes G J, J.K. Norskov, 2007 *J. Electroanal. Chem.* **607** 83-89
[9] Vazquez-zcuchillo O, Manzo-Robeldo A, Zanella Elizondo-Villareal R N and Cruz-Lopez A *Ultrason. Sonochem.* **20** 498-501
[10] Li X and Zang J 2011 *Catal. Commun.* **12** 1380-83
[11] Yi X, Li J, 2010 *J. Sol-Gel Sci. Technol.* **53** 480-484

[12] Lin W-H, Cheng C, Hu C-C and Teng H, 2006 *Appl. Phys. Lett.* **89** 211904-906
[13] Kato H and Kudo A,2001 *J. Phys. Chem B* **105** 4285-92
[14] Kato H, Kiyotaka K and Kudo A, 2003 *J. Am. Chem. Soc.* **125** 3082-89
[15] Mohamed R M and Aazam E S, 2014 *J. Ind. Eng. Chem.* **20** 4377-81
[16] Amano F, Nogami K, Tanaka M and Ohtani B, 2010 *Langmuir* **26** 7174-80
[17] Ryu S Y, Balcerski W, Lee T K and Hoffmann M R, 2007 *J. Phys. Chem. C* **111** 18195-203
[18] Zhang K, Yuan E-H, Xu L-L, Xue Q-S, Luo C, Albela B and Bonneviot L 2002 *Eur. J. Inorg. Chem.* **2012** 4183-89
[19] Peng R, Zhao D, Baltrusatis J, Wu C-M and Koodali R T 2012 *RSC Adv.* **2** 5754-67
[20] Aleksandrovic V and Djonlagic J 2001 *J. Serb. Chem. Soc.* **66** 139-152
[21] Peng R, Zhao D, Dimitrijevic N M, Rajh T and Koodali R T 2012 *J. Phys. Chem. C* **116** 1605-13
[22] Hu C-C and Teng H 2007 *Appl. Catal. A: General* **331** 44-50
[23] Tüysüz H and Chan C K 2013 *Nano Energy* **2** 116-123
[24] Senevirathne K, Burns A, Bussell M E and Brock S 2007 *Adv. Funct. Mater.* **17** 3933-39
[25] Ravikovitch P I, Vishnyakov A, Russo R and Neimark A V, 2000 *Langmuir* **16** 2311-20

Phase diagram of the ferrimagnetic mixed-spin Blume-Capel model with four-spin and next-nearest neighbor interactions

M. Azhari, N. Benayad*, M. Mouhib

Laboratory of High Energy Physics and Scientific Computing,
Hassan II University–Casablanca, Faculty of Sciences-Aïn Chock,
B.P : 5366 Maarif, Casablanca 20100, Morocco.

*Corresponding author: E-mail: n.benayad@fsac.ac.ma; noureddine_benayad@yahoo.fr
Phone: + 212 5 22 23 06 84, Fax: + 212 5 22 23 06 74

Abstract. A finite cluster approximation based on a single cluster theory is developed for a ferrimagnetic mixed spin-1/2 and spin-1 Blume-Capel model with four-spin interaction J_4 and next-nearest neighbor couplings J'. The state equations are derived for the two-dimensional square lattice and the phase diagram is explored. We found that the system can exhibits both critical and tricritical behavior according to appropriate values of J', D and J_4.

1. Introduction

The synthesis of new stable crystalline room-temperature magnets, that are able to keep magnetization in the absence of an applied magnetic field, is a major issue for many industrial applications. This is not only because of their potential device application, including those in areas such as thermomagnetic recording, electronic and computer technologies [1], as well as biomedical materials [2] and catalysts [3], but also due to their common uses as research tools in multi-interdisciplinary areas of chemistry [4], geology, biology and medicine [5]. In recent years both chemists and physicists started to collaborate very closely on chemistry and material sciences in order to produce materials soluble in organic solvents, biocompatible, optically transparent, with spontaneous moments at room temperature [1][4][6]. Generally, the ferrimagnetic ordering seems often to play a prominent relevance in these materials. There is currently a great deal of interest in the synthesis of single-chain magnets (SCM) with ferrimagnetic arrangement, such as $Co^{II}(hfac)_2(NITPhOMe)$ [7], Co^{II} – Cu^{II} bimetallic single-chain magnet [8] and $[Fe^{II}(ClO_4)_2\ \{Fe^{III}(bpca)_2\}](ClO_4)$ [9], because they can act as elementary binary units (bits) used for information storage, providing potential applications in high-density information storage [10] or as qubits in quantum computers [11]. As purely one-dimensional systems are known to have a long-range order only at $T = 0\ K$, these SCM materials promote long relaxation times and the system can behave like a magnet [12]. In the search for new and improved molecular materials, synthesis has been expanded toward 2-D and 3-D dimensional ferrimagnets, such as 2-D organometallic ferrimagnets [12]. 2-D networks of the mixed-metal material {[P(Phenyl)4][MnCr(oxalate)3]}n [13], 3-D ferrimagnets with T_C=240K and T_C=190K [14] and the amorphous $V(TCNE)_x.y(solvent)$ with temperatures ordering as high as 400*K* [15]. For all these purposes, intensive efforts are required in the theoretical study of these materials in order to clarify their very interesting and sometimes unusual behaviors.

Mixed-spin Ising systems provide good models to study ferrimagnetism. The magnetic properties of these systems on the square lattice have been investigated by various methods. In particular, the mixed spin-1/2 and spin-1 Ising model with only nearest neighbor coupling and crystal field, has been studied by mean field theory [16] and high(low)-temperature series expansions [17]. Both approaches report tricritical point while compensation points were only detected using mean field approximation. As a result the phase diagram and the magnetic properties of the system have been reinvestigated using numerical transfer matrix techniques, supplemented by Monte Carlo simulations by Buendia and Novotny [18], renormalization-group technique [19] and also using extensive Monte Carlo simulations by Selke and Oitmaa [20]. The resulting papers found no evidence of either a compensation point or a tricritical point. On the other hand, they showed that compensation points are induced by the presence of an interaction between the next nearest neighbor of the spin-1/2. The inclusion of further-neighbor interactions would allow for a better modeling of real magnetic systems [21] and of course of all other systems that can be mapped onto the Ising models such as models of microemulsions [22].

Recently, there has been considerable interest in experimental and theoretical researches of Ising model with multispin interactions. These models are interesting because they found their theoretical explanation in the theories of super exchange interaction, the magnetoelastic effect and the spin-phonon coupling [23]. Moreover, it was pointed out that the models with the higher-order exchange interactions may exhibit rich phase diagrams and can describe phase transition in some physical systems. Additionally, they show physical behavior not detected in the usual spin systems. For example, the non-universal critical phenomena [24] and deviation from $T^{3/2}$ Block law at low temperature [25].

From the theoretical point of view, the monoatomic Ising models with multispin interactions have been investigated in detail within different methods, such as mean field approximation [26], effective field theory [27], some more accurate treatments such as series expansion [28], renormalization group methods [29], Monte Carlo simulations [30], and also exact calculations [31]. Experimentally, an interesting fact for the models with multispin interactions has been reported. Indeed, it can be used to describe various physical systems such as classical fluid [32], solid ^3He [33], lipid bilayers [34], and rare gases [35]. Moreover for some materials it has been shown that the multispin interactions play a significant role; and they are comparable or even much important than the bilinear ones. The models with pair and quartet interaction have been applied successfully to study and explain the existence of first order phase transition in squaric acid crystal $H_2C_2O_4$ [36]. Such models have been also used to describe thermodynamical properties of hydrogen-bonded ferroelectric $PbHPO_4$, $PbDPO_4$ [37], some copolymers [38] and optical conductivity [39] observed in cuprate ladder $La_xCa_{14-X}Cu_{24}O_{41}$. On the other hand, some experimental studies on $La_6Ca_8Cu_{24}O_{41}$ [40] and $La_4Sr_{10}Cu_{24}O_{41}$ [41] reveal that they could be explained by the introduction of the four-spin interaction. It is worthy to note here that this later plays an important role in the two dimensional antiferromagnet La_2CuO_2 [42], the parent material of high-T_C superconductors.

Intense interest has been directed to study the magnetic properties of two-sublattices mixed spin Ising system. They have less translational symmetry than their single spin counterparts, and are well adapted to study a certain type of ferrimagnetism [43]. Experimentally, it has been shown that the

MnNi(EDTA)-6H_2O complex [44] is a good example of a mixed system. The thermodynamic properties of the ferromagnetic mixed Ising model consisting of spin-1/2 and spin-1 with four-spin and next nearest neighbor (NNN) interactions have been studied, very recently, by Monte Carlo simulations (MCS) [45] and finite cluster approximation (FCA) [46]. According to the value of the NNN couplings it has been shown that the system behaves qualitatively and quantitatively different from that obtained when the interaction between NNN is ignored. In the present paper we examine the same system using FCA in the context of the ferrimagnetism and compensation points. We will focus our attention principally on the behavior of the phase diagrams.

The ferrimagnetic mixed-spin Ising system, we are interested in, is shown in Fig.1. Such system can be described by the following Hamiltonian:

$$H \quad J_2 \sum_{\langle ij \rangle} \sigma_i S_j - J_4 \sum_{\{i,j,k,l\}} \sigma_i \sigma_k S_j S_l - J' \sum_{\langle ik \rangle} \sigma_i \sigma_k + D \sum_{j} (S_j)^2 \qquad (1)$$

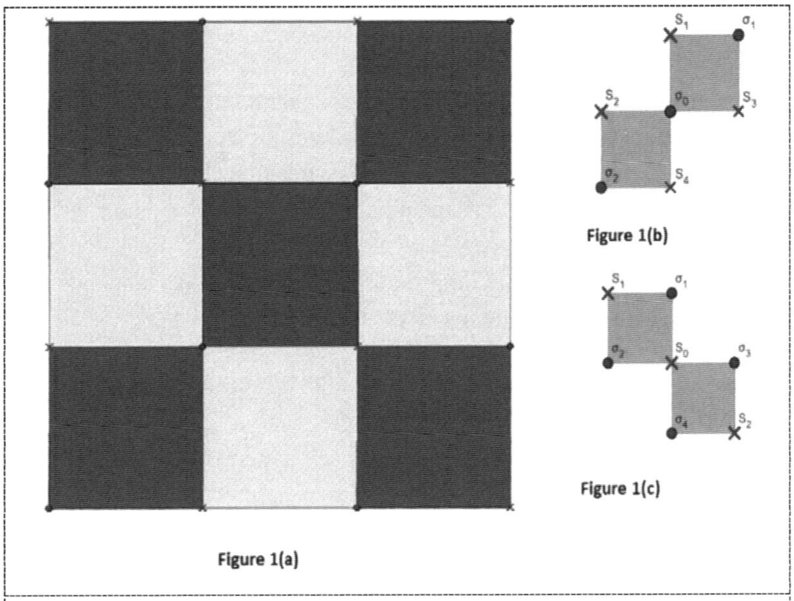

Figure 1. (a) Part of the square lattice. • and × correspond to σ and S-sublattice sites, respectively. (b) Neighbors of spin σ_0 with which directly interacts. (c) Neighbors of spin S_0 with which directly interacts.

The underlying lattice is composed of two interpenetrating sublattices. One occupied by spins with spin moment $\sigma=\pm1/2$ and the other one is occupied by spins with moment S= 0,±1 which interact with one another with ferrimagnetic parameter J_2 ($J_2>0$). The first summation carried out only over nearest-neighbor pair of spins. The second and the third summations represent the four-spin and NNN interactions, respectively, where the summations concern all alternate squares shaded in Fig.1. In the present work, our study is focused on ferrimagnetic state with $J'>0$.

This paper is organized as fellows. The second section is devoted to the theoretical framework and the state equations whereas the third section focuses on the numerical result and the discussion. Finally, some conclusions are drawn in the fourth section.

2. Theoretical framework

The theoretical framework that we adopt in the study of the mixed spin-1/2 and spin-1 Ising model with four-spin and NNN interactions, described by the Hamiltonian (1), is the finite cluster approximation (FCA) [47] based on a single-site cluster theory. We have to mention that this method has been successfully applied to a number of interesting pure and disordered spin Ising systems [48]. It has also been used for transverse Ising models [49] and semi-infinite Ising systems [50]. In all these applications, it was shown that the FCA improves qualitatively and quantitatively the results obtained in the frame of the mean-field theory. In this approach, attention is focused on a cluster comprising just a single selected spin $\sigma_o(S_o)$ and its neighbour spins $\{\sigma_1,\sigma_2,S_1,S_2,S_3,S_4\}$ ($\{S_1,S_2,\sigma_1,\sigma_2,\sigma_3,\sigma_4\}$) with which it directly interacts (see Figs. 1-(a) & (b)).

We split the total Hamiltonian (1) into two parts, $H=H_0+H'$ where H_o includes all parts of H associated with the lattice site 0. In the present system, H_o takes the form

$$H_{0\sigma} \quad -\left[-J_2\sum_{i=1}^{4} S_i + J_4(S_1S_2\sigma_1 + S_3S_4\sigma_2) + J'(\sigma_1+\sigma_2)\right]\sigma_0 \tag{2}$$

$$H_{0S} \quad -\left[-J_2\sum_{i=1}^{4} \sigma_i + J_4(\sigma_1\sigma_2S_1 + \sigma_3\sigma_4S_2)\right]S_0 + D(S_0)^2 \tag{3}$$

whether the lattice site 0 belongs to σ or S-sublattice, respectively.

The problem consists in evaluating the sublattice magnetizations and the quadrupolar moment. To this end, we denote by $<\sigma_o>_c$ and, $<S_o^n>_c$ (n=1, 2), respectively the mean value of σ_o and S_o^n for a given configuration c of all other spins (i.e. when all other spin σ_i and S_j (i, j≠0) are kept fixed). $<\sigma_o>_c$ and $<S_o^n>_c$ are given by

$$\langle\sigma_0\rangle_C \quad \frac{\underset{\sigma_0}{Tr}\,\sigma_0 exp(-\beta H_{0\sigma})}{\underset{\sigma_0}{Tr}\,exp(-\beta H_{0\sigma})} \tag{4}$$

$$\langle S_0^n\rangle_C \quad \frac{\underset{S_0}{Tr}\,S_0^n exp(-\beta H_{0S})}{\underset{S_0}{Tr}\,exp(-\beta H_{0S})} \tag{5}$$

where $Tr_{\sigma o}$ (or Tr_{So}) means the trace performed over σ_o (or S_o) only. As usual $\beta=1/T$ where T is the absolute temperature. The sublattice magnetizations μ, m and the quadrupolar moment q are then given by

$$\mu \equiv \langle\langle\sigma_0\rangle_C\rangle = \langle\frac{\underset{\sigma_0}{Tr}\,\sigma_0 exp(-\beta H_{0\sigma})}{\underset{\sigma_0}{Tr}\,exp(-\beta H_{0\sigma})}\rangle \tag{6}$$

$$m \equiv \langle\langle S_0\rangle_C\rangle = \langle\frac{Tr_{S_0} S_0 exp(-\beta H_{0S})}{Tr_{S_0} exp(-\beta H_{0S})}\rangle \tag{7}$$

$$q \equiv \langle\langle S_0^n\rangle_C\rangle = \langle\frac{Tr_{S_0} S_0^n exp(-\beta H_{0S})}{Tr_{S_0} exp(-\beta H_{0S})}\rangle \tag{8}$$

where <···> denotes the average over all spin configurations. Performing the inner traces in (6), (7) and (8) over the states of the selected spin σ_o (S_o), we obtain the following exact relations

$$\mu \quad \langle\frac{1}{2}\tanh\left[\frac{K}{2}\{-(S_1+S_2+S_3+S_4)+\alpha(S_1S_2\sigma_1+S_3S_4\sigma_2)+\gamma(\sigma_1+\sigma_2)\}\right]\rangle \tag{9}$$

$$m \quad \langle\frac{\sinh[K\{-(\sigma_1+\sigma_2+\sigma_3+\sigma_4)+\alpha(\sigma_1\sigma_2S_1+\sigma_3\sigma_4S_2)\}]}{\exp(\beta D)+\cosh[K\{-(\sigma_1+\sigma_2+\sigma_3+\sigma_4)+\alpha(\sigma_1\sigma_2S_1+\sigma_3\sigma_4S_2)\}]}\rangle \tag{10}$$

$$q \quad \langle\frac{\cosh[K\{-(\sigma_1+\sigma_2+\sigma_3+\sigma_4)+\alpha(\sigma_1\sigma_2S_1+\sigma_3\sigma_4S_2)\}]}{\exp(\beta D)+\cosh[K\{-(\sigma_1+\sigma_2+\sigma_3+\sigma_4)+\alpha(\sigma_1\sigma_2S_1+\sigma_3\sigma_4S_2)\}]}\rangle \tag{11}$$

where $K=\beta J_2$, $\alpha=J_4/J_2$ and $\gamma=J'/J_2$.

It is a formidable task to calculate the average on the right-hand sides of Eqs.(9) to (11) over all spin configurations. We can easily observe that any function such as $f(\sigma,S)$ of σ and S can be written as the linear superposition

$$f(\sigma,S) \quad f_1+f_2\sigma+f_3S+f_4S^2+f_5\sigma S+f_6\sigma S^2 \tag{12}$$

which appropriate coefficients f_i (i=1,….,6) . After applying this to all spins σ_i and S_j in expressions between brackets in equations (9) to (11), we average over all spin configurations. In this paper we use the simplest approximation in which we treat all spin self-correlations exactly while still neglecting correlations between quantities pertaining to different sites. This leads to the following coupled equations.

$$\begin{aligned}\mu \quad & \mu[2A_1+4(A_2+A_3)q+(2A_4+8A_5+2A_6)q^2+4(A_7+A_8)q^3+2A_9q^4]+\\ & m[4A_{10}+(4A_{11}+8A_{12})q+(4A_{13}+8A_{14})q^2+4A_{15}q^3]+\\ & m\mu^2[4A_{16}+[4A_{17}+8A_{18})q+(4A_{19}+8A_{20})q^2+4A_{21}q^3]+\\ & \mu m^2[2A_{22}+8A_{23}+2A_{24}+(4A_{25}+8A_{26}+4A_{27}+8A_{28})q+(2A_{29}+8A_{30}+2A_{31})q^2]+\\ & m^3[4A_{32}+4A_{33}q]+\mu m^4[2A_{34}]+\mu^2m^3[4A_{35}+4A_{36}q]\end{aligned} \tag{13}$$

$$\begin{aligned}m \quad & \mu[4B_1+8B_2q+4B_3q^2]+m[2B_4+2B_5q]+\mu^3[4B_6+4B_7q+4B_8q^2]+\\ & m\mu^2[2B_9+2B_{10}q]+\mu m^2[4B_{11}]+m\mu^4[2B_{12}+2B_{13}q]+m^2\mu^3[4B_{14}]\end{aligned} \tag{14}$$

$$\begin{aligned}q \quad & [C_1+2C_2q+C_3q^2]+\mu^2[6C_4+12C_5q+6C_6q^2]+m^2[C_7]+\mu m[4C_8+4C_9q]+\\ & \mu^4[C_{10}+2C_{11}q+C_{12}q^2]+m\mu^3[4C_{13}+4C_{14}q]+\mu^2m^2[6C_{15}]+m^2\mu^4[C_{16}]\end{aligned} \tag{15}$$

Journal of Physics: Conference Series **758** (2016) 012004 doi:10.1088/1742-6596/758/1/012004

The non-zero coefficients quoted in Eq. (13) are listed in the Appendix while those quoted in Eqs. (14) and (15) can be taken from Appendix in [51], where the definition of the functions g(x) and h(x) must be changed, respectively, to

$$g(x) \equiv \frac{2\sinh(Kx)}{\exp(\beta D) + 2\cosh(Kx)} \quad , \quad h(x) \equiv \frac{2\cosh(Kx)}{\exp(\beta D) + 2\cosh(Kx)}$$

and take into account that the bilinear interaction J_2 is ferrimagnetic.

If we replace m and q in (13) by their expressions taken from (14) and (15), we obtain an equation for μ of the form

$$\mu \quad a\mu + b\mu^3 + \cdots \tag{16}$$

In order to obtain the second-order transition temperature, we neglect higher- order terms in the magnetizations in Eqs. (13)-(15). Therefore, the critical temperature is analytically obtained through a determinantal equation, i.e.

$$1 \quad 2A_1 + 4(A_2 + A_3)q_0 + (2A_4 + 8A_5 + 2A_6)q_0^2 + 4(A_7 + A_8)q_0^3 + 2A_9q_0^4 + \frac{4B_1 + 8B_2q_0 + 4B_3q_0^2}{1 - (2B_4 + 2B_5q_0)}[4A_{10} + (4A_{11} + 8A_{12})q_0 + (4A_{13} + 8A_{14})q_0^2 + 4A_{15}q_0^3] \tag{17}$$

where q_0 is the solution of

$$q_0 \quad C_1 + 2C_2q_0 + C_3q_0^2 \tag{18}$$

Equation (17) means that its right-hand side corresponds to the coefficient *a* in (16).

In the vicinity of the second-order transition, the sublattice magnetization μ is given by

$$\mu^2 \quad \frac{1-a}{b} \tag{19}$$

The right-hand side of (19) is positive since we are in the long-ranged ordered regime. This means that the signs of *1-a* and b are the same.

When *b* changes sign (*1-a* keeping its sign), μ^2 becomes negative. It means that we are not in the vicinity of a second-order transition line. So, the transition is of the first order. Therefore the point at which

$$a(K, \alpha, \gamma) \quad 1 \text{ and } b(K, \alpha, \gamma) \quad 0 \tag{20}$$

characterizes the tricritical points.

To obtain the expression for *b*, one has to solve (13)-(15) for small μ and *m*. The solution is of the form

$$q \quad q_0 + q_1\mu^2 + q_2m^2 + q_3\mu m \tag{21}$$

where q_1, q_2 and q_3 are given by

$$q_1 \quad \frac{6C_4 + 12C_5 q_0 + 6C_6 q_0^2}{1 - 2C_2 - 2C_3 q_0}, \quad q_2 \quad \frac{C_7}{1 - 2C_2 - 2C_3 q_0}, \quad q_3 \quad \frac{4C_8 + 4C_9 q_0}{1 - 2C_2 - 2C_3 q_0}.$$

After some algebraic manipulations, (14) and (21) can be written in the following forms

$$m \quad \frac{A}{B}\mu + (\frac{8B_2 q_4}{B} + \frac{4B_3 q_0 q_4}{B} + \frac{C}{B} + \frac{DA}{B^2} + \frac{2B_5 q_4 A}{B^2} + \frac{4B_{11} A^2}{B^3})\mu^3 + \cdots \tag{22}$$

$$q \quad q_0 + q_4 \mu^2 \tag{23}$$

where

$$A \quad 4B_1 + 8B_2 q_0 + 4B_3 q_0^2, B \quad 1 - (2B_4 + 2B_5 q_0), C \quad 4B_6 + 4B_7 q_0 + 4B_8 q_0^2$$

$$D \quad 2B_9 + 2B_{10} q_0 \ , \quad q_4 \quad q_1 + \left(\frac{A}{B}\right)^2 q_2 + \left(\frac{A}{B}\right) q_3$$

By substituting m and q in Eq. (13), with their expressions taken from Eqs. (22) and (23), we obtain the Eq. (16), where b is given by

$$b \quad [4(A_2 + A_3)q_4 + 2(2A_4 + 8A_5 + 2A_6)q_0 q_4 + 3(4A_7 + 4A_8)q_0^2 q_4 + 8A_9 q_0^3 q_4] +$$

$$\frac{A}{B}[(4A_{11} + 8A_{12})q_4 + 2(4A_{13} + 8A_{14})q_0 q_4 + 12A_{15} q_0^2 q_4] + \frac{A}{B}[4A_{16} + 4(A_{17} + 8A_{18})q_0 +$$

$$(4A_{19} + 8A_{20})q_0^2 + 4A_{21} q_0^3] + \frac{EA^2}{B^2} + \frac{FA^3}{B^3} + [\frac{8B_2 q_4 + 8B_3 q_0 q_4 + C}{B} + \frac{AD + 2B_5 q_4 A}{B^2} + \frac{4B_{11} A^2}{B^3}]$$

$$[4A_{10} + (4A_{11} + 8A_{12})q_0 + (4A_{13} + 8A_{14})q_0^2 + 4A_{15} q_0^3] \tag{24}$$

with

$$E \quad 2A_{22} + 8A_{23} + 2A_{24} + (4A_{25} + 8A_{26} + 4A_{27} + 8A_{28})q_0 + (2A_{29} + 8A_{30} + 2A_{31})q_0^2$$

And

$$F \quad 4A_{32} + 4A_{33} q_0$$

3. Results and discussion

In this section, let us examine numerically the phase diagrams of the ferrimagnetic mixed spin-1 and spin-1/2 Blume-Capel model with four-spin interaction J_4 and next-nearest neighbor coupling J', described by the Hamiltonian (1), using the formulation of the FCA given in the section 3. Indeed, we study the effect of the crystal field *D*, the four-spin interactions J_4 as well as the next nearest neighbor couplings on the phase diagram of the system under study.

Journal of Physics: Conference Series **758** (2016) 012004 doi:10.1088/1742-6596/758/1/012004

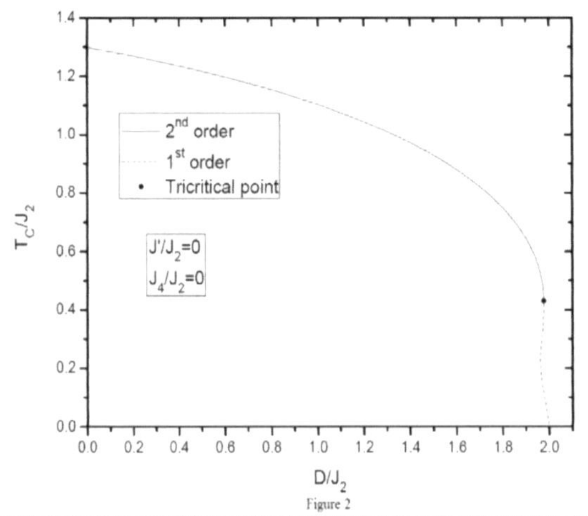

Figure 2. The phase diagram in $(D/J_2, T_C/J_2)$ plane for the mixed spin-1/2 and spin-1 Blume-Capel model.

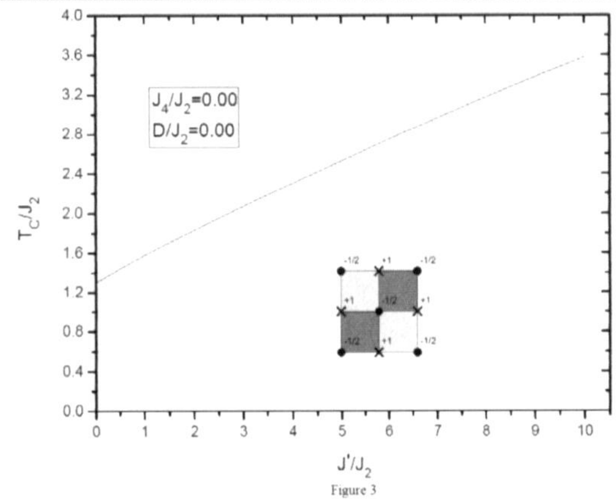

Figure 3. The phase diagram of the mixed spin-1/2 and spin-1 Ising model with next nearest neighbor interaction. No tricritical point is detected in this case.

First of all, let us take $J_4=J'=0$ and consider the system with only bilinear interaction and crystal field. The phase diagram in the $(T_c/J_2, D/J_2)$ plane depicted in Fig. 2 shows a tricritical point at $(D/J_2)_{Tri.}=1.976$ which can be compared with the mean-field treatment $(D/J_2)_{Tri.}=1.860$ [55]. In order to elucidate the role of the next nearest neighbor interaction on the system, we plot in Fig. 3 the phase diagram for $J_4/J_2=D/J_2=0$. From this latter (Fig. 3), on one hand we note that no tricritical point can be found which means that the system exhibits second order transition for any positive value of J'. On the other hand, a positive value of J' strengthens the ferrimagnetic order at low temperature. In Fig.4, we plot the phase diagram in the $(T_c/J_2, D/J_2)$ plane for selected values of the ratio J_4/J_2. From this figure, we point out that the T-component of the tricritical point decreases with decreasing values of the four-spin coupling. The insert plots in Fig.4 show that the latter behavior (tricriticality) disappears, and all transitions are of second-order kind. One also notes that for $J'\leq-0.5$, the phase diagrams show bulges suggesting the occurrence of the reentrant phenomenon in which competing pair, quartet and crystal field interactions may occur. In order to have an idea on the effects of the four-spin interaction on the mixed spin Ising model with next nearest neighbor interaction, we have plotted in Fig.5 sections of the critical surface $T_C(J', J_4)$ with a plan of fixed values of four-spin coupling. We note that negative values of J_4 decrease the critical temperature as J' decreases. As seen from this figure, the effects become dramatic when J_4 approaches -4 and the critical temperature vanishes exactly for $J_4/J_2=-4$ and $J'/J_2=0$.

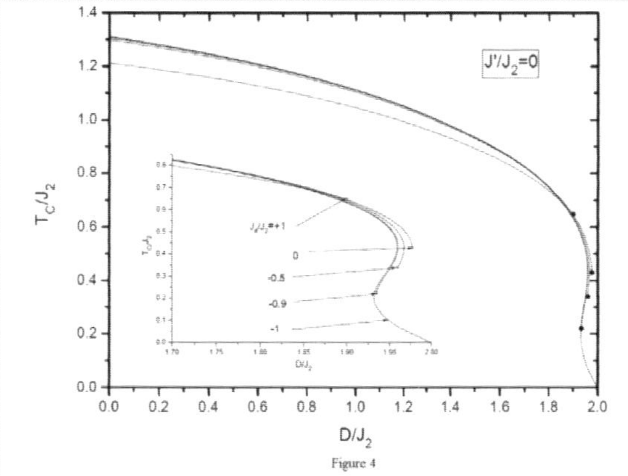

Figure 4. The phase diagram in *T-D* plane. Lines give critical temperature versus single ion anisotropy for various strength of J_4/J_2. The black point is the tricritical point.

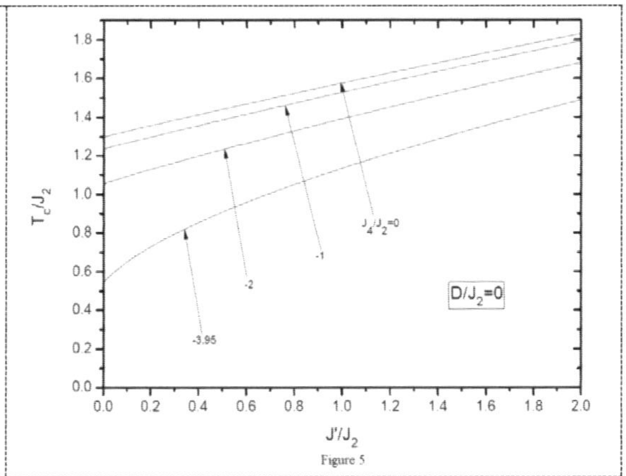

Figure 5. The phase diagram in *T-J'* plane when the four-spin interaction J_4/J_2 approaches the value -4.

4. Conclusion

In this work, we have investigated the phase diagram of the ferrimagnetic mixed spin Blume-Capel model with four-spin J_4 and next nearest neighbor J' interactions on the square lattice by the use of the finite cluster approximation within the framework of a single-site cluster theory. We have discussed the effect of J_4, J' and D, normalized with the interaction strength J_2, on the phase diagrams. We have shown that the tricritical behavior arises when the four-spin or the crystal field interaction is included. This latter feature disappears when J_4 and D take appropriate values and all transitions are of second-order. We have also shown that no tricritical behavior is detected when including next nearest neighbor interaction J' between the sites of the σ-sublattice and the critical temperature vanishes when $J_4/J_2=-4$ and $J'/J_2=0$.

Appendix

The coefficients appearing in Eq.(13) are given by:
For abbreviation we define new function:

$$f(x) \quad \frac{1}{2}\tanh\left(\frac{kx}{2}\right)$$

$$A_1 \quad f(\gamma)$$

$$A_2 \quad A_3 \quad -f(\gamma) + \frac{1}{2}\{f(1+\gamma) + f(-1+\gamma)\}$$

$$A_4 \quad \{f(\gamma) - f(1+\gamma) - f(-1+\gamma)\} - \frac{1}{2}\{f\left(\frac{\alpha}{2}\right) + f\left(\frac{\alpha}{2} - \gamma\right)\} + \frac{1}{4}\{f\left(-2+\frac{\alpha}{2}+\gamma\right) + f\left(-2+\frac{\alpha}{2}\right)$$
$$+f\left(2+\frac{\alpha}{2}+\gamma\right) + f\left(2+\frac{\alpha}{2}\right)\}$$

$$A_5 \quad \frac{3}{2}f(\alpha') - \{f(1+\gamma) + f(-1+\gamma)\} + \frac{1}{4}\{f(-2+\gamma) + f(2+\gamma)\}$$

$$A_6 \quad \frac{1}{2}\left\{f\left(\frac{\alpha}{2}\right) - f\left(\frac{\alpha}{2} - \gamma\right)\right\} + \{f(\gamma) - f(1+\gamma) - f(-1+\gamma)\} + \frac{1}{4}\{f\left(-2+\frac{\alpha}{2}+\gamma\right) - f\left(-2+\frac{\alpha}{2}\right)$$
$$+f\left(2+\frac{\alpha}{2}+\gamma\right) - f\left(2+\frac{\alpha}{2}\right)\}$$

A_7 $-2f(\alpha') + \frac{3}{2}\{f(-1+\gamma) + f(1-\gamma)\} + \frac{1}{2}\left\{f\left(\frac{\alpha}{2}\right) + f\left(\frac{\alpha}{2}-\gamma\right) - f(-2+\gamma) - f(2+\gamma)\right\}$
$-\frac{1}{4}\left\{f\left(2+\frac{\alpha}{2}\right) + f\left(-2+\frac{\alpha}{2}\right) + f\left(2+\frac{\alpha}{2}+\gamma\right) + f\left(-2+\frac{\alpha}{2}+\gamma\right) + f\left(-1+\frac{\alpha}{2}-\gamma\right) + f\left(1+\frac{\alpha}{2}-\gamma\right)\right\}$
$+\frac{1}{8}\{f\left(3+\frac{\alpha}{2}+\gamma\right) - f\left(-1+\frac{\alpha}{2}\right) - f\left(1+\frac{\alpha}{2}\right) + f\left(1+\frac{\alpha}{2}+\gamma\right) + f\left(-3+\frac{\alpha}{2}\right) + f\left(-3+\frac{\alpha}{2}+\gamma\right)$
$+f\left(3+\frac{\alpha}{2}\right) + f\left(-1+\frac{\alpha}{2}+\gamma\right)\}$

A_8 $-2f(\alpha') + \frac{3}{2}\{f(-1+\gamma) + f(1+\gamma)\} + \frac{1}{2}\left\{-f\left(\frac{\alpha}{2}\right) + f\left(\frac{\alpha}{2}-\gamma\right) - f(-2+\gamma) - f(2+\gamma)\right\}$
$+\frac{1}{4}\left\{f\left(-2+\frac{\alpha}{2}\right) + f\left(2+\frac{\alpha}{2}\right) - f\left(2+\frac{\alpha}{2}+\gamma\right) - f\left(-2+\frac{\alpha}{2}+\gamma\right) - f\left(-1+\frac{\alpha}{2}-\gamma\right) - f\left(1+\frac{\alpha}{2}-\gamma\right)\right\}$
$+\frac{1}{8}\{f\left(-1+\frac{\alpha}{2}\right) + f\left(1+\frac{\alpha}{2}\right) + f\left(3+\frac{\alpha}{2}+\gamma\right) + f\left(1+\frac{\alpha}{2}+\gamma\right) + f\left(-1+\frac{\alpha}{2}+\gamma\right) + f\left(-3+\frac{\alpha}{2}+\gamma\right)$
$-f\left(3+\frac{\alpha}{2}\right) - f\left(-3+\frac{\alpha}{2}\right)\}$

A_9 $3f(\alpha') - 2f(-1+\gamma) - 2f(1+\gamma) + \frac{1}{8}f(\alpha+\gamma) + \left\{f\left(-1+\frac{\alpha}{2}-\gamma\right) + f\left(1+\frac{\alpha}{2}-\gamma\right) - f\left(\frac{\alpha}{2}-\gamma\right)\right\}$
$+\frac{1}{16}\{f(-4+\alpha+\gamma) + f(4+\alpha+\gamma)\} + \frac{1}{4}\{5f(-2+\gamma) + 5f(2+\gamma) - f(\alpha-\gamma)\} + \frac{1}{2}\{f\left(2+\frac{\alpha}{2}+\gamma\right)$
$+f\left(-2+\frac{\alpha}{2}+\gamma\right) - f\left(3+\frac{\alpha}{2}+\gamma\right) - f\left(1+\frac{\alpha}{2}+\gamma\right) - f\left(-1+\frac{\alpha}{2}+\gamma\right) - f\left(-3+\frac{\alpha}{2}+\gamma\right)\}$

A_{10} $\frac{1}{2}f(-1) + \frac{1}{4}\{f(-1+\gamma) - f(1+\gamma)\}$

A_{11} $-\frac{1}{2}f(-1) + \frac{1}{4}\{f(1+\gamma) - f(-1+\gamma)\} + \frac{1}{8}\{f\left(-2+\frac{\alpha}{2}\right) - f\left(2+\frac{\alpha}{2}\right) + f\left(-2+\frac{\alpha}{2}+\gamma\right)$
$-f\left(2+\frac{\alpha}{2}+\gamma\right)\}$

A_{12} $\frac{1}{4}\{f(-2) - f(-1)\} + \frac{1}{8}\{f(-2+\gamma) + 3f(1+\gamma) - f(2+\gamma) - 3f(-1+\gamma)\}$

A_{13} $\frac{1}{2}\{f(-1+\gamma) - f(1+\gamma) - f(-2)\} + \frac{1}{4}\{f(2+\gamma) - f(-2+\gamma)\} + \frac{1}{8}\{f\left(-1+\frac{\alpha}{2}-\gamma\right)$
$-f\left(1+\frac{\alpha}{2}-\gamma\right)\} + \frac{1}{16}\{f\left(-1+\frac{\alpha}{2}\right) - f\left(1+\frac{\alpha}{2}\right) - f\left(3+\frac{\alpha}{2}+\gamma\right) + f\left(1+\frac{\alpha}{2}+\gamma\right) - f\left(-1+\frac{\alpha}{2}+\gamma\right)$
$+f\left(-3+\frac{\alpha}{2}+\gamma\right) + f\left(-3+\frac{\alpha}{2}\right) - f\left(3+\frac{\alpha}{2}\right)\}$

A_{14} $\frac{1}{4}\{f(-1) - f(-2)\} + \frac{1}{8}\{3f(-1+\gamma) - 3f(1+\gamma) + f\left(2+\frac{\alpha}{2}\right) - f\left(-2+\frac{\alpha}{2}\right) + f\left(2+\frac{\alpha}{2}+\gamma\right)$
$-f\left(-2+\frac{\alpha}{2}+\gamma\right) - f(-2+\gamma) + f(2+\gamma)\} + \frac{1}{16}\{f\left(-1+\frac{\alpha}{2}\right) - f\left(1+\frac{\alpha}{2}\right) - f\left(3+\frac{\alpha}{2}+\gamma\right)$
$-f\left(1+\frac{\alpha}{2}+\gamma\right) + f\left(-1+\frac{\alpha}{2}+\gamma\right) + f\left(-3+\frac{\alpha}{2}+\gamma\right) - f\left(3+\frac{\alpha}{2}\right) + f\left(-3+\frac{\alpha}{2}\right)\}$

A_{15} $\frac{1}{32}\{f(-4+\alpha+\gamma) - f(4+\alpha+\gamma)\} + \frac{1}{2}\{f(-2) + f(1+\gamma) - f(-1+\gamma)\} + \frac{1}{16}\{f(-2+\alpha)$
$+f(-4) - f(2+\alpha) + f\left(1+\frac{\alpha}{2}+\gamma\right) - f\left(-1+\frac{\alpha}{2}+\gamma\right) - 3f\left(-1+\frac{\alpha}{2}\right) + 3f\left(1+\frac{\alpha}{2}\right)$
$-3f\left(-3+\frac{\alpha}{2}+\gamma\right) + 3f\left(3+\frac{\alpha}{2}\right) - 3f\left(-3+\frac{\alpha}{2}\right) + 3f\left(3+\frac{\alpha}{2}+\gamma\right) + 5f(-2+\gamma) - 5f(2+\gamma)\}$
$+\frac{1}{8}\{f\left(1+\frac{\alpha}{2}-\gamma\right) - f\left(-1+\frac{\alpha}{2}-\gamma\right) - f\left(2+\frac{\alpha}{2}\right) + f\left(-2+\frac{\alpha}{2}\right) - f\left(2+\frac{\alpha}{2}+\gamma\right) + f\left(-2+\frac{\alpha}{2}+\gamma\right)\}$

A_{16} $\{f(-1+\gamma) - f(1+\gamma)\} - 2f(-1)$

A_{17} $2f(-1) + \{f(1+\gamma) - f(-1+\gamma)\} + \frac{1}{2}\{f\left(-2+\frac{\alpha}{2}+\gamma\right) - f\left(2+\frac{\alpha}{2}+\gamma\right) - f\left(-2+\frac{\alpha}{2}\right)$
$+f\left(2+\frac{\alpha}{2}\right)\}$

A_{18} $\{f(-1)-f(-2)\}+\frac{1}{2}\{f(1+\gamma)-f(2+\gamma)-f(-1+\gamma)+f(-2+\gamma)\}$

A_{19} $2f(-2)+\{f(2+\gamma)-f(-2+\gamma)\}+\frac{1}{2}\{f\left(-1+\frac{\alpha}{2}-\gamma\right)-f\left(1+\frac{\alpha}{2}-\gamma\right)\}+\frac{1}{4}\{f\left(1+\frac{\alpha}{2}\right)$
$-f\left(-1+\frac{\alpha}{2}\right)+f\left(-3+\frac{\alpha}{2}+\gamma\right)-f\left(3+\frac{\alpha}{2}+\gamma\right)+f\left(1+\frac{\alpha}{2}+\gamma\right)-f\left(-1+\frac{\alpha}{2}+\gamma\right)+f\left(3+\frac{\alpha}{2}\right)$
$-f\left(-3+\frac{\alpha}{2}\right)\}$

A_{20} $\{f(-2)-f(-1)\}+\frac{1}{2}\{f(-1+\gamma)-f(1+\gamma)-f(-2+\gamma)+f\left(-2+\frac{\alpha}{2}\right)-f\left(2+\frac{\alpha}{2}\right)$
$+f\left(2+\frac{\alpha}{2}+\gamma\right)-f\left(-2+\frac{\alpha}{2}+\gamma\right)+f(2+\gamma)\}+\frac{1}{4}\{f\left(1+\frac{\alpha}{2}\right)-f\left(-1+\frac{\alpha}{2}\right)-f\left(3+\frac{\alpha}{2}+\gamma\right)$
$-f\left(1+\frac{\alpha}{2}+\gamma\right)+f\left(-1+\frac{\alpha}{2}+\gamma\right)+f\left(-3+\frac{\alpha}{2}+\gamma\right)+f\left(3+\frac{\alpha}{2}\right)-f\left(-3+\frac{\alpha}{2}\right)\}$

A_{21} $-2f(-2)+\frac{1}{2}\left\{f\left(-2+\frac{\alpha}{2}+\gamma\right)-f\left(2+\frac{\alpha}{2}+\gamma\right)-f\left(-1+\frac{\alpha}{2}-\gamma\right)+f\left(1+\frac{\alpha}{2}-\gamma\right)+f\left(2+\frac{\alpha}{2}\right)\right\}$
$+\frac{1}{8}\{f(-4+\alpha+\gamma)-f(4+\alpha+\gamma)\}+\frac{1}{4}\{f(2+\alpha)-f(-2+\alpha)-f(-4)+f\left(1+\frac{\alpha}{2}+\gamma\right)$
$-f\left(-1+\frac{\alpha}{2}+\gamma\right)+3f\left(-1+\frac{\alpha}{2}\right)-3f\left(1+\frac{\alpha}{2}\right)+3f\left(3+\frac{\alpha}{2}+\gamma\right)-3f\left(-3+\frac{\alpha}{2}+\gamma\right)-3f\left(3+\frac{\alpha}{2}\right)$
$+3f\left(-3+\frac{\alpha}{2}\right)+5f(-2+\gamma)-5f(2+\gamma)\}$

A_{22} $\frac{1}{4}\{f\left(-2+\frac{\alpha}{2}\right)+f\left(2+\frac{\alpha}{2}+\gamma\right)+f\left(2+\frac{\alpha}{2}\right)+f\left(-2+\frac{\alpha}{2}+\gamma\right)\}+\frac{1}{4}\{f\left(\frac{\alpha}{2}\right)+f\left(\frac{\alpha}{2}-\gamma\right)\}$

A_{23} $-\frac{1}{2}f(\gamma)+\frac{1}{4}\{f(-2+\gamma)+f(2+\gamma)\}$

A_{24} $\frac{1}{2}\{f\left(\frac{\alpha}{2}-\gamma\right)-f\left(\frac{\alpha}{2}\right)\}+\frac{1}{4}\{f\left(-2+\frac{\alpha}{2}+\gamma\right)-f\left(-2+\frac{\alpha}{2}\right)+f\left(2+\frac{\alpha}{2}+\gamma\right)-f\left(2+\frac{\alpha}{2}\right)\}$

A_{25} $-\frac{1}{2}\{f\left(\frac{\alpha}{2}\right)+f\left(\frac{\alpha}{2}-\gamma\right)\}+\frac{1}{4}\{f\left(-1+\frac{\alpha}{2}-\gamma\right)-f\left(-2+\frac{\alpha}{2}\right)+f\left(1+\frac{\alpha}{2}-\gamma\right)-f\left(2+\frac{\alpha}{2}+\gamma\right)$
$-f\left(-2+\frac{\alpha}{2}+\gamma\right)-f\left(2+\frac{\alpha}{2}\right)\}+\frac{1}{8}\{3f\left(-1+\frac{\alpha}{2}\right)+3f\left(1+\frac{\alpha}{2}\right)+f\left(3+\frac{\alpha}{2}+\gamma\right)+f\left(1+\frac{\alpha}{2}+\gamma\right)$
$+f\left(-1+\frac{\alpha}{2}+\gamma\right)+f\left(-3+\frac{\alpha}{2}+\gamma\right)+f\left(3+\frac{\alpha}{2}\right)+f\left(-3+\frac{\alpha}{2}\right)\}$

A_{26} $\frac{1}{2}f(\alpha')-\frac{1}{4}\{f(-2+\gamma)+f(2+\gamma)\}+\frac{1}{8}\{f\left(-1+\frac{\alpha}{2}\right)+f\left(-3+\frac{\alpha}{2}+\gamma\right)+f\left(3+\frac{\alpha}{2}+\gamma\right)$
$-f\left(1+\frac{\alpha}{2}+\gamma\right)-f\left(-1+\frac{\alpha}{2}+\gamma\right)-f\left(3+\frac{\alpha}{2}\right)-f\left(-3+\frac{\alpha}{2}\right)+f\left(1+\frac{\alpha}{2}\right)\}$

A_{27} $\frac{1}{2}\{f\left(\frac{\alpha}{2}\right)-f\left(\frac{\alpha}{2}-\gamma\right)\}+\frac{1}{4}\{f\left(2+\frac{\alpha}{2}\right)+f\left(-2+\frac{\alpha}{2}\right)-f\left(2+\frac{\alpha}{2}+\gamma\right)-f\left(-2+\frac{\alpha}{2}+\gamma\right)$
$+f\left(-1+\frac{\alpha}{2}-\gamma\right)+f\left(1+\frac{\alpha}{2}-\gamma\right)\}+\frac{1}{8}\{f\left(3+\frac{\alpha}{2}+\gamma\right)-f\left(-3+\frac{\alpha}{2}\right)+f\left(1+\frac{\alpha}{2}+\gamma\right)+f\left(-1+\frac{\alpha}{2}+\gamma\right)$
$+f\left(-3+\frac{\alpha}{2}+\gamma\right)-f\left(3+\frac{\alpha}{2}\right)-3f\left(-1+\frac{\alpha}{2}\right)-3f\left(1+\frac{\alpha}{2}\right)\}$

A_{28} $\frac{1}{2}f(\gamma)-\frac{1}{4}\{f(-2+\gamma)+f(2+\gamma)\}+\frac{1}{8}\{-f\left(-1+\frac{\alpha}{2}\right)+f\left(-3+\frac{\alpha}{2}+\gamma\right)+f\left(3+\frac{\alpha}{2}+\gamma\right)$
$-f\left(1+\frac{\alpha}{2}+\gamma\right)-f\left(-1+\frac{\alpha}{2}+\gamma\right)+f\left(3+\frac{\alpha}{2}\right)+f\left(-3+\frac{\alpha}{2}\right)-f\left(1+\frac{\alpha}{2}\right)\}$

A_{29} $\frac{1}{8}f(\alpha+\gamma)+\frac{1}{16}\{f(-4+\alpha+\gamma)+f(4+\alpha+\gamma)\}+\frac{1}{2}\{f\left(\frac{\alpha}{2}-\gamma\right)-f\left(-1+\frac{\alpha}{2}-\gamma\right)-f\left(1+\frac{\alpha}{2}-\gamma\right)$
$+f\left(\frac{\alpha}{2}\right)\}+\frac{1}{4}\{f(-2+\alpha)+f(\alpha-\gamma)+f(2+\alpha)-f\left(3+\frac{\alpha}{2}+\gamma\right)-f\left(1+\frac{\alpha}{2}+\gamma\right)-f\left(-1+\frac{\alpha}{2}+\gamma\right)$

$-f\left(-3+\frac{\alpha}{2}+\gamma\right)-f\left(3+\frac{\alpha}{2}\right)-f\left(-3+\frac{\alpha}{2}\right)+f\left(-2+\frac{\alpha}{2}\right)+f\left(2+\frac{\alpha}{2}\right)+f\left(2+\frac{\alpha}{2}+\gamma\right)+$
$f\left(2+\frac{\alpha}{2}+\gamma\right)\ -3f\left(-1+\frac{\alpha}{2}\right)-3f\left(1+\frac{\alpha}{2}\right)\}$

A_{30} $-\frac{1}{2}f(\gamma)-\frac{1}{8}f(\alpha+\gamma)+\frac{1}{16}\{f(4+\alpha+\gamma)+f(-4+\alpha+\gamma)\}+\frac{1}{4}\{f\left(1+\frac{\alpha}{2}+\gamma\right)$
$-f\left(3+\frac{\alpha}{2}+\gamma\right)+f\left(-1+\frac{\alpha}{2}+\gamma\right)-f\left(-3+\frac{\alpha}{2}+\gamma\right)+f(-2+\gamma)+f(2+\gamma)\}$

A_{31} $\frac{1}{8}f(-\alpha+\gamma)+\frac{1}{16}\{f(-4+\alpha+\gamma)+f(4+\alpha+\gamma)\}+\frac{1}{2}\{f\left(\frac{\alpha}{2}-\gamma\right)-f\left(-1+\frac{\alpha}{2}-\gamma\right)$
$-f\left(1+\frac{\alpha}{2}-\gamma\right)-f\left(\frac{\alpha}{2}\right)\}+\frac{1}{4}\{f\left(-3+\frac{\alpha}{2}\right)-f(-2+\alpha)+f(\alpha-\gamma)-f(2+\alpha)-f\left(2+\frac{\alpha}{2}\right)$
$-f\left(-2+\frac{\alpha}{2}\right)+f\left(3+\frac{\alpha}{2}\right)+f\left(2+\frac{\alpha}{2}+\gamma\right)+f\left(-2+\frac{\alpha}{2}+\gamma\right)-f\left(3+\frac{\alpha}{2}+\gamma\right)-f\left(1+\frac{\alpha}{2}+\gamma\right)$
$-f\left(-1+\frac{\alpha}{2}+\gamma\right)-f\left(-3+\frac{\alpha}{2}+\gamma\right)+3f\left(-1+\frac{\alpha}{2}\right)+3f\left(1+\frac{\alpha}{2}\right)\}$

A_{32} $\frac{1}{16}\{3f\left(1+\frac{\alpha}{2}\right)-3f\left(-1+\frac{\alpha}{2}\right)+f\left(-3+\frac{\alpha}{2}+\gamma\right)-f\left(3+\frac{\alpha}{2}+\gamma\right)+f\left(1+\frac{\alpha}{2}+\gamma\right)$
$-f\left(-1+\frac{\alpha}{2}+\gamma\right)-f\left(3+\frac{\alpha}{2}\right)+f\left(-3+\frac{\alpha}{2}\right)\}+\frac{1}{8}\{f\left(1+\frac{\alpha}{2}-\gamma\right)-f\left(-1+\frac{\alpha}{2}-\gamma\right)\}$

A_{33} $\frac{1}{32}\{f(-4+\alpha+\gamma)-f(4+\alpha+\gamma)\}+\frac{1}{16}\{f(-4)-f(-2+\alpha)+f(2+\alpha)+f\left(3+\frac{\alpha}{2}+\gamma\right)$
$-f(-2+\gamma)-f(-2-\gamma)-f\left(-3+\frac{\alpha}{2}+\gamma\right)+f\left(3+\frac{\alpha}{2}\right)-f\left(-3+\frac{\alpha}{2}\right)+3f\left(-1+\frac{\alpha}{2}\right)-f\left(1+\frac{\alpha}{2}+\gamma\right)$
$-3f\left(1+\frac{\alpha}{2}\right)+f\left(-1+\frac{\alpha}{2}+\gamma\right)\}+\frac{1}{8}\{f\left(-1+\frac{\alpha}{2}-\gamma\right)-f\left(1+\frac{\alpha}{2}-\gamma\right)\}$

A_{34} $\frac{1}{8}f(\alpha+\gamma)+\frac{1}{16}\{f(-4+\alpha+\gamma)+f(4+\alpha+\gamma)\}+\frac{1}{4}\{f(-2-\gamma)-f(-2+\gamma)-f(\alpha+\gamma)\}$

A_{35} $\frac{1}{2}\left\{f\left(1+\frac{\alpha}{2}-\gamma\right)-f\left(-1+\frac{\alpha}{2}-\gamma\right)\right\}+\frac{1}{4}\{3f\left(-1+\frac{\alpha}{2}\right)-3f\left(1+\frac{\alpha}{2}\right)-f\left(3+\frac{\alpha}{2}+\gamma\right)$
$+f\left(-3+\frac{\alpha}{2}+\gamma\right)+f\left(1+\frac{\alpha}{2}+\gamma\right)-f\left(-1+\frac{\alpha}{2}+\gamma\right)+f\left(3+\frac{\alpha}{2}\right)-f\left(-3+\frac{\alpha}{2}\right)\}$

A_{36} $\frac{1}{8}\{f(-4+\alpha+\gamma)-f(4+\alpha+\gamma)\}+\frac{1}{4}\{f(-2+\alpha)-f(2+\alpha)-f(-4)+f\left(3+\frac{\alpha}{2}+\gamma\right)$
$-f\left(-3+\frac{\alpha}{2}+\gamma\right)-f\left(1+\frac{\alpha}{2}+\gamma\right)+f\left(-1+\frac{\alpha}{2}+\gamma\right)-f\left(3+\frac{\alpha}{2}\right)+f\left(-3+\frac{\alpha}{2}\right)-f(-2+\gamma)$
$-f(-2-\gamma)-3f\left(-1+\frac{\alpha}{2}\right)+3f\left(1+\frac{\alpha}{2}\right)\}+\frac{1}{2}\left\{f\left(-1+\frac{\alpha}{2}-\gamma\right)-f\left(1+\frac{\alpha}{2}-\gamma\right)\right\}$

References

[1] Magnetic Molecular Materials, NATO ASI Series, edited by D.Gatteshi, O.Kahn, J.S. Miller, and Palacio (Kluwe Academic, Dordrecht, 1991).

[2] A. T. Apostolov, I. N. Apostolova, and J. M. Wesselinowa, Eur. Phys. J. B **86** (2013) 483.

[3] M. G. Whitesides, L. H. Craig, J-C. Brunie, Ind. Eng. Chem. Process Des. Dev., **15** (1976) 226–227.

[4] Mark M. Turnbull , Toyonari Sugimoto , Lawrence K. Thompson Molecule-Based Magnetic Materials: Theory, Techniques, and Applications (ACS Symposium Series) Hardcover – May 5, 1996.

[5] R.H Kodama J. Magn. Magn. Mater. **200** (1999) 359.

[6] T. Mallah, S. Tiebaut, M. Verdager, P. Veillet, Science **262** (1993) 1554;

H. Iwamura, J. S. Miller, Proceedings of the conference on Ferrimagnetic and High Spin Molecular-based Materials, Mol. Cryst. Liq. Crys. **233** (1993) ;

H. Ohawa, N. Matsumoto, M. Obba, Mol. Cryst. Liq. Crys. **233** (1993) 257;

M. Trunbull, C. P. landee, T. C. Soesbe, R. D. willet, Mol. Cryst. Liq. Crys. **233** (1993) 269.

[7] A. Caneschi, D. Gatteschi, N. Lalioti, C. Sangregorio, R. Sessoli, G. Venturi, A. Vindigni, A. Rettori, M-G Pini, M-A. Novak, Angew. Chem. Int. Ed. **40** (2001) 1760.

[8] E. Pardo, R. Ruiz-Garcia, F. Lloret, J. Faus, M. Julve, Y. Journaux, F. Delgado, C. Ruiz-Perez, Adv. Mater. **16** (2004) 1597.

[9] T. Kajiwara, M. Nakano, Y. Kaneko, S. Takaishi, T. Ito, M. Yamashita, A. Igashira-Kamiyama, H. Nojiri, Y. Ono, N. Kojima, J. Am. Chem. Soc. **127** (2005) 10150.

[10] O. Kahn & C. J. Martinez, Science **279** (1998) 44 ;
A. Bousseksou, G. Molna´r, , P. Demont, J.Menegotto, J. Mater. Chem. **13** (2003) 2069.

[11] M-N. Leuenberger, D. Loss, Nature **410** (2001) 789.

[12] Single-Molecule Magnets and Related Phenomena Editors: R. Winpenny, Springer (2006).

[13] S. Decurtins, H. W. Schmalle, H. R. Oswald, A. Linden, J. Ensling, P. Gütlich, A. Hauser, Inorg. Chi. Acta **65** (1994) 216.

[14] T. Mallah, S. Thiébaut, M. Verdaguer, P. Veillet, Science **262** (1993) 1554.

[15] G. DU, J. Joo, A. J. Epstein, J. Appl. Phys. **73** (1993) 6566.

[16] T. Kaneyoshi, J.C. Chen, , J. Magn. Magn. Mater. **98** (1991) 201.

[17] J. Oitmaa, I. G. Enting, J. Phys.: Condens. Matter **18** (2006) 10931.

[18] G M Buendía, M A Novotny, J. Phys.: Condens. Matter **9** (1997) 5951.

[19] B. Beatriz, R.A. Filgueiras, C. Claudette, N.S. Branco, Physica A **304** 2002 429.

[20] W. Selke, C. Ekiz, J. Phys.: Condens. Matter **23** (2011) 496002.

[21] L.J. de Jongh, A.R. Miedema, Experiments on Simple Magnetic Models Systems, Taylor and Francis, London, 1974.

[22] B. Widom, J. Chem. Phys. 84 (1986) 6943; M. Schick, W. Shih, Phys. Rev. 34 (1986) 1797.

[23] T. Iwashita, N. Uryu, J. phys. C **17** (1984) 855–868;

P. Gluck, O. Wohlman-Entin, Phys. Status Solidi B **52** (1972) 323–333.

[24] F.Y. Wu, Phys. Rev. B **4** (1971) 2312–2314.

[25] U. Köbler, R. Mueller, L. Smardz, D. Maier, K. Fischer, B. Olefs, W. Zinn, Z. Phys. B **100** (1996) 497–506;

E. Müller-Hartmann, U. Köbler, L. Smardz, J. Magn. Magn. Mater. **173** (1997) 133–140.

[26] C.L. Wang, Z.K. Qin, D.L. Lin, J. Magn. Magn. Mater. **88** (1990) 87–92 ;

K.G. Chakraborty, J. Magn. Magn. Mater. **114** (1992) 155–160.

[27] T. Kaneyoshi, T. Aoyama, J. Magn. Magn. Mater. **96** (1991) 67–76 ;

B. Laaboudi, M. Kerouad, Physica A **241** (1997) 729–736.

[28] H.P. Griffiths, D.W. Wood, J. Phys. C **7** (1974) L54–L58;

H.P. Griffiths, D.W. Wood, J. Phys. C **7** (1974) 4021–4036.

J. Oitmaa, R.W. Gibberd, J. Phys. C **6** (1973) 2077–2088.

[29] F. Lee, H.H. Chen, F.Y. Wu, Phys. Rev. B **40** (1989) 4871–4876;

M. Gitterman, Mikulinsky, J. Phys. C **10** (1977) 4073–4078;

M.P. Nightingale, Phys. Lett. A **59** (1977) 468.

[30] T. Iwashita, K. Uragami, A. Shimizu, A. Nagaki, T. Kasama, T. Idogaki, J. Magn. Magn. Mater. **310** (2007) e435–e437;

T. Iwashita, K. Uragami, K. Goto, M. Arao, T. Kasama, T. Idogaki, J. Magn. Magn. Mater. **272-276** (2004) 672–673;

Guo-Ming Zhang, Chuan-Zhang Yang, Phys. Status Solidi B **175** (1993) 459–463.

[31] S. Lacková, T. Horiguchi, Physica A **319** (2003) 311–318;

F.Y. Wu, Phys. Lett. A **38** (1972) 77–78;

F. Wang, M. Suzuki, Physica A **230** (1996) 639–650;

A. Lipowski, Physica A **248** (1998) 207–212 ;

Lucia. Gàlisová, Phys. Status Solidi B **250** (2013) 187.

[32] M. Grimsditch, P. Loubeyre, A. Polian, Phys. Rev. B **33** (1986) 7192–7200.

[33] M. Roger, J.H. Hetherington, J.M. Delrieu, Rev. Mod. Phys. **55** (1983) 1–64.

[34] H.L. Scott, Phys. Rev. A **37** (1988) 263–268.

[35] J.A. Barker, Phys. Rev. Lett. **57** (1986) 230–233.

[36] C.L. Wang, Z.K. Qin, D.L. Lin, Solid State Commun. **71** (1989) 45–48;

C.L. Wang, Z.K. Qin, D.L. Lin, Phys. Rev. B **40** (1980) 680–685.

[37] W. Chunlei, Q. Zikai, Z. Jingbo, Ferroelectrics **77** (1988) 21–29.

[38] P.R. Silva, B.V. Costa, R.L. Moreira, Polymer **34** (1993) 3107–3108.

[39] T.S. Nunner, P. Brune, T. Kopp, M. Windt, M. Grueninger, Acta. Phys. Pol. B **34** (2003) 1545–1548.

[40] S. Brehmer, H.-J. Mikeska, M. Müller, N. Nagaosa, S. Uchida, Phys. Rev. B **60** (1999) 329;

M. Matsuda, K. Katsumata, R.S. Eccleston, S. Brehmer, H.J. Mikeska, Phys. Rev. B **60** (2000) 8903.

[41] S. Notbohm, P. Ribeiro, B. Lake, D.A. Tennant, K.P. Schmidt, G.S. Uhrig, C. Hess, R. Klingeler, G. Behr, B. Büchner, M. Reehuis, R.I. Bewley, C.D. Frost, P. Manuel, R.S. Eccleston, Phys. Rev. Lett. **98** (2007) 027403.

[42] R. Coldea, S.M. Hayden, G. Aeppli, T.G. Perring, C.D. Frost, T.E. Mason, S.W. Cheong, Z. Fisk, Phys. Rev. Lett. **86** (2001) 5377–5380.

[43] L. Néel, Ann. Phys: (Paris) **3** (1948) 137.

[44] M. Drillon, E. Coronado, D. Beltran, R. Georges, J. Chem. Phys. **79** (1983) 449.

[45] M. Azhari, N. Benayad, M. Mouhib, Superlattices Microstruct. **79** (2015) 96–107.

[46] M. Mouhib, M. Azhari, N. Benayad, Submitted to M.J. CONDENSED MATTER.

[47] N. Boccara, Phys. Lett. A. **94** (1983) 185–187.

A. Benyoussef, N. Boccara, J. Phys. **44** (1983) 1143–1147.

[48] N. Benayad, A. Klümper, J. Zittartz, A. Benyoussef, Z. Phys. B:. **77** (1989) 339–341.

N. Benayad, A. Dakhama, A. Klümper, J. Zittartz, Ann. Phys. **508** (1996) 387–398.

N. Benayad, L. Khaya, A. Fathi, J. Phys: Condens. Matter. **14** (2002) 9667–9685.

N. Benayad, A. Fathi, L. Khaya, Physica A. **300** (2001) 225–244.

[49] M. Ghliyem, N. Benayad, M. Azhari, Physica A. **402** (2014) 14–29.

N. Benayad, A. Fathi, L. Khaya, J. Magn. Magn. Mater. **278** (2004) 407–429.

N. Benayad, A. Fathi, R. Zerhouni, J. Magn. Magn. Mater. **222** (2000) 355–367.

N. Benayad, R. Zerhouni, A. Klümper, Eur. Phys. J. B. **5** (1998) 687–695.

A. Benyoussef, H. Ez-Zahraouy, J. Phys: Condens. Matter. **6** (1994) 3411–3420.

[50] T. Lahcini, N. Benayad, Phase Transitions **82** (2009) 197–210.

L. Khaya, N. Benayad, A. Dakhama, Phys. Status Solidi. **241** (2004) 1078–1087.

N. Benayad, A. Dakhama, J. Magn. Magn. Mater. **168** (1997) 105–120.

N. Benayad, A. Dakhama, Phys. Rev. B. **55** (1997) 12276–12289.

A. Benyoussef, N. Boccara, M. Saber, J. Phys. C:. **18** (1985) 4275–4289.

[51] N. Benayad, M. Ghliyem, Physica B. **407** (2012) 6–13.

[52] T.Kaneyoshi and J. C. Chen, J. Magn. Magn. Mater. **98** (1991) 201.

[53] S. Ohkoshi, Y. Abe, A. Fujishima, K. Hashimoto, Phys. Rev. Lett. **82** (1999) 1285.

Magnetic properties of mixed spin (1/2,1) Ising nanoparticles

M Mouhib[1], N Benayad[1] and M Azhari[1]

[1]Laboratory of High Energy Physics and Scientific Computing,
HASSAN II University–Casablanca, Faculty of Sciences-Aïn Chock,
B.P : 5366 Maarif , Casablanca 20100, Morocco.

E-mail: majid.mouhib83@gmail.com ; noureddine_benayad@yahoo.fr

Abstract. Mixed spin 2D-nanoparticles (circle and square) with core – shell structure consisting of spin 1/2 and spin 1 distributed in concentric and two alternate rings are investigated by the use of the finite cluster approximation (FCA). In this approach, the state equations are derived. The effects of the exchange interactions on the phase diagrams are systematically discussed. A number of interesting phenomena have been found, in particular, the transition temperature is not too sensitive of shell exchange interaction where the spins S are strongly correlated.

1. Introduction

During the last few years, the research on magnetic nanomaterials such as nanoparticles, nanorods, nanobelts, nanowires and nanotubes, have been attracting considerable attention, and considered the most actively studied topics in statistical mechanics and condensed matter physics. The reason is due to their promising applications in nanotechnology [1,2] such as permanent magnets [3], information storage devices [4,5], biosensors [6] and some medical applications [7]. At present, the scientist can produce such kinds of fine nanoscaled materials [8,9], and the magnetization of certain nanomaterials such as γ-Fe_2O_3 nanoparticles has been experimentally measured [10]. In particular, magnetic nanowires and nanotubes such as ZnO [11], FePt, and Fe_3O_4 [12] can be synthesized by various experimental techniques and they have many applications in nanotechnology [13,14]. They are also utilized as raw materials in fabrication of ultra-high density magnetic recording media [15-17]. On the other hand, the magnetic properties of a nanometric scale material depend highly on the size and dimensionality of the nanomaterial.

From the theoretical point of view, Ising nanomagnetic systems have been studied by a variety of techniques, including mean field theory [18-20], effective field theory with correlations [21-24], Monte Carlo simulation [25-28], variational cumulant expansion method [29,30], and Green's function techniques [31].

Recently, the study of two sublattice mixed spin systems have become one of the actively studied subject in the statistical physics. They are of interest for the following main reasons. The first one is that they have less symmetry than their single-spin counterparts. The second one is that they are well adapted to study a certain type of ferrimagnetism [32], and finally many new phenomena are observed in these systems, which cannot be seen in the single-spin Ising model [33].

The aim of this paper is to investigate the mixed spin Ising nanoparticles. These latters, consisting of spin 1/2 and spin 1, have two different shapes (circle and square). In particular, we study the effect of the core and shell exchange interactions on the phase diagrams and the magnetizations. To this end, we use the finite cluster approximation (FCA) [34,35] within the framework of a single-site cluster theory.

The outline of this work is as follows: in section 2, we present the model and the theoretical framework. The state equations are derived. The main results and discussion are arranged in section 3. Finally, the last section is devoted to a brief conclusion.

2. Theoretical framework

We consider two kinds of 2D-nanoparticles (circle and square), as depicted in figure 1. They are consisted of the surface shell and the core. The surface shell surrounds the core. Each site is occupied by an Ising spin where the σ and S spins are distributed in concentric and alternate rings.

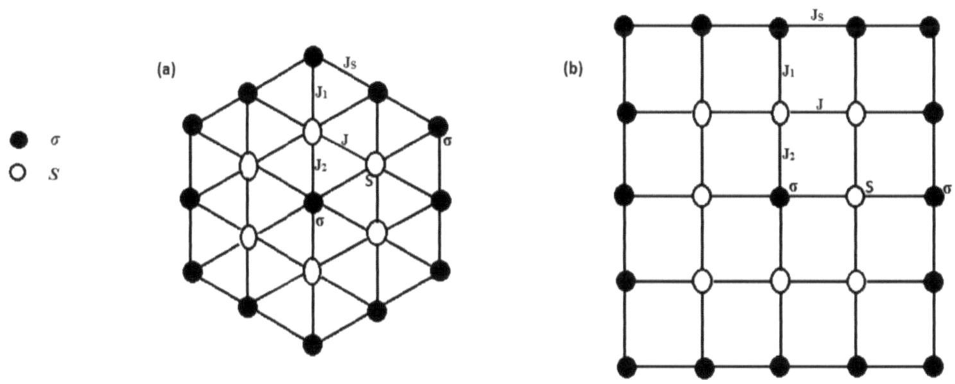

Figure 1. Schematic representations of two 2D-nanoparticles. They represent the circle nanoparticle (a) and the square nanoparticle (b). Atoms are denoted by σ_i (spin 1/2) and S_i (spin 1).

The Hamiltonian of the system is given by

$$H = -J_{S}.\sum_{\langle i,j\rangle} \sigma_i\sigma_j - J.\sum_{\langle m,n\rangle} S_m S_n - J_1.\sum_{\langle i,m\rangle} \sigma_i S_m - J_2.\sum_{\langle k,m\rangle} \sigma_k S_m \tag{1}$$

S_i and σ_j are the Pauli spin operators with $S=\pm 1,0$ and $\sigma=\pm 1/2$. J_s is the exchange interaction between two nearest neighbor magnetic atoms at the surface shell. This latter is coupled to the next shell in the core with the exchange interaction J_1. J is the exchange interaction between spins S_i in the core. J_2 is the exchange interaction between the central site σ_i and the sites S_i belonging to the first ring.

The method that we adopt in the study of the mixed spin Ising of nanoparticles (circle and square) consisting of spin 1/2 and spin 1 described by the Hamiltonian (1) is the finite cluster approximation (FCA) [34,35] based on a single-site cluster theory. We have to mention that this method has been successfully applied to a number of interesting pure and disordered spin Ising systems [36,39]. It has also been used for transverse Ising models [40-43] and semi-infinite Ising systems [44-48].

Within the core-shell concept, for the nanoparticle with a circle shape depicted in figure 1 (a), we have to determine the magnetizations at each site in the core and at the surface shell. Indeed, we denote by $\mu_3=\langle\sigma_3\rangle$, $m=\langle S\rangle$ the magnetizations per site in the core corresponding to the central site σ and the site belonging to the ring consisting by spin S, respectively. At the surface shell of the system, we have to define two magnetizations $\mu_1=<\sigma_1>$ and $\mu_2=<\sigma_2>$ which correspond to the coordination numbers z=3 and z=4, respectively.

We split the total Hamiltonian (1) into two parts, $H=H_0+H'$ where H_0 includes all parts of H associated with the lattice site 0. In the present system, H_0 takes the form

$$H_{0\sigma_1} = -[J_S.(\sigma_{2,1} + \sigma_{2,2}) + J.S_1].\sigma_{1,0} \tag{2}$$

$$H_{0\sigma_2} = -[J_S.(\sigma_{1,1} + \sigma_{1,2}) + J_1.(S_1 + S_2)].\sigma_{2,0} \tag{3}$$

$$H_{0\sigma_3} = -[J_2.\sum_{i=1}^{6} S_i].\sigma_{3,0} \tag{4}$$

$$H_{0S} = -[J.(S_1 + S_2) + J_1.(\sigma_{1,1}+\sigma_{2,1} + \sigma_{2,2}) + J_2.\sigma_3].S_0 \tag{5}$$

Whether the lattice site 0 belongs to σ or S-sublattice, respectively.

Using the framework of the FCA, we evaluate $<\sigma_{1,0}>_c$, $<\sigma_{2,0}>_c$, $<\sigma_{3,0}>_c$ and $<(S_0)^n>_c$ (n=1, 2) the mean values of $\sigma_{1,0}$,$\sigma_{2,0}$, $\sigma_{3,0}$, $(S_0)^n$ for a given configuration c of all other spins (i.e when all other spins σ_i and S_j are kept fixed).

$<\sigma_{1,0}>_c$, $<\sigma_{2,0}>_c$, $<\sigma_{3,0}>_c$ and $<(S_0)^n>_c$ are given by

$$\langle\sigma_{1,0}\rangle_c = \frac{Tr_{\sigma_{1,0}}(\sigma_{1,0}.\exp(-\beta.H_{0\sigma_1}))}{Tr_{\sigma_{1,0}}(\exp(-\beta.H_{0\sigma_1}))} \tag{6}$$

$$\langle\sigma_{2,0}\rangle_c = \frac{Tr_{\sigma_{2,0}}(\sigma_{2,0}.\exp(-\beta.H_{0\sigma_2}))}{Tr_{\sigma_{2,0}}(\exp(-\beta.H_{0\sigma_2}))} \tag{7}$$

$$\langle\sigma_{3,0}\rangle_c = \frac{Tr_{\sigma_{3,0}}(\sigma_{3,0}.\exp(-\beta.H_{0\sigma_3}))}{Tr_{\sigma_{3,0}}(\exp(-\beta.H_{0\sigma_3}))} \tag{8}$$

$$\langle S_0^n\rangle_c = \frac{Tr_{S_0}(S_0^n.\exp(-\beta.H_{0S}))}{Tr_{S_0}(\exp(-\beta.H_{0S}))} \tag{9}$$

Here, $Tr_{\sigma 0}$ (Tr_{S0} or) means the trace performed over σ_0 (or S_0). As usual $\beta=1/T$, where T is the absolute temperature. n=1,2 correspond to the magnetization and the quadrupolar moment respectively. Therefore, the magnetizations at the shell and in the core are then given by

$$\mu_1 = \langle\langle\sigma_{1,0}\rangle_c\rangle = \langle\frac{Tr_{\sigma_{1,0}}(\sigma_{1,0}.\exp(-\beta.H_{0\sigma_1}))}{Tr_{\sigma_{1,0}}(\exp(-\beta.H_{0\sigma_1}))}\rangle \tag{10}$$

$$\mu_2 = \langle\langle\sigma_{2,0}\rangle_c\rangle = \langle\frac{Tr_{\sigma_{2,0}}(\sigma_{2,0}.\exp(-\beta.H_{0\sigma_2}))}{Tr_{\sigma_{2,0}}(\exp(-\beta.H_{0\sigma_2}))}\rangle \tag{11}$$

$$\mu_3 = \langle\langle\sigma_{3,0}\rangle_c\rangle = \langle\frac{Tr_{\sigma_{3,0}}(\sigma_{3,0}.\exp(-\beta.H_{0\sigma_3}))}{Tr_{\sigma_{3,0}}(\exp(-\beta.H_{0\sigma_3}))}\rangle \tag{12}$$

$$m = \langle\langle S_0\rangle_c\rangle = \langle\frac{Tr_{S_0}(S_0.\exp(-\beta.H_{0S}))}{Tr_{S_0}(\exp(-\beta.H_{0S}))}\rangle \tag{13}$$

$$q = \langle\langle S_0^2\rangle_c\rangle = \langle\frac{Tr_{S_0}(S_0^2.\exp(-\beta.H_{0S}))}{Tr_{S_0}(\exp(-\beta.H_{0S}))}\rangle \tag{14}$$

where <...> denotes the average over all spin configurations. q is the quadrupolar moment. Performing the inner traces in (10)-(14) over the states of the selected σ_0 spin (S_0), we obtain the following exact relations

$$\mu_1 = \langle \frac{1}{2}\tanh\left[\frac{K}{2}\{\alpha_2.(\sigma_{2,1} + \sigma_{2,2}) + \alpha_1.S_1\}\right]\rangle \tag{15}$$

$$\mu_2 = \langle \frac{1}{2}\tanh\left[\frac{K}{2}\{\alpha_2.(\sigma_{1,1} + \sigma_{1,2}) + \alpha_1.(S_1 + S_2)\}\right]\rangle \tag{16}$$

$$\mu_3 = \langle \frac{1}{2}\tanh\left[\frac{K}{2}\{S_1 + S_2 + S_3 + S_4 + S_5 + S_6\}\right]\rangle \tag{17}$$

$$m = \langle \frac{2\sinh[K\{\alpha_1.(\sigma_1+\sigma_{2,1}+\sigma_{2,2})+\sigma_3+\alpha_3.(S_1+S_2)\}]}{1+2\cosh[K\{\alpha_1.(\sigma_1+\sigma_{2,1}+\sigma_{2,2})+\sigma_3+\alpha_3.(S_1+S_2)\}]}\rangle \tag{18}$$

$$q = \langle \frac{2\cosh[K\{\alpha_1.(\sigma_1+\sigma_{2,1}+\sigma_{2,2})+\sigma_3+\alpha_3.(S_1+S_2)\}]}{1+2\cosh[K\{\alpha_1.(\sigma_1+\sigma_{2,1}+\sigma_{2,2})+\sigma_3+\alpha_3.(S_1+S_2)\}]}\rangle \tag{19}$$

where $K=\beta.J_2$, $\alpha 1=J_1/J_2$, $\alpha_2=J_s/J_2$ and $\alpha_3=J/J_2$.

Now, we have to average the right-hand sides of Eqs. (15)-(19) over all spin configurations, within the FCA framework. This latter has been designed to treat all spin self-correlations exactly while still neglecting correlations between different spins. We can easily observe that any functions such as $f(\sigma,S)$and $g(S)$ of σ and S can be written as the linear superposition

$$f(\sigma,S) = f_1 + f_2.\sigma + f_3.S + f_4.S^2 + f_5.\sigma.S + f_6.\sigma.S^2 \tag{20}$$

$$g(S) = g_1 + g_2.S + g_3.S^2 \tag{21}$$

with appropriate coefficients $f_i(i=1,\ldots,6)$ and $g_j(j=1,2,3)$. Applying this to all spins σ_i and S_j in expressions between brackets in Eqs. (15)-(19), we obtain

$$\langle\sigma_{1,0}\rangle_c = \sum_{p=0}^{2}\sum_{n=0}^{1}\sum_{\ell=0}^{1-n}\sum_{i=1}^{2} A_i^{p,n,\ell}[\sigma_i S_m S_m^2] \tag{22}$$

$$\langle\sigma_{2,0}\rangle_c = \sum_{p=0}^{2}\sum_{n=0}^{2}\sum_{\ell=0}^{2-n}\sum_{i=1}^{2}\sum_{j=1}^{2}\sum_{\substack{k=1\\(k\neq j)}}^{2} B_{i,j,k}^{p,n,\ell}[\sigma_i S_j S_k^2] \tag{23}$$

$$\langle\sigma_{3,0}\rangle_c = \sum_{n=0}^{6}\sum_{\ell=0}^{6-n}\sum_{i=1}^{6}\sum_{\substack{j=1\\j\neq i}}^{6} C_{i,j}^{n,\ell}[S_i S_j^2] \tag{24}$$

$$\langle S_0\rangle_c = \sum_{p=0}^{4}\sum_{n=0}^{2}\sum_{\ell=0}^{2-n}\sum_{i=1}^{4}\sum_{j=1}^{2}\sum_{\substack{k=1\\k\neq j}}^{2} D_{i,j,k}^{p,n,\ell}[\sigma_i S_j S_k^2] \tag{25}$$

$$\langle S_0^2\rangle_c = \sum_{p=0}^{4}\sum_{n=0}^{2}\sum_{\ell=0}^{2-n}\sum_{i=1}^{4}\sum_{j=1}^{2}\sum_{\substack{k=1\\k\neq j}}^{2} E_{i,j,k}^{p,n,\ell}[\sigma_i S_j S_k^2] \tag{26}$$

Journal of Physics: Conference Series **758** (2016) 012005 doi:10.1088/1742-6596/758/1/012005

where $[\sigma_i S_j (S_k)^2]$ denotes the term containing p different factors of σ_i, n different factors of S_j, and ℓ different factors of $(S_k)^2$ with ($k \neq j$). $[S_j(S_k)^2]$ denotes the term containing n different factors of S_j, and ℓ different factors of $(S_k)^2$. These factors are selected from sets $\{\sigma_1,\sigma_2,S_m\}$, $\{\sigma_1,\sigma_2,S_1,S_2\}$, $\{S_1,S_2, S_3,S_4, S_5,S_6\}$ and $\{\sigma_1,\sigma_2,\sigma_3,\sigma_4,S_1,S_2\}$ for $\langle\sigma_{1,0}\rangle_c$, $\langle\sigma_{2,0}\rangle_c$, $\langle\sigma_{3,0}\rangle_c$ and $\langle(S_0)^n\rangle_c$, respectively.

For example:

- p=1, n=0, ℓ=1: $A_i^{p,n,\ell}\sigma_i S_m S_m^2 = A_1^{1,0,1}\sigma_1 S_m^2$
- p=2 , n=1 , ℓ=1: $B_{i,j,k}^{p,n,\ell}\sigma_i S_j S_k^2 = B_{(1,2),1,2}^{2,1,1}\sigma_1\sigma_2 S_1 S_2^2$
- n=3 , ℓ=1: $C_{i,j}^{n,\ell} S_i S_j^2 = C_{(1,2,3),5}^{3,1} S_1 S_2 S_3 S_5^2$

The magnetizations at the shell and in the core are given by Eqs. (15)-(19) using the expansions (22)-(26). They constitute a set of exact relations according to which we can study the nanoparticle with a circle shape. However, in order to carry out the thermal average over all spin configurations, we have to deal with multispin correlations appearing via the right-hand side of Eqs. (22)-(26). The problem becomes mathematically intractable if we try to treat them exactly in the spirit of the FCA. In this paper, we use the simplest approximation in which we treat all spin self-correlations exactly while still neglecting correlations between quantities pertaining to different sites. This leads to the following coupled equations

$$\mu_1 = \mu_2.(2A_1 + 2A_3.q) + m.A_2 + \mu_2^2.m.A_4 \quad (27)$$

$$\mu_2 = \mu_1.(2B_1 + 4B_3.q + 2B_7.q^2) + m.(2B_2 + 2B_6.q) + m.\mu_1^2.(2B_4 + 2B_8.q) + 2.B_5.m^2.\mu_1 \quad (28)$$

$$\mu_3 = m.(6C_1 + 30C_2.q + 60C_3.q^2 + 60C_4.q^3 + 30C_5.q^4 + 6C_6.q^5) + m^3.(20C_7 + 60C_8.q + 60C_9.q^2 + 20C_{10}.q^3) + m^5.(6C_{11} + 6C_{12}.q) \quad (29)$$

$$m = \mu_1.(D_1 + 2D_{10}.q + D_{21}.q^2) + \mu_2.(2D_1 + 4D_{10}.q + 2D_{21}.q^2) + \mu_3.(D_2 + 2D_{11}.q + D_{22}.q^2) + m.(2D_3 + 2D_4.q) + \mu_1.\mu_2^2.(D_5 + 2D_{12}.q + D_{24}.q^2) + \mu_3.\mu_2^2.(D_6 + 2D_{13}.q + D_{23}.q^2) + m.\mu_2^2.(2D_7 + 2D_{18}.q) + \mu_1.m^2.(D_{14}) + \mu_2.m^2.(2D_{14}) + \mu_3.m^2.(D_{15}) + \mu_1.\mu_2.\mu_3.(2D_6 + 4D_{13}.q + 2D_{23}.q^2) + m.\mu_1.\mu_2.(4D_7 + 4D_{18}.q) + m.\mu_1.\mu_3.(2D_8 + 2D_{19}.q) + m.\mu_2.\mu_3.(4D_8 + 4D_{19}.q) + m.\mu_1.\mu_3.\mu_2^2.(2D_9 + 2D_{20}.q) + \mu_1.m^2.\mu_2^2.(D_{16}) + \mu_3.m^2.\mu_2^2.(D_{17}) + \mu_1.\mu_2.\mu_3.m^2.(2D_{17}) \quad (30)$$

$$q = (E_1 + 2E_5.q + E_7.q^2) + \mu_2^2.(E_2 + 2E_{12}.q + E_{22}.q^2) + \mu_1.\mu_2.(2E_2 + 4E_{12}.q + 2E_{22}.q^2) + \mu_1.\mu_3.(E_3 + 2E_{13}.q + E_{23}.q^2) + \mu_2.\mu_3.(2E_3 + 4E_{13}.q + 2E_{23}.q^2) + m^2.(E_6) + m.\mu_1.(2E_8 + 2E_{18}.q) + m.\mu_2.(4E_8 + 4E_{18}.q) + m.\mu_3.(2E_9 + 2E_{19}.q) + \mu_1.\mu_3.\mu_2^2.(E_4 + 2E_{14}.q + E_{24}.q^2) + \mu_1.\mu_2.\mu_3.m.(4E_{10} + 4E_{21}.q) + m.\mu_1.\mu_2^2.(2E_{11} + 2E_{20}.q) + m.\mu_3.\mu_2^2.(2E_{10} + 2E_{21}.q) + \mu_1.\mu_2.m^2.(2E_{16}) + \mu_1.\mu_3.m^2.(E_{17}) + \mu_2.\mu_3.m^2.(2E_{17}) + m^2.\mu_2^2.(E_{16}) + m^2.\mu_2^2.\mu_1.\mu_3.(E_{15}) \quad (31)$$

The non-zero coefficients quoted in Eqs. (27)-(31), are listed in the appendix.

If we substitute μ_i (i=1,2,3) and q into (30) with their expressions taken from (27), (28), (29) and (31), we obtain an equation for m of the form

$$m=a.m+ b.m^3 +.... \quad (32)$$

The second-order transition is determined by the condition a=1. For this reason, we neglect higher-order terms in the magnetizations in (27)-(31). Therefore, the transition temperature is analytically determined by the equation

$$1 = 2D_3 + 2D_4.q_0 + (D_1 + 2D_{10}.q_0 + D_{21}.q_0^2).(A + 2B) + C.(D_2 + 2D_{11}.q_0 + D_{22}.q_0^2) \quad (33)$$

where q_0 is the solution of

$$q_0 = E_1 + 2E_5.q_0 + E_7.q_0^2 \quad (34)$$

Eq. (33) means that the right-hand-side corresponds to the coefficient a in Eq. (32).

The magnetization m in the vicinity of the second-order transition is given by

$$m^2 = \frac{1-a}{b} \quad (35)$$

The right-hand side of (35) must be positive. If this is not the case, the transition is of the first-order and the point at which a=1 and b=0 characterizes the tricritical point.

To obtain the expression for b, one has to solve Eqs (27)-(31) for small μ_1, μ_2, μ_3 and m. The solution is of the form

$$q = q_0 + q_1.\mu_2^2 + q_2.m^2 + q_3.\mu_1.m + q_4.\mu_2.m + q_5.\mu_3.m + q_6.\mu_1.\mu_2 + q_7.\mu_1.\mu_3 + q_8.\mu_2.\mu_3 \quad (36)$$

After some algebraic manipulations, Eqs (27), (28), (29) and (36) can be written in the following forms

μ_1=A.m+ D.m^3+...... (37)

μ_2=B.m+ Em^3+...... (38)

μ_3=C.m+ F.m^3+...... (39)

$q=q_0+q_9.m^2$ (40)

By substituting μ_1, μ_2, μ_3 and q in Eq.(30), with their expressions taken from (37)-(40), we obtain the equation (32), where b is given by

$$b = D_1.(2E + D) + D_2.F + 2D_4.q_9 + D_5.A.B^2 + D_6.(C.B^2 + 2A.B.C) + D_7.(4A.B + 2B^2) + D_8.(4B.C + 2A.C) + D_{10}.(4q_0.E + 4q_9.B + 2q_0.D + 2q_9.A) + 2D_{11}.(q_0.F + q_9.C) + 2D_{12}.q_0.A.B^2 + D_{13}.(4q_0.A.B.C + 2q_0.B^2.C) + D_{14}.(2B + A) + D_{15}.C + D_{18}.(2q_0.B^2 + 4q_0.A.B) + D_{19}.(4q_0.B.C + 2q_0.A.C) + D_{21}.(2q_0.q_9.A + q_0^2.D + 2q_0^2.E + 4q_0.q_9.B) + D_{22}.(q_0^2.F + 2q_0.q_9.C) + D_{23}.(q_0^2.B^2.C + 2q_0^2.A.B.C) + D_{24}.q_0^2.A.B^2 \quad (41)$$

For the nanoparticle with a square shape depicted in figure 1(b), we define three magnetizations μ_1, μ_2, μ_3 at the surface shell, three magnetizations (m_1 and m_2 at the second shell and μ_4 at the center) in the core, and two quadrupolar moments q_1 and q_2. Using FCA, we determine all these quantities from coupled equations, similar to those (Eqs. (27)-(31)) written for the circle nanoparticle. They will not be presented here.

3. Results and discussion

Let us introduce the magnetizations m_{sh} and m_C at the shell and in the core as well as the total magnetization per site m_T. Their expressions are given by

$$m_{sh} = \frac{\mu_1+\mu_2}{2}, \quad m_C = \frac{\mu_3+6m}{7}, \quad m_T = \frac{6\mu_1+6\mu_2+\mu_3+6m}{19} \quad (42)$$

for circle nanoparticle

$$m_{sh} = \frac{\mu_1+2\mu_2+\mu_3}{4}, \quad m_C = \frac{\mu_4+4m_1+4m_2}{9}, \quad m_T = \frac{4\mu_1+8\mu_2+4\mu_3+\mu_4+4m_1+4m_2}{25} \quad (43)$$

Journal of Physics: Conference Series **758** (2016) 012005 doi:10.1088/1742-6596/758/1/012005

for square nanoparticle

In this section, we study the effects of core and shell couplings on the phase diagrams of the circle and square 2D-nanoparticles. In the following discussions, all interactions as well as the temperature are normalized with exchange interaction J_2 between the central site and the sites belonging to the outer boundary of the core. In the present work, we consider that the shells at the surface and in the core interact with each other via the same coupling (J_1=J_2).

In order to examine the transition temperature of the shell and the core, we investigate the thermal dependence of the magnetizations. Firstly, we can see from figure 2 and figure 3 that all magnetizations (μ_1, μ_2, μ_3 and m) decrease to zero continuously as the temperature increases; therefore, a second-order phase transition occurs. On the other hand, by analyzing the coupled equations (27)-(31), we find that all magnetizations vanish at a unique transition temperature, which is solution of Eq. (33). This behavior is observed for any strength of the Hamiltonian parameters as is shown in figure 2 and figure 3. Therefore, the shell and the core undergo a transition at the same temperature. Further, we note from these figures that the core exchange interaction J has the effect of increasing the critical temperature. In other words, the long-range ferromagnetic order domain becomes more and more wider with increasing values of exchange interaction J. The same behaviors are also observed for the nanoparticle with square shape.

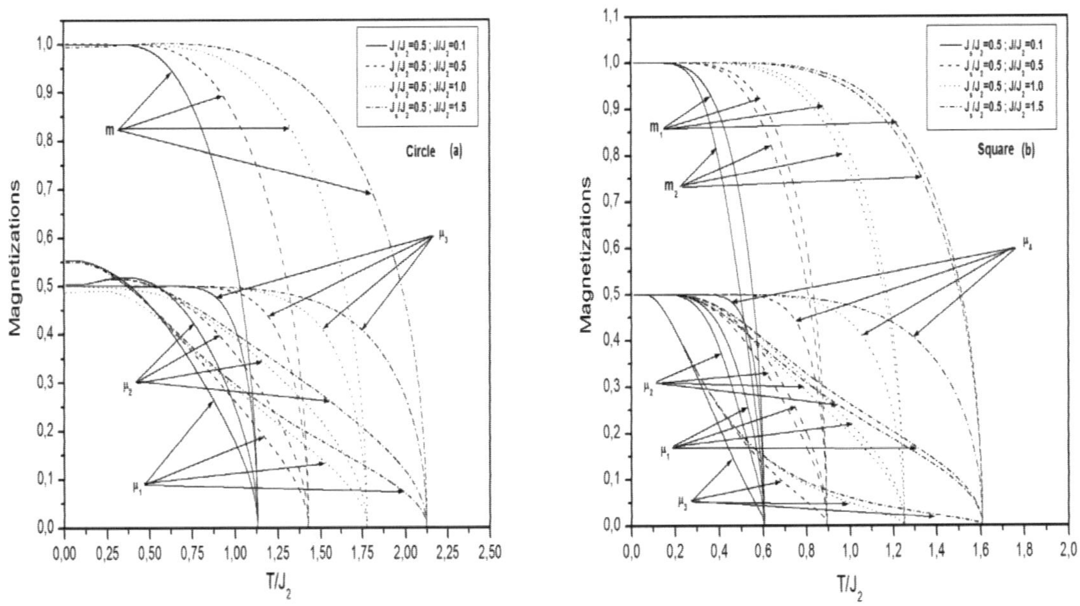

Figure 2. The thermal dependence of the magnetizations (μ and m) for selected values of Js and J. (a) circle nanoparticle and (b) square nanoparticle.

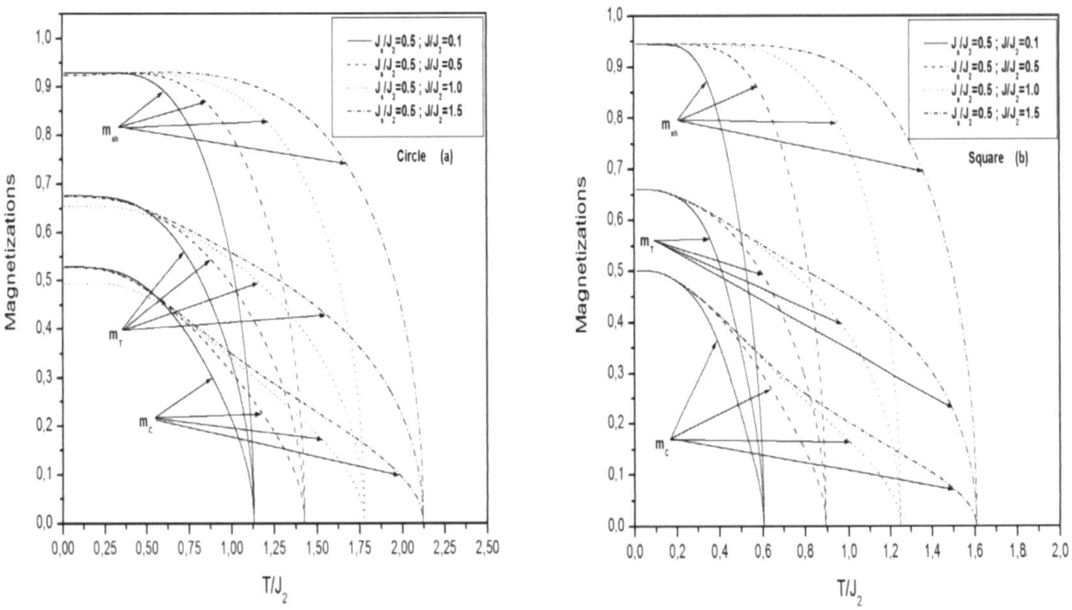

Figure 3. The thermal dependence of the magnetizations at the shell m_{sh}, in the core m_C and the total magnetization m_T, for selected values of J_s and J. (a) circle nanoparticle and (b) square nanoparticle.

In figure 4, we present in (T_c/J_2,J/J_2) plane the phase diagrams for circle and square nanoparticles for selected values of the shell coupling J_s. As seen in this figure, for fixed value of the shell coupling J_s, the transition temperature increases gradually with the increase of J. For relatively low strength of this latter interaction, the transition temperature increases with J_s. This dependence becomes less and less sensitive for high values of the surface exchange interaction. In other word, for important values of J the ferromagnetic order of the core coupling are dominant against the shell coupling J_s, hence the transition temperature of the system becomes independent of shell coupling J_s. We note that this behavior is more pronounced for the nanoparticle with a square shape. Indeed, the positive value of the exchange interaction J rises the absolute value of lattice energy coming from the spin1-spin1 interaction, and then reinforce the longitudinal magnetization. Therefore it acts in favor of the order. Thus the domain of the order state becomes wider and the transition temperature increases with J. On the other hand, one should notice that the transition temperature of circle nanoparticle is greater than the one of square nanoparticle. This is because the difference of the atomic arrangement, namely the atomic arrangement in the circle is more closely packed than the one in the square. The similar phenomena have also been observed in the phase diagrams of a nanowire and a nanotube [22].

Journal of Physics: Conference Series **758** (2016) 012005 doi:10.1088/1742-6596/758/1/012005

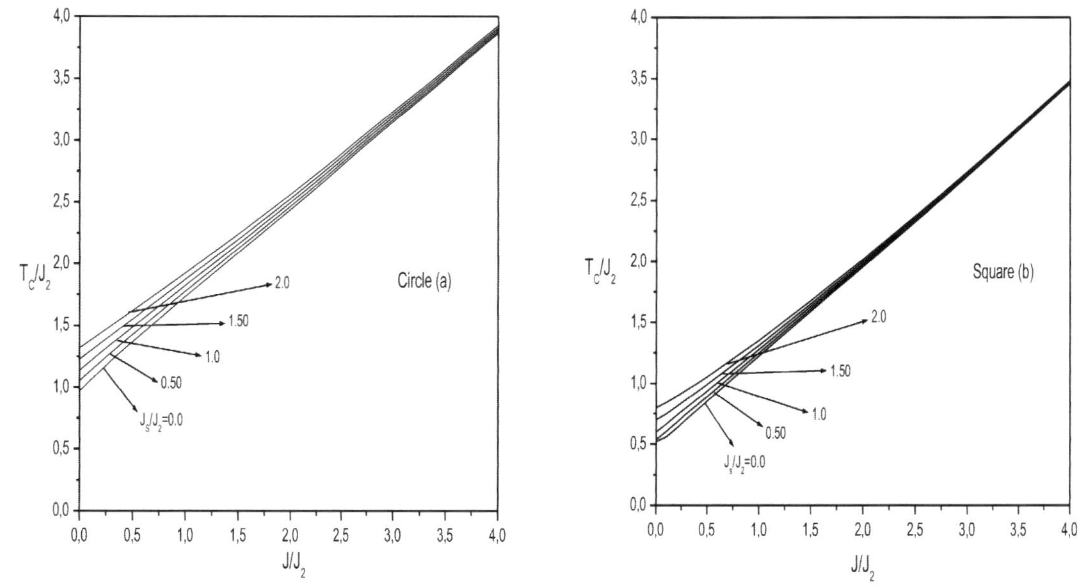

Figure 4. The phase diagrams in the (T_c/J_2,J/J_2) plane of the two nanoparticles, (a) the circle nanoparticle and (b) the square nanoparticle. The number accompanying each curve denotes the value of J_s/J_2.

4. Conclusion

In this work, we have studied the magnetic properties of the mixed spin Ising 2D-nanoparticles (circle and square) formed by alternate layers of σ and S spins using the finite cluster approximation (FCA) within the single-cluster theory. Let summarize by stating the main results of this investigation.

Firstly, we have reported the effects of the exchange interaction between spins S and the shell coupling on the phase diagrams, it has been shown that the increasing values of J_s and J interaction increase the transition temperature. This latter becomes not so sensitive of shell exchange interaction where the spins S are strongly correlated.

Secondly, the thermal behaviours of the longitudinal magnetizations have also been examined. It has been found that all sublattice magnetizations vanish at a unique transition temperature for any strength of the hamiltonian parameters Therefore the shell and the core undergo a transition at the same temperature.

References

[1] Kim T Y, Yamazaki Y and Hirano T 2004 *Phys. Status Solidi B* **241** 1601
[2] Kodama R H 1999 *J. Magn. Magn. Mater.* **200** 359
[3] Zeng H, Li J, Liu J P, Wang Z L and Sun S 2002 *Nature* **420** 395
[4] Hayashi T, Hirono S, Tomita M and Umemura S 1996 *Nature* **381** 772
[5] Kodama R H, BerkowitzJ A E, McNiff E J and Foner J S 1996 *Phys. Rev. Lett.* **774** 394
[6] Kurlyandskaya G V, Sanchez M L, Hernando B, Prida V M, Gorria P and M. Tejedor 2003 *Appl. Phys. Lett.* **82** 3053
[7] Alexiou C, Schmidt A, Klein R, Hullin P, Bergemann C and Arnold W 2002 *J. Magn. Magn. Mater.* **252** 363
[8] Ruhrig M, Khamsehpour B, Kirk K J, Chapman J N, Aitchison P, McVitie S and Wilkinsons C D W 1996 *IEEE Trans. Magn.* **32** 4452
[9] Scherfl T, Fidler J, Kirk K J and Chapman J N 1997 *J. Magn. Magn. Mater.* **175** 193
[10] Martinez B, Obradors X, Balcells L, Rouanet A and Monty C 1998 *Phys. Rev. Lett.* **80** 181
[11] Fan Z and LU J G 2006 *Int. J. High speed Electron. Systems* **16** 883
[12] Su Y C, Skomski R, Sorge K D and Sellmyer D J 2004 *Appl. Phys. Left.* **84** 1525
[13] Skomski R 2003 *J. Phys. Condens, Matter* **15** R841.
[14] Schlörb H, Haehnel V, Khatri M S, Srivastav A, Kumar A, Schultz L and Fähler S 2010 *Phys. Status solidi B* **247** 2364
[15] Wegrove J E, Kelly D, Jaccard Y, Guittienne Ph and Ansermet J –Ph 1999 *Europhys. Lett.* **45** 626.
[16] Fert A and Piraux L 1999 *J. Magn. Magn. Mater.* **200** 338.
[17] Bader S D 2006 *Rev. Mod. Phys.* **78** 1
[18] Leite V S and Figueiredo W 2005 *Physica A* **350** 379
[19] Kaneyoshi T 2009 *Phys. Status Solidi B* **246** 2359
[20] Michael F, Gonzalez C, Mujica V, Marquez M and Ratner A M 1996 *Phys. Rev. B* **76** 224409
[21] Kaneyoshi T 2011 *J. Magn. Magn. Mater.* **323** 1145
[22] Kaneyoshi T 2011 *Phys. Status Solidi B* **248** 250
[23] Sarli N and Keskin M 2012 *Solid State Commun.* **152** 354
[24] Canko O, Erdinç A, Taskin F and Yildirim A F 2012 *J. Magn. Magn. Mater.* **324** 508
[25] Vasilakaki M and Trohidou K N 2009 *Phys. Rev. B* **79** 144402
[26] Zaim A, Kerouad M and Y. El Amraoui Y 2009 *J. Magn. Magn. Mater.* **321** 1083
[27] Eglesias O, Battle X and Labarta A 2007 *J. Phys. Condens. Mater* **19** 406232
[28] Eglesias O, Battle X and Labarta A 2005 *Phys. Rev. B* **72** 212401
[29] Wang H, Zhou Y, Lin D L and Wang C 2002 *Phys. Status Solidi B* **232** 254
[30] Wang H, Zhou Y, Wang E and Lin D L 2001 *Chin. J. Phys.* **39** 85
[31] Garanin D A and Kachkachi H 2003 *Phys. Rev. Lett.* **90** 65504
[32] Néel L 1948 *Ann. Phys. (Paris)* **3** 137
[33] Albayrak E and Keskin M 2003 *J. Magn. Magn. Mater.* **261** 196
[34] Boccara N 1983 *Phys. Lett A* **94** 185
[35] Benyoussef A and Boccara N 1983 *J. Phys.* **44** 1143.
[36] Benayad N, Clümper A, Zittartz J and Benyoussef A 1989 *Z. Phys. B* **77** 339
[37] Benayad N, Dakhama A, Clümper A and Zittartz J 1996 *Ann. Phys. Lpz.* **5** 387
[38] Benayad N, Khaya L and Fathi A 2002 *J. Phys. Condens. Mater.* **14** 9667
[39] Benayad N, Fathi A and Khaya L 2001 *Physica A* **300** 225
[40] Benayad N, Zerhouni R and Clümper A 1998 *Eur. Phys. J. B* **5** 687
[41] Benayad N, Fathi A and Khaya L 2004 *J. Magn. Magn. Mater.* **278** 407
[42] Benayad N, Fathi A and Zerhouni R 2000 *J. Magn. Magn. Mater.* **222** 355
[43] Benyoussef A and Ez-Zahhraoui H 1994 *J. Phys. Condens. Mater.* **6** 3411
[44] Lahcini T and Benayad N 2009 Phase Transitions **82** 197

[45] Khaya L, Benayad N and Dakhama A 2004 *Phys Status Solidi B* **241** 1078
[46] Benayad N and Dakhama A 1997 *Phys. Rev B* **55** 12276
[47] Benayad N and Dakhama A 1997 *J. Magn. Magn. Mater.* **168** 105
[48] Benyoussef A, Boccara N and Saber M 1985 *J. Phys C* **18** 4275

Journal of Physics: Conference Series **758** (2016) 012005 doi:10.1088/1742-6596/758/1/012005

Appendix

The coefficients A_i(i=1,…,4), B_i(i=1,…,8), C_i(i=1,…,12), D_i(i=1,…,24) and E_i(i=1,…,24) appearing respectively in eqs. (27), (28), (29), (30) and (31), also the coefficients A, B, C, D, E, F, q_1, q_2, q_3, q_4, q_5, q_6, q_7, q_8 and q_9 are given by:

For abbreviation we consider new functions

$$f(x) = \frac{1}{2} tanh\left(\frac{\beta}{2}x\right)$$

$$F_1(x) = \frac{2sinh(\beta x)}{1+2cosh(\beta x)}$$

$$F_2(x) = \frac{2cosh(\beta x)}{1+2cosh(\beta x)}$$

$$A_1 = f(J_s)$$

$$A_2 = \frac{1}{4}f(J_s + J_1) - \frac{1}{4}f(J_s - J_1) + \frac{1}{2}f(J_1)$$

$$A_3 = \frac{1}{2}f(J_s + J_1) + \frac{1}{2}f(J_s - J_1) - f(J_s)$$

$$A_4 = f(J_s + J_1, \Omega_s) - f(J_s - J_1) - 2f(J_1)$$

$$B_1 = f(J_s,)$$

$$B_2 = \frac{1}{4}f(J_s + J_1) - \frac{1}{4}f(J_s - J_1) + \frac{1}{2}f(J_1)$$

$$B_3 = \frac{1}{2}f(J_s + J_1) + \frac{1}{2}f(J_s - J_1) - f(J_s)$$

$$B_4 = f(J_s + J_1) - f(J_s - J_1) - 2. f(J_1)$$

$$B_5 = \frac{1}{4}f(J_s + 2J_1) + \frac{1}{4}f(J_s - 2J_1) - \frac{1}{2}f(J_s)$$

$$B_6 = \frac{1}{8}f(J_s + 2J_1, \Omega_s) - \frac{1}{8}f(J_s - 2J_1) - \frac{1}{4}f(J_s + J_1) + \frac{1}{4}f(J_s - J_1) + \frac{1}{4}f(2J_1) - \frac{1}{2}f(J_1,)$$

$$B_7 = \frac{1}{4}f(J_s + 2J_1) + \frac{1}{4}f(J_s - 2J_1) - f(J_s + J_1) - f(J_s - J_1) + \frac{3}{2}f(J_s)$$

$$B_8 = \frac{1}{2}f(J_s +. J_1) - \frac{1}{2}f(J_s - 2J_1) - f(J_s + J_1) + f(J_s - J_1, \Omega_s) - f(2J_1, \Omega_s) + 2f(J_1)$$

$$C_1 = f(J_2)$$

$$C_2 = -f(J_2) + \frac{1}{2}f(2J_2)$$

$$C_3 = \frac{5}{4}f(J_2) - f(2J_2) + \frac{1}{4}f(3J_2)$$

Journal of Physics: Conference Series **758** (2016) 012005 doi:10.1088/1742-6596/758/1/012005

$$C_4 = -\frac{7}{4}f(J_2) + \frac{7}{4}f(2J_2) - \frac{3}{4}f(3J_2) + \frac{1}{8}f(4J_2)$$

$$C_5 = \frac{21}{8}f(J_2) - 3f(2J_2) + \frac{27}{16}f(3J_2) - \frac{1}{2}f(4J_2) + \frac{1}{16}f(5J_2)$$

$$C_6 = -\frac{33}{8}f(J_2) + \frac{165}{32}f(2J_2) - \frac{55}{16}f(3J_2) + \frac{11}{8}f(4J_2) - \frac{5}{16}f(5J_2) + \frac{1}{32}(6J_2)$$

$$C_7 = -\frac{3}{4}f(J_2) + \frac{1}{4}(3J_2)$$

$$C_8 = \frac{3}{4}f(J_2) - \frac{1}{4}f(2J_2) + \frac{1}{4}f(3J_2) + \frac{1}{8}f(4J_2)$$

$$C_9 = -\frac{7}{8}f(J_2) + \frac{1}{2}f(2J_2) + \frac{3}{16}f(3J_2) - \frac{1}{4}f(4J_2) + \frac{1}{16}f(5J_2)$$

$$C_{10} = \frac{9}{8}f(J_2) - \frac{27}{32}f(2J_2) - \frac{1}{16}f(3J_2) + \frac{3}{8}f(4J_2) - \frac{3}{16}f(5J_2) + \frac{1}{32}f(6J_2)$$

$$C_{11} = \frac{5}{8}f(J_2) - \frac{5}{16}f(3J_2) + \frac{1}{16}f(5J_2)$$

$$C_{12} = -\frac{5}{8}f(J_2) + \frac{5}{32}f(2J_2) + \frac{5}{16}f(3J_2) - \frac{1}{8}f(4J_2) - \frac{1}{16}f(5J_2) + \frac{1}{32}f(6J_2)$$

$$D_1 = \frac{1}{4}F_1\left(\frac{3}{2}J_1 + \frac{1}{2}J_2\right) + \frac{1}{4}F_1\left(\frac{3}{2}J_1 - \frac{1}{2}J_2\right) + \frac{1}{4}F_1\left(\frac{1}{2}J_1 + \frac{1}{2}J_2\right) + \frac{1}{4}F_1\left(\frac{1}{2}J_1 - \frac{1}{2}J_2\right)$$

$$D_2 = \frac{1}{4}F_1\left(\frac{3}{2}J_1 + \frac{1}{2}J_2\right) - \frac{1}{4}F_1\left(\frac{3}{2}J_1 - \frac{1}{2}J_2\right) + \frac{3}{4}F_1\left(\frac{1}{2}J_1 + \frac{1}{2}J_2\right) - \frac{3}{4}F_1\left(\frac{1}{2}J_1 - \frac{1}{2}J_2\right)$$

$$D_3 = \frac{1}{16}.\left[F_1\left(J + \frac{3}{2}J_1 + \frac{1}{2}J_2\right) + F_1\left(J + \frac{3}{2}J_1 - \frac{1}{2}J_2\right) + 3F_1\left(J + \frac{1}{2}J_1 + \frac{1}{2}J_2\right) + 3F_1\left(J + \frac{1}{2}J_1 - \frac{1}{2}J_2\right) - F_1\left(-J + \frac{3}{2}J_1 + \frac{1}{2}J_2\right) - F_1\left(-J + \frac{3}{2}J_1 - \frac{1}{2}J_2\right) - 3F_1\left(-J + \frac{1}{2}J_1 + \frac{1}{2}J_2\right) - 3F_1\left(-J + \frac{1}{2}J_1 - \frac{1}{2}J_2\right)\right]$$

$$D_4 = \frac{1}{32}.\left[F_1\left(2.J + \frac{3}{2}.J_1 + \frac{1}{2}.J_2\right) + F_1\left(2.J + \frac{3}{2}.J_1 - \frac{1}{2}.J_2\right) + 3F_1\left(2.J + \frac{1}{2}.J_1 + \frac{1}{2}.J_2\right) + 3F_1\left(2.J + \frac{1}{2}.J_1 - \frac{1}{2}.J_2\right) - F_1\left(-2.J + \frac{3}{2}.J_1 + \frac{1}{2}.J_2\right) - F_1\left(-2.J + \frac{3}{2}.J_1 - \frac{1}{2}.J_2\right) - 3F_1\left(-2.J + \frac{1}{2}.J_1 + \frac{1}{2}.J_2\right) - 3F_1\left(-2.J + \frac{1}{2}.J_1 - \frac{1}{2}.J_2\right) - 2F_1\left(J + \frac{3}{2}.J_1 + \frac{1}{2}.J_2\right) - 2F_1\left(J + \frac{3}{2}.J_1 - \frac{1}{2}.J_2\right) - 6F_1\left(J + \frac{1}{2}.J_1 + \frac{1}{2}.J_2\right) - 6F_1\left(J + \frac{1}{2}.J_1 - \frac{1}{2}.J_2\right) + 2F_1\left(-J + \frac{3}{2}.J_1 + \frac{1}{2}.J_2\right) + 2F_1\left(-J + \frac{3}{2}.J_1 - \frac{1}{2}.J_2\right) + 6F_1\left(-J + \frac{1}{2}.J_1 + \frac{1}{2}.J_2\right) + 6F_1\left(-J + \frac{1}{2}.J_1 - \frac{1}{2}.J_2\right)\right]$$

$$D_5 = F_1\left(\frac{3}{2}.J_1 + \frac{1}{2}.J_2\right) + F_1\left(\frac{3}{2}.J_1 - \frac{1}{2}.J_2\right) - 3F_1\left(\frac{1}{2}.J_1 + \frac{1}{2}.J_2\right) - 3F_1\left(\frac{1}{2}.J_1 - \frac{1}{2}.J_2\right)$$

$$D_6 = F_1\left(\frac{3}{2}.J_1 + \frac{1}{2}.J_2\right) - F_1\left(\frac{3}{2}.J_1 - \frac{1}{2}.J_2\right) - F_1\left(\frac{1}{2}.J_1 + \frac{1}{2}.J_2\right) + F_1\left(\frac{1}{2}.J_1 - \frac{1}{2}.J_2\right)$$

$$D_7 = \frac{1}{4}.\left[F_1\left(J + \frac{3}{2}.J_1 + \frac{1}{2}.J_2\right) + F_1\left(J + \frac{3}{2}.J_1 - \frac{1}{2}.J_2\right) - F_1\left(J + \frac{1}{2}.J_1 + \frac{1}{2}.J_2\right) - F_1\left(J + \frac{1}{2}.J_1 - \frac{1}{2}.J_2\right) - F_1\left(-J + \frac{3}{2}.J_1 + \frac{1}{2}.J_2\right) - F_1\left(-J + \frac{3}{2}.J_1 - \frac{1}{2}.J_2\right) + F_1\left(-J + \frac{1}{2}.J_1 + \frac{1}{2}.J_2\right) + F_1\left(-J + \frac{1}{2}.J_1 - \frac{1}{2}.J_2\right)\right]$$

$$D_8 = \frac{1}{4}.\left[F_1\left(J + \frac{3}{2}.J_1 + \frac{1}{2}.J_2\right) - F_1\left(J + \frac{3}{2}.J_1 - \frac{1}{2}.J_2\right) + F_1\left(J + \frac{1}{2}.J_1 + \frac{1}{2}.J_2\right) - F_1\left(J + \frac{1}{2}.J_1 - \frac{1}{2}.J_2\right) - F_1\left(-J + \frac{3}{2}.J_1 + \frac{1}{2}.J_2\right) + F_1\left(-J + \frac{3}{2}.J_1 - \frac{1}{2}.J_2\right) - F_1\left(-J + \frac{1}{2}.J_1 + \frac{1}{2}.J_2\right) + F_1\left(-J + \frac{1}{2}.J_1 - \frac{1}{2}.J_2\right)\right]$$

$$D_9 = F_1\left(J + \frac{3}{2}.J_1 + \frac{1}{2}.J_2\right) - F_1\left(J + \frac{3}{2}.J_1 - \frac{1}{2}.J_2\right) - 3F_1\left(J + \frac{1}{2}.J_1 + \frac{1}{2}.J_2\right) - 3F_1\left(J + \frac{1}{2}.J_1 - \frac{1}{2}.J_2\right) - F_1\left(-J + \frac{3}{2}.J_1 + \frac{1}{2}.J_2\right) + F_1\left(-J + \frac{3}{2}.J_1 - \frac{1}{2}.J_2\right) + 3F_1\left(-J + \frac{1}{2}.J_1 + \frac{1}{2}.J_2\right) - 3F_1\left(-J + \frac{1}{2}.J_1 - \frac{1}{2}.J_2\right)$$

$$D_{10} = \frac{1}{8}.\left[F_1\left(J + \frac{3}{2}.J_1 + \frac{1}{2}.J_2\right) + F_1\left(J + \frac{3}{2}.J_1 - \frac{1}{2}.J_2\right) + F_1\left(J + \frac{1}{2}.J_1 + \frac{1}{2}.J_2\right) + F_1\left(J + \frac{1}{2}.J_1 - \frac{1}{2}.J_2\right) + F_1\left(-J + \frac{3}{2}.J_1 + \frac{1}{2}.J_2\right) + F_1\left(-J + \frac{3}{2}.J_1 - \frac{1}{2}.J_2\right) + F_1\left(-J + \frac{1}{2}.J_1 + \frac{1}{2}.J_2\right) + F_1\left(-J + \frac{1}{2}.J_1 - \frac{1}{2}.J_2\right) - 2F_1\left(\frac{3}{2}.J_1 + \frac{1}{2}.J_2\right) - 2F_1\left(\frac{3}{2}.J_1 - \frac{1}{2}.J_2\right) - 2F_1\left(\frac{1}{2}.J_1 + \frac{1}{2}.J_2\right) - 2F_1\left(\frac{1}{2}.J_1 - \frac{1}{2}.J_2\right)\right]$$

$$D_{11} = \frac{1}{8}.\left[F_1\left(J + \frac{3}{2}.J_1 + \frac{1}{2}.J_2\right) - F_1\left(J + \frac{3}{2}.J_1 - \frac{1}{2}.J_2\right) + 3F_1\left(J + \frac{1}{2}.J_1 + \frac{1}{2}.J_2\right) - 3F_1\left(J + \frac{1}{2}.J_1 - \frac{1}{2}.J_2\right) + F_1\left(-J + \frac{3}{2}.J_1 + \frac{1}{2}.J_2\right) - F_1^z\left(-J + \frac{3}{2}.J_1 - \frac{1}{2}.J_2\right) + 3F_1\left(-J + \frac{1}{2}.J_1 + \frac{1}{2}.J_2\right) - 3F_1\left(-J + \frac{1}{2}.J_1 - \frac{1}{2}.J_2\right) - 2F_1\left(\frac{3}{2}.J_1 + \frac{1}{2}.J_2\right) + 2F_1\left(\frac{3}{2}.J_1 - \frac{1}{2}.J_2\right) - 6F_1\left(\frac{1}{2}.J_1 + \frac{1}{2}.J_2\right) + 6F_1\left(\frac{1}{2}.J_1 - \frac{1}{2}.J_2\right)\right]$$

$$D_{12} = \frac{1}{2}.\left[F_1\left(J + \frac{3}{2}.J_1 + \frac{1}{2}.J_2\right) + F_1\left(J + \frac{3}{2}.J_1 - \frac{1}{2}.J_2\right) - 3F_1\left(J + \frac{1}{2}.J_1 + \frac{1}{2}.J_2\right) - 3\,F_1\left(J + \frac{1}{2}.J_1 - \frac{1}{2}.J_2\right) + F_1\left(-J + \frac{3}{2}.J_1 + \frac{1}{2}.J_2\right) + F_1\left(-J + \frac{3}{2}.J_1 - \frac{1}{2}.J_2\right) - 3F_1\left(-J + \frac{1}{2}.J_1 + \frac{1}{2}.J_2\right) - 3F_1\left(-J + \frac{1}{2}.J_1 - \frac{1}{2}.J_2\right) - 2F_1\left(\frac{3}{2}.J_1 + \frac{1}{2}.J_2\right) - 2F_1\left(\frac{3}{2}.J_1 - \frac{1}{2}.J_2\right) + 6F_1\left(\frac{1}{2}.J_1 + \frac{1}{2}.J_2\right) + 6F_1\left(\frac{1}{2}.J_1 - \frac{1}{2}.J_2\right)\right]$$

$$D_{13} = \frac{1}{2}.\left[F_1\left(J + \frac{3}{2}.J_1 + \frac{1}{2}.J_2\right) - F_1\left(J + \frac{3}{2}.J_1 - \frac{1}{2}.J_2\right) - F_1\left(J + \frac{1}{2}.J_1 + \frac{1}{2}.J_2\right) + F_1\left(J + \frac{1}{2}.J_1 - \frac{1}{2}.J_2\right) + F_1^z\left(-J + \frac{3}{2}.J_1 + \frac{1}{2}.J_2\right) - F_1\left(-J + \frac{3}{2}.J_1 - \frac{1}{2}.J_2\right) - F_1\left(-J + \frac{1}{2}.J_1 + \frac{1}{2}.J_2\right) + F_1\left(-J + \frac{1}{2}.J_1 - \frac{1}{2}.J_2\right) - 2F_1\left(\frac{3}{2}.J_1 + \frac{1}{2}.J_2\right) + 2F_1\left(\frac{3}{2}.J_1 - \frac{1}{2}.J_2\right) + 2F_1\left(\frac{1}{2}.J_1 + \frac{1}{2}.J_2\right) - 2F_1\left(\frac{1}{2}.J_1 - \frac{1}{2}.J_2\right)\right]$$

$$D_{14} = \frac{1}{16}.\Big[F_1\left(2.J + \frac{3}{2}.J_1 + \frac{1}{2}.J_2\right) + F_1\left(2.J + \frac{3}{2}.J_1 - \frac{1}{2}.J_2\right) + F_1\left(2.J + \frac{1}{2}.J_1 + \frac{1}{2}.J_2\right) + F_1\left(2.J + \frac{1}{2}.J_1 - \frac{1}{2}.J_2\right) + F_1\left(-2.J + \frac{3}{2}.J_1 + \frac{1}{2}.J_2\right) + F_1\left(-2.J + \frac{3}{2}.J_1 - \frac{1}{2}.J_2\right) + F_1\left(-2.J + \frac{1}{2}.J_1 + \frac{1}{2}.J_2\right) + F_1\left(-2.J + \frac{1}{2}.J_1 - \frac{1}{2}.J_2\right) - 2F_1\left(\frac{3}{2}.J_1 + \frac{1}{2}.J_2\right) - 2F_1\left(\frac{3}{2}.J_1 - \frac{1}{2}.J_2\right) - 2F_1\left(\frac{1}{2}.J_1 + \frac{1}{2}.J_2\right) - 2F_1\left(\frac{1}{2}.J_1 - \frac{1}{2}.J_2\right)\Big]$$

$$D_{15} = \frac{1}{16}.\Big[F_1\left(2.J + \frac{3}{2}.J_1 + \frac{1}{2}.J_2\right) - F_1\left(2.J + \frac{3}{2}.J_1 - \frac{1}{2}.J_2\right) + 3F_1\left(2.J + \frac{1}{2}.J_1 + \frac{1}{2}.J_2\right) - 3F_1\left(2.J + \frac{1}{2}.J_1 - \frac{1}{2}.J_2\right) + F_1\left(-2.J + \frac{3}{2}.J_1 + \frac{1}{2}.J_2\right) - F_1\left(-2.J + \frac{3}{2}.J_1 - \frac{1}{2}.J_2\right) + 3F_1\left(-2.J + \frac{1}{2}.J_1 + \frac{1}{2}.J_2\right) - 3F_1\left(-2.J + \frac{1}{2}.J_1 - \frac{1}{2}.J_2\right) - 2F_1\left(\frac{3}{2}.J_1 + \frac{1}{2}.J_2\right) + 2F_1\left(\frac{3}{2}.J_1 - \frac{1}{2}.J_2\right) - 6F_1\left(\frac{1}{2}.J_1 + \frac{1}{2}.J_2\right) + 6F_1\left(\frac{1}{2}.J_1 - \frac{1}{2}.J_2\right)\Big]$$

$$D_{16} = \frac{1}{4}.\Big[F_1\left(2.J + \frac{3}{2}.J_1 + \frac{1}{2}.J_2\right) + F_1\left(2.J + \frac{3}{2}.J_1 - \frac{1}{2}.J_2\right) - 3F_1\left(2.J + \frac{1}{2}.J_1 + \frac{1}{2}.J_2\right) - 3F_1\left(2.J + \frac{1}{2}.J_1 - \frac{1}{2}.J_2\right) + F_1\left(-2.J + \frac{3}{2}.J_1 + \frac{1}{2}.J_2\right) + F_1\left(-2.J + \frac{3}{2}.J_1 - \frac{1}{2}.J_2\right) - 3F_1\left(-2.J + \frac{1}{2}.J_1 + \frac{1}{2}.J_2\right) - 3F_1\left(-2.J + \frac{1}{2}.J_1 - \frac{1}{2}.J_2\right) - 2F_1\left(\frac{3}{2}.J_1 + \frac{1}{2}.J_2\right) - 2F_1\left(\frac{3}{2}.J_1 - \frac{1}{2}.J_2\right) + 6F_1\left(\frac{1}{2}.J_1 + \frac{1}{2}.J_2\right) + 6F_1\left(\frac{1}{2}.J_1 - \frac{1}{2}.J_2\right)\Big]$$

$$D_{17} = \frac{1}{4}.\Big[F_1\left(2.J + \frac{3}{2}.J_1 + \frac{1}{2}.J_2\right) - F_1\left(2.J + \frac{3}{2}.J_1 - \frac{1}{2}.J_2\right) - F_1\left(2.J + \frac{1}{2}.J_1 + \frac{1}{2}.J_2\right) + F_1\left(2.J + \frac{1}{2}.J_1 - \frac{1}{2}.J_2\right) + F_1\left(-2.J + \frac{3}{2}.J_1 + \frac{1}{2}.J_2\right) - F_1\left(-2.J + \frac{3}{2}.J_1 - \frac{1}{2}.J_2\right) - F_1\left(-2.J + \frac{1}{2}.J_1 + \frac{1}{2}.J_2\right) + F_1\left(-2.J + \frac{1}{2}.J_1 - \frac{1}{2}.J_2\right) - 2F_1\left(\frac{3}{2}.J_1 + \frac{1}{2}.J_2\right) + 2F_1\left(\frac{3}{2}.J_1 - \frac{1}{2}.J_2\right) + 2F_1\left(\frac{1}{2}.J_1 + \frac{1}{2}.J_2\right) - 2F_1\left(\frac{1}{2}.J_1 - \frac{1}{2}.J_2\right)\Big]$$

$$D_{18} = \frac{1}{8}.\Big[F_1\left(2.J + \frac{3}{2}.J_1 + \frac{1}{2}.J_2\right) + F_1\left(2.J + \frac{3}{2}.J_1 - \frac{1}{2}.J_2\right) - F_1\left(2.J + \frac{1}{2}.J_1 + \frac{1}{2}.J_2\right) - F_1\left(2.J + \frac{1}{2}.J_1 - \frac{1}{2}.J_2\right) - F_1\left(-2.J + \frac{3}{2}.J_1 + \frac{1}{2}.J_2\right) - F_1\left(-2.J + \frac{3}{2}.J_1 - \frac{1}{2}.J_2\right) + F_1\left(-2.J + \frac{1}{2}.J_1 + \frac{1}{2}.J_2\right) + F_1\left(-2.J + \frac{1}{2}.J_1 - \frac{1}{2}.J_2\right) - 2F_1\left(J + \frac{3}{2}.J_1 + \frac{1}{2}.J_2\right) - 2F_1\left(J + \frac{3}{2}.J_1 - \frac{1}{2}.J_2\right) + 2F_1\left(J + \frac{1}{2}.J_1 + \frac{1}{2}.J_2\right) + 2F_1\left(J + \frac{1}{2}.J_1 - \frac{1}{2}.J_2\right) + 2F_1\left(-J + \frac{3}{2}.J_1 + \frac{1}{2}.J_2\right) + 2F_1\left(-J + \frac{3}{2}.J_1 - \frac{1}{2}.J_2\right) - 2F_1\left(-J + \frac{1}{2}.J_1 + \frac{1}{2}.J_2\right) - 2F_1\left(-J + \frac{1}{2}.J_1 - \frac{1}{2}.J_2\right)\Big]$$

$$D_{19} = \frac{1}{8}.\Big[F_1\left(2.J + \frac{3}{2}.J_1 + \frac{1}{2}.J_2\right) - F_1\left(2.J + \frac{3}{2}.J_1 - \frac{1}{2}.J_2\right) + F_1\left(2.J + \frac{1}{2}.J_1 + \frac{1}{2}.J_2\right) - F_1\left(2.J + \frac{1}{2}.J_1 - \frac{1}{2}.J_2\right) - F_1\left(-2.J + \frac{3}{2}.J_1 + \frac{1}{2}.J_2\right) + F_1\left(-2.J + \frac{3}{2}.J_1 - \frac{1}{2}.J_2\right) - F_1\left(-2.J + \frac{1}{2}.J_1 + \frac{1}{2}.J_2\right) + F_1\left(-2.J + \frac{1}{2}.J_1 - \frac{1}{2}.J_2\right) - 2F_1\left(J + \frac{3}{2}.J_1 + \frac{1}{2}.J_2\right) + 2F_1\left(J + \frac{3}{2}.J_1 - \frac{1}{2}.J_2\right) - 2F_1\left(J + \frac{1}{2}.J_1 + \frac{1}{2}.J_2\right) + 2F_1\left(J + \frac{1}{2}.J_1 - \frac{1}{2}.J_2\right) + 2F_1\left(-J + \frac{3}{2}.J_1 + \frac{1}{2}.J_2\right) - 2F_1\left(-J + \frac{3}{2}.J_1 - \frac{1}{2}.J_2\right) + 2F_1\left(-J + \frac{1}{2}.J_1 + \frac{1}{2}.J_2\right) - 2F_1\left(-J + \frac{1}{2}.J_1 - \frac{1}{2}.J_2\right)\Big]$$

$$D_{20}=\frac{1}{2}.\Big[F_1\left(2.J+\tfrac{3}{2}.J_1+\tfrac{1}{2}.J_2\right)-F_1\left(2.J+\tfrac{3}{2}.J_1-\tfrac{1}{2}.J_2\right)-3F_1\left(2.J+\tfrac{1}{2}.J_1+\tfrac{1}{2}.J_2\right)+3F_1\left(2.J+\tfrac{1}{2}.J_1-\tfrac{1}{2}.J_2\right)-F_1\left(-2.J+\tfrac{3}{2}.J_1+\tfrac{1}{2}.J_2\right)+F_1\left(-2.J+\tfrac{3}{2}.J_1-\tfrac{1}{2}.J_2\right)+3F_1\left(-2.J+\tfrac{1}{2}.J_1+\tfrac{1}{2}.J_2\right)-3F_1\left(-2.J+\tfrac{1}{2}.J_1-\tfrac{1}{2}.J_2\right)-2F_1\left(J+\tfrac{3}{2}.J_1+\tfrac{1}{2}.J_2\right)+2F_1\left(J+\tfrac{3}{2}.J_1-\tfrac{1}{2}.J_2\right)+6F_1\left(J+\tfrac{1}{2}.J_1+\tfrac{1}{2}.J_2\right)-6F_1\left(J+\tfrac{1}{2}.J_1-\tfrac{1}{2}.J_2\right)+2F_1\left(-J+\tfrac{3}{2}.J_1+\tfrac{1}{2}.J_2\right)-2F_1\left(-J+\tfrac{3}{2}.J_1-\tfrac{1}{2}.J_2\right)-6F_1\left(-J+\tfrac{1}{2}.J_1+\tfrac{1}{2}.J_2\right)+6F_1\left(-J+\tfrac{1}{2}.J_1-\tfrac{1}{2}.J_2\right)\Big]$$

$$D_{21}=\frac{1}{16}.\Big[F_1\left(2.J+\tfrac{3}{2}.J_1+\tfrac{1}{2}.J_2\right)+F_1\left(2.J+\tfrac{3}{2}.J_1-\tfrac{1}{2}.J_2\right)+F_1\left(2.J+\tfrac{1}{2}.J_1+\tfrac{1}{2}.J_2\right)+F_1\left(2.J+\tfrac{1}{2}.J_1-\tfrac{1}{2}.J_2\right)+F_1\left(-2.J+\tfrac{3}{2}.J_1+\tfrac{1}{2}.J_2\right)+F_1\left(-2.J+\tfrac{3}{2}.J_1-\tfrac{1}{2}.J_2\right)+F_1\left(-2.J+\tfrac{1}{2}.J_1+\tfrac{1}{2}.J_2\right)+F_1\left(-2.J+\tfrac{1}{2}.J_1-\tfrac{1}{2}.J_2\right)-4F_1\left(J+\tfrac{3}{2}.J_1+\tfrac{1}{2}.J_2\right)-4F_1\left(J+\tfrac{3}{2}.J_1-\tfrac{1}{2}.J_2\right)-4F_1\left(J+\tfrac{1}{2}.J_1+\tfrac{1}{2}.J_2\right)-4F_1\left(J+\tfrac{1}{2}.J_1-\tfrac{1}{2}.J_2\right)-4F_1\left(-J+\tfrac{3}{2}.J_1+\tfrac{1}{2}.J_2\right)-4F_1\left(-J+\tfrac{3}{2}.J_1-\tfrac{1}{2}.J_2\right)-4F_1\left(-J+\tfrac{1}{2}.J_1+\tfrac{1}{2}.J_2\right)-4F_1\left(-J+\tfrac{1}{2}.J_1-\tfrac{1}{2}.J_2\right)+6F_1\left(\tfrac{3}{2}.J_1+\tfrac{1}{2}.J_2\right)+6F_1\left(\tfrac{3}{2}.J_1-\tfrac{1}{2}.J_2\right)+6F_1\left(\tfrac{1}{2}.J_1+\tfrac{1}{2}.J_2\right)+6F_1\left(\tfrac{1}{2}.J_1-\tfrac{1}{2}.J_2\right)\Big]$$

$$D_{22}=\frac{1}{16}.\Big[F_1\left(2.J+\tfrac{3}{2}.J_1+\tfrac{1}{2}.J_2\right)-F_1\left(2.J+\tfrac{3}{2}.J_1-\tfrac{1}{2}.J_2\right)+3F_1\left(2.J+\tfrac{1}{2}.J_1+\tfrac{1}{2}.J_2\right)-3\,F_1\left(2.J+\tfrac{1}{2}.J_1-\tfrac{1}{2}.J_2\right)+F_1\left(-2.J+\tfrac{3}{2}.J_1+\tfrac{1}{2}.J_2\right)-F_1\left(-2.J+\tfrac{3}{2}.J_1-\tfrac{1}{2}.J_2\right)+3F_1\left(-2.J+\tfrac{1}{2}.J_1+\tfrac{1}{2}.J_2\right)-3F_1\left(-2.J+\tfrac{1}{2}.J_1-\tfrac{1}{2}.J_2\right)-4F_1\left(J+\tfrac{3}{2}.J_1+\tfrac{1}{2}.J_2\right)+4F_1\left(J+\tfrac{3}{2}.J_1-\tfrac{1}{2}.J_2\right)-12F_1\left(J+\tfrac{1}{2}.J_1+\tfrac{1}{2}.J_2\right)+12F_1\left(J+\tfrac{1}{2}.J_1-\tfrac{1}{2}.J_2\right)-4F_1\left(-J+\tfrac{3}{2}.J_1+\tfrac{1}{2}.J_2\right)+4F_1\left(-J+\tfrac{3}{2}.J_1-\tfrac{1}{2}.J_2\right)-12F_1\left(-J+\tfrac{1}{2}.J_1+\tfrac{1}{2}.J_2\right)+12F_1\left(-J+\tfrac{1}{2}.J_1-\tfrac{1}{2}.J_2\right)+6F_1\left(\tfrac{3}{2}.J_1+\tfrac{1}{2}.J_2\right)-6F_1\left(\tfrac{3}{2}.J_1-\tfrac{1}{2}.J_2\right)-18F_1\left(\tfrac{1}{2}.J_1+\tfrac{1}{2}.J_2\right)+18F_1\left(\tfrac{1}{2}.J_1-\tfrac{1}{2}.J_2\right)\Big]$$

$$D_{23}=\frac{1}{4}.\Big[F_1\left(2.J+\tfrac{3}{2}.J_1+\tfrac{1}{2}.J_2\right)-F_1\left(2.J+\tfrac{3}{2}.J_1-\tfrac{1}{2}.J_2\right)-F_1\left(2.J+\tfrac{1}{2}.J_1+\tfrac{1}{2}.J_2\right)+F_1\left(2.J+\tfrac{1}{2}.J_1-\tfrac{1}{2}.J_2\right)+F_1\left(-2.J+\tfrac{3}{2}.J_1+\tfrac{1}{2}.J_2\right)-F_1\left(-2.J+\tfrac{3}{2}.J_1-\tfrac{1}{2}.J_2\right)-F_1\left(-2.J+\tfrac{1}{2}.J_1+\tfrac{1}{2}.J_2\right)+F_1\left(-2.J+\tfrac{1}{2}.J_1-\tfrac{1}{2}.J_2\right)-4F_1\left(J+\tfrac{3}{2}.J_1+\tfrac{1}{2}.J_2\right)+4F_1\left(J+\tfrac{3}{2}.J_1-\tfrac{1}{2}.J_2\right)+4F_1^z\left(J+\tfrac{1}{2}.J_1+\tfrac{1}{2}.J_2\right)-4F_1\left(J+\tfrac{1}{2}.J_1-\tfrac{1}{2}.J_2\right)-4F_1\left(-J+\tfrac{3}{2}.J_1+\tfrac{1}{2}.J_2\right)+4F_1\left(-J+\tfrac{3}{2}.J_1-\tfrac{1}{2}.J_2\right)+4F_1\left(-J+\tfrac{1}{2}.J_1+\tfrac{1}{2}.J_2\right)-4F_1\left(-J+\tfrac{1}{2}.J_1-\tfrac{1}{2}.J_2\right)+6F_1\left(\tfrac{3}{2}.J_1+\tfrac{1}{2}.J_2\right)-6F_1\left(\tfrac{3}{2}.J_1-\tfrac{1}{2}.J_2\right)-6F_1\left(\tfrac{1}{2}.J_1+\tfrac{1}{2}.J_2\right)+6F_1\left(\tfrac{1}{2}.J_1-\tfrac{1}{2}.J_2\right)\Big]$$

$$D_{24}=\frac{1}{4}.\Big[F_1\left(2.J+\tfrac{3}{2}.J_1+\tfrac{1}{2}.J_2\right)+F_1\left(2.J+\tfrac{3}{2}.J_1-\tfrac{1}{2}.J_2\right)-3F_1\left(2.J+\tfrac{1}{2}.J_1+\tfrac{1}{2}.J_2\right)-3F_1\left(2.J+\tfrac{1}{2}.J_1-\tfrac{1}{2}.J_2\right)+F_1\left(-2.J+\tfrac{3}{2}.J_1+\tfrac{1}{2}.J_2\right)+F_1\left(-2.J+\tfrac{3}{2}.J_1-\tfrac{1}{2}.J_2\right)-3F_1\left(-2.J+\tfrac{1}{2}.J_1+\tfrac{1}{2}.J_2\right)-3F_1\left(-2.J+\tfrac{1}{2}.J_1-\tfrac{1}{2}.J_2\right)-4F_1\left(J+\tfrac{3}{2}.J_1+\tfrac{1}{2}.J_2\right)-4F_1\left(J+\tfrac{3}{2}.J_1-\tfrac{1}{2}.J_2\right)+12F_1\left(J+\tfrac{1}{2}.J_1+\tfrac{1}{2}.J_2\right)+12F_1\left(J+\tfrac{1}{2}.J_1-\tfrac{1}{2}.J_2\right)-$$

$$4F_1\left(-J+\frac{3}{2}.J_1+\frac{1}{2}.J_2\right)-4F_1\left(-J+\frac{3}{2}.J_1-\frac{1}{2}.J_2\right)+12F_1\left(-J+\frac{1}{2}.J_1+\frac{1}{2}.J_2\right)+ 12F_1\left(-J+\frac{1}{2}.J_1-\frac{1}{2}.J_2\right)+6F_1\left(\frac{3}{2}.J_1+\frac{1}{2}.J_2\right)+6F_1\left(\frac{3}{2}.J_1-\frac{1}{2}.J_2\right)-18F_1\left(\frac{1}{2}.J_1+\frac{1}{2}.J_2\right)- 18F_1\left(\frac{1}{2}.J_1-\frac{1}{2}.J_2\right)\Big]$$

$$E_1=\frac{1}{8}F_2\left(\frac{3}{2}.J_1+\frac{1}{2}.J_2\right)+\frac{1}{8}F_2\left(\frac{3}{2}.J_1-\frac{1}{2}.J_2\right)+\frac{3}{8}F_2\left(\frac{1}{2}.J_1+\frac{1}{2}.J_2\right)+\frac{3}{8}F_2\left(\frac{1}{2}.J_1-\frac{1}{2}.J_2\right)$$

$$E_2=\frac{1}{2}F_2\left(\frac{3}{2}.J_1+\frac{1}{2}.J_2\right)+\frac{1}{2}F_2\left(\frac{3}{2}.J_1-\frac{1}{2}.J_2\right)-\frac{1}{2}F_2\left(\frac{1}{2}.J_1+\frac{1}{2}.J_2\right)-\frac{1}{2}F_2\left(\frac{1}{2}.J_1-\frac{1}{2}.J_2\right)$$

$$E_3=\frac{1}{2}F_2\left(\frac{3}{2}.J_1+\frac{1}{2}.J_2\right)-\frac{1}{2}F_2\left(\frac{3}{2}.J_1-\frac{1}{2}.J_2\right)+\frac{1}{2}F_2\left(\frac{1}{2}.J_1+\frac{1}{2}.J_2\right)-\frac{1}{2}F_2\left(\frac{1}{2}.J_1-\frac{1}{2}.J_2\right)$$

$$E_4=2F_2\left(\frac{3}{2}.J_1+\frac{1}{2}.J_2\right)-2F_2\left(\frac{3}{2}.J_1-\frac{1}{2}.J_2\right)-6F_2\left(\frac{1}{2}.J_1+\frac{1}{2}.J_2\right)+6F_2\left(\frac{1}{2}.J_1-\frac{1}{2}.J_2\right)$$

$$E_5=\frac{1}{16}.\Big[F_2\left(J+\frac{3}{2}.J_1+\frac{1}{2}.J_2\right)+F_2\left(J+\frac{3}{2}.J_1-\frac{1}{2}.J_2\right)+3F_2\left(J+\frac{1}{2}.J_1+\frac{1}{2}.J_2\right)+3F_2\left(J+\frac{1}{2}.J_1-\frac{1}{2}.J_2\right)+F_2\left(-J+\frac{3}{2}.J_1+\frac{1}{2}.J_2\right)+F_2\left(-J+\frac{3}{2}.J_1-\frac{1}{2}.J_2\right)+3F_2\left(-J+\frac{1}{2}.J_1+\frac{1}{2}.J_2\right)+ 3F_2\left(-J+\frac{1}{2}.J_1-\frac{1}{2}.J_2\right)-2F_2\left(\frac{3}{2}.J_1+\frac{1}{2}.J_2\right)-2F_2\left(\frac{3}{2}.J_1-\frac{1}{2}.J_2\right)-6F_2\left(\frac{1}{2}.J_1+\frac{1}{2}.J_2\right)- 6F_2\left(\frac{1}{2}.J_1-\frac{1}{2}.J_2\right)\Big]$$

$$E_6=\frac{1}{32}.\Big[F_2\left(2.J+\frac{3}{2}.J_1+\frac{1}{2}.J_2\right)+F_2\left(2.J+\frac{3}{2}.J_1-\frac{1}{2}.J_2\right)+3F_2\left(2.J+\frac{1}{2}.J_1+\frac{1}{2}.J_2\right)+ 3F_2\left(2.J+\frac{1}{2}.J_1-\frac{1}{2}.J_2\right)+F_2\left(-2.J+\frac{3}{2}.J_1+\frac{1}{2}.J_2\right)+F_2\left(-2.J+\frac{3}{2}.J_1-\frac{1}{2}.J_2\right)+ 3F_2\left(-2.J+\frac{1}{2}.J_1+\frac{1}{2}.J_2\right)+3F_2\left(-2.J+\frac{1}{2}.J_1-\frac{1}{2}.J_2\right)-2F_2\left(\frac{3}{2}.J_1+\frac{1}{2}.J_2\right)- 2F_2\left(\frac{3}{2}.J_1-\frac{1}{2}.J_2\right)-6F_2\left(\frac{1}{2}.J_1+\frac{1}{2}.J_2\right)-6F_2\left(\frac{1}{2}.J_1-\frac{1}{2}.J_2\right)\Big]$$

$$E_7=\frac{1}{32}.\Big[F_2\left(2.J+\frac{3}{2}.J_1+\frac{1}{2}.J_2\right)+F_2\left(2.J+\frac{3}{2}.J_1-\frac{1}{2}.J_2\right)+3F_2\left(2.J+\frac{1}{2}.J_1+\frac{1}{2}.J_2\right)+ 3F_2\left(2.J+\frac{1}{2}.J_1-\frac{1}{2}.J_2\right)+F_2\left(-2.J+\frac{3}{2}.J_1+\frac{1}{2}.J_2\right)+F_2\left(-2.J+\frac{3}{2}.J_1-\frac{1}{2}.J_2\right)+ 3F_2\left(-2.J+\frac{1}{2}.J_1+\frac{1}{2}.J_2\right)+3F_2\left(-2.J+\frac{1}{2}.J_1-\frac{1}{2}.J_2\right)-4F_2\left(J+\frac{3}{2}.J_1+\frac{1}{2}.J_2\right)- 4F_2\left(J+\frac{3}{2}.J_1-\frac{1}{2}.J_2\right)-12F_2\left(J+\frac{1}{2}.J_1+\frac{1}{2}.J_2\right)-12F_2\left(J+\frac{1}{2}.J_1-\frac{1}{2}.J_2\right)-4F_2\left(-J+\frac{3}{2}.J_1+\frac{1}{2}.J_2\right)-4F_2\left(-J+\frac{3}{2}.J_1-\frac{1}{2}.J_2\right)-12F_2\left(-J+\frac{1}{2}.J_1+\frac{1}{2}.J_2\right)-12F_2\left(-J+\frac{1}{2}.J_1-\frac{1}{2}.J_2\right)+6F_2\left(\frac{3}{2}.J_1+\frac{1}{2}.J_2\right)+6F_2\left(\frac{3}{2}.J_1-\frac{1}{2}.J_2\right)+18F_2\left(\frac{1}{2}.J_1+\frac{1}{2}.J_2\right)+18F_2\left(\frac{1}{2}.J_1-\frac{1}{2}.J_2\right)\Big]$$

$$E_8=\frac{1}{8}.\Big[F_2\left(J+\frac{3}{2}.J_1+\frac{1}{2}.J_2\right)+F_2\left(J+\frac{3}{2}.J_1-\frac{1}{2}.J_2\right)+F_2\left(J+\frac{1}{2}.J_1+\frac{1}{2}.J_2\right)+F_2\left(J+\frac{1}{2}.J_1-\frac{1}{2}.J_2\right)-F_2\left(-J+\frac{3}{2}.J_1+\frac{1}{2}.J_2\right)-F_2\left(-J+\frac{3}{2}.J_1-\frac{1}{2}.J_2\right)-F_2\left(-J+\frac{1}{2}.J_1+\frac{1}{2}.J_2\right)- F_2\left(-J+\frac{1}{2}.J_1-\frac{1}{2}.J_2\right)\Big]$$

$$E_9 = \frac{1}{8}.\Big[F_2\left(J + \frac{3}{2}.J_1 + \frac{1}{2}.J_2\right) - F_2\left(J + \frac{3}{2}.J_1 - \frac{1}{2}.J_2\right) + 3F_2\left(J + \frac{1}{2}.J_1 + \frac{1}{2}.J_2\right) - 3F_2\left(J + \frac{1}{2}.J_1 - \frac{1}{2}.J_2\right) - F_2\left(-J + \frac{3}{2}.J_1 + \frac{1}{2}.J_2\right) + F_2\left(-J + \frac{3}{2}.J_1 - \frac{1}{2}.J_2\right) - 3F_2\left(-J + \frac{1}{2}.J_1 + \frac{1}{2}.J_2\right) + 3F_2\left(-J + \frac{1}{2}.J_1 - \frac{1}{2}.J_2\right)\Big]$$

$$E_{10} = \frac{1}{2}.\Big[F_2\left(J + \frac{3}{2}.J_1 + \frac{1}{2}.J_2\right) - F_2\left(J + \frac{3}{2}.J_1 - \frac{1}{2}.J_2\right) - F_2\left(J + \frac{1}{2}.J_1 + \frac{1}{2}.J_2\right) + F_2\left(J + \frac{1}{2}.J_1 - \frac{1}{2}.J_2\right) - F_2\left(-J + \frac{3}{2}.J_1 + \frac{1}{2}.J_2\right) + F_2\left(-J + \frac{3}{2}.J_1 - \frac{1}{2}.J_2\right) + F_2\left(-J + \frac{1}{2}.J_1 + \frac{1}{2}.J_2\right) - F_2\left(-J + \frac{1}{2}.J_1 - \frac{1}{2}.J_2\right)\Big]$$

$$E_{11} = \frac{1}{2}.\Big[F_2\left(J + \frac{3}{2}.J_1 + \frac{1}{2}.J_2\right) + F_2\left(J + \frac{3}{2}.J_1 - \frac{1}{2}.J_2\right) - 3F_2\left(J + \frac{1}{2}.J_1 + \frac{1}{2}.J_2\right) - 3F_2\left(J + \frac{1}{2}.J_1 - \frac{1}{2}.J_2\right) - F_2\left(-J + \frac{3}{2}.J_1 + \frac{1}{2}.J_2\right) - F_2\left(-J + \frac{3}{2}.J_1 - \frac{1}{2}.J_2\right) + 3F_2\left(-J + \frac{1}{2}.J_1 + \frac{1}{2}.J_2\right) + 3F_2\left(-J + \frac{1}{2}.J_1 - \frac{1}{2}.J_2\right)\Big]$$

$$E_{12} = \frac{1}{4}.\Big[F_2\left(J + \frac{3}{2}.J_1 + \frac{1}{2}.J_2\right) + F_2\left(J + \frac{3}{2}.J_1 - \frac{1}{2}.J_2\right) - F_2\left(J + \frac{1}{2}.J_1 + \frac{1}{2}.J_2\right) - F_2\left(J + \frac{1}{2}.J_1 - \frac{1}{2}.J_2\right) + F_2\left(-J + \frac{3}{2}.J_1 + \frac{1}{2}.J_2\right) + F_2\left(-J + \frac{3}{2}.J_1 - \frac{1}{2}.J_2\right) - F_2\left(-J + \frac{1}{2}.J_1 + \frac{1}{2}.J_2\right) - F_2\left(-J + \frac{1}{2}.J_1 - \frac{1}{2}.J_2\right) - 2F_2\left(\frac{3}{2}.J_1 + \frac{1}{2}.J_2\right) - 2F_2\left(\frac{3}{2}.J_1 - \frac{1}{2}.J_2\right) + 2F_2\left(\frac{1}{2}.J_1 + \frac{1}{2}.J_2\right) + 2F_2\left(\frac{1}{2}.J_1 - \frac{1}{2}.J_2\right)\Big]$$

$$E_{13} = \frac{1}{4}.\Big[F_2\left(J + \frac{3}{2}.J_1 + \frac{1}{2}.J_2\right) - F_2\left(J + \frac{3}{2}.J_1 - \frac{1}{2}.J_2\right) + F_2\left(J + \frac{1}{2}.J_1 + \frac{1}{2}.J_2\right) - \left(J + \frac{1}{2}.J_1 - \frac{1}{2}.J_2\right) + F_2\left(-J + \frac{3}{2}.J_1 + \frac{1}{2}.J_2\right) - F_2\left(-J + \frac{3}{2}.J_1 - \frac{1}{2}.J_2\right) + F_2\left(-J + \frac{1}{2}.J_1 + \frac{1}{2}.J_2\right) - F_2\left(-J + \frac{1}{2}.J_1 - \frac{1}{2}.J_2\right) - 2F_2\left(\frac{3}{2}.J_1 + \frac{1}{2}.J_2\right) + 2F_2\left(\frac{3}{2}.J_1 - \frac{1}{2}.J_2\right) - 2F_2\left(\frac{1}{2}.J_1 + \frac{1}{2}.J_2\right) + 2F_2\left(\frac{1}{2}.J_1 - \frac{1}{2}.J_2\right)\Big]$$

$$E_{14} = F_2\left(J + \frac{3}{2}.J_1 + \frac{1}{2}.J_2\right) - F_2\left(J + \frac{3}{2}.J_1 - \frac{1}{2}.J_2\right) - 3F_2\left(J + \frac{1}{2}.J_1 + \frac{1}{2}.J_2\right) + 3F_2\left(J + \frac{1}{2}.J_1 - \frac{1}{2}.J_2\right) + F_2\left(-J + \frac{3}{2}.J_1 + \frac{1}{2}.J_2\right) - F_2\left(-J + \frac{3}{2}.J_1 - \frac{1}{2}.J_2\right) - 3F_2\left(-J + \frac{1}{2}.J_1 + \frac{1}{2}.J_2\right) + 3F_2\left(-J + \frac{1}{2}.J_1 - \frac{1}{2}.J_2\right) - 2F_2\left(\frac{3}{2}.J_1 + \frac{1}{2}.J_2\right) + 2F_2\left(\frac{3}{2}.J_1 - \frac{1}{2}.J_2\right) + 6F_2\left(\frac{1}{2}.J_1 + \frac{1}{2}.J_2\right) - 6F_2\left(\frac{1}{2}.J_1 - \frac{1}{2}.J_2\right)$$

$$E_{15} = \frac{1}{2}.\Big[F_2\left(J + \frac{3}{2}.J_1 + \frac{1}{2}.J_2\right) - F_2\left(J + \frac{3}{2}.J_1 - \frac{1}{2}.J_2\right) - 3F_2\left(J + \frac{1}{2}.J_1 + \frac{1}{2}.J_2\right) + 3F_2\left(J + \frac{1}{2}.J_1 - \frac{1}{2}.J_2\right) + F_2\left(-J + \frac{3}{2}.J_1 + \frac{1}{2}.J_2\right) - F_2\left(-J + \frac{3}{2}.J_1 - \frac{1}{2}.J_2\right) - 3F_2\left(-J + \frac{1}{2}.J_1 + \frac{1}{2}.J_2\right) + 3F_2\left(-J + \frac{1}{2}.J_1 - \frac{1}{2}.J_2\right) - 2F_2\left(\frac{3}{2}.J_1 + \frac{1}{2}.J_2\right) + 2F_2\left(\frac{3}{2}.J_1 - \frac{1}{2}.J_2\right) + 6F_2\left(\frac{1}{2}.J_1 + \frac{1}{2}.J_2\right) - 6F_2\left(\frac{1}{2}.J_1 - \frac{1}{2}.J_2\right)\Big]$$

$$E_{16} = \frac{1}{8}.\Big[F_2\left(J + \frac{3}{2}.J_1 + \frac{1}{2}.J_2\right) + F_2\left(J + \frac{3}{2}.J_1 - \frac{1}{2}.J_2\right) - F_2\left(J + \frac{1}{2}.J_1 + \frac{1}{2}.J_2\right) - F_2\left(J + \frac{1}{2}.J_1 - \frac{1}{2}.J_2\right) + F_2\left(-J + \frac{3}{2}.J_1 + \frac{1}{2}.J_2\right) + F_2\left(-J + \frac{3}{2}.J_1 - \frac{1}{2}.J_2\right) - F_2\left(-J + \frac{1}{2}.J_1 + \frac{1}{2}.J_2\right) -$$

$$F_2\left(-J+\frac{1}{2}.J_1-\frac{1}{2}.J_2\right)-2F_2\left(\frac{3}{2}.J_1+\frac{1}{2}.J_2\right)-2F_2\left(\frac{3}{2}.J_1-\frac{1}{2}.J_2\right)+2F_2\left(\frac{1}{2}.J_1+\frac{1}{2}.J_2\right)+2F_2\left(\frac{1}{2}.J_1-\frac{1}{2}.J_2\right)\Big]$$

$$E_{17}=\frac{1}{8}.\Big[F_2\left(J+\frac{3}{2}.J_1+\frac{1}{2}.J_2\right)-F_2\left(J+\frac{3}{2}.J_1-\frac{1}{2}.J_2\right)+F_2\left(J+\frac{1}{2}.J_1+\frac{1}{2}.J_2\right)-F_2\left(J+\frac{1}{2}.J_1-\frac{1}{2}.J_2\right)+F_2\left(-J+\frac{3}{2}.J_1+\frac{1}{2}.J_2\right)-F_2\left(-J+\frac{3}{2}.J_1-\frac{1}{2}.J_2\right)+F_2\left(-J+\frac{1}{2}.J_1+\frac{1}{2}.J_2\right)-F_2\left(-J+\frac{1}{2}.J_1-\frac{1}{2}.J_2\right)-2F_2\left(\frac{3}{2}.J_1+\frac{1}{2}.J_2\right)+2F_2\left(\frac{3}{2}.J_1-\frac{1}{2}.J_2\right)-2F_2\left(\frac{1}{2}.J_1+\frac{1}{2}.J_2\right)+2F_2\left(\frac{1}{2}.J_1-\frac{1}{2}.J_2\right)\Big]$$

$$E_{18}=\frac{1}{16}.\Big[F_2\left(2.J+\frac{3}{2}.J_1+\frac{1}{2}.J_2\right)+F_2\left(2.J+\frac{3}{2}.J_1-\frac{1}{2}.J_2\right)+F_2\left(2.J+\frac{1}{2}.J_1+\frac{1}{2}.J_2\right)+F_2\left(2.J+\frac{1}{2}.J_1-\frac{1}{2}.J_2\right)-F_2\left(-2.J+\frac{3}{2}.J_1+\frac{1}{2}.J_2\right)-F_2\left(-2.J+\frac{3}{2}.J_1-\frac{1}{2}.J_2\right)-F_2\left(-2.J+\frac{1}{2}.J_1+\frac{1}{2}.J_2\right)-F_2\left(-2.J+\frac{1}{2}.J_1-\frac{1}{2}.J_2\right)-2F_2\left(J+\frac{3}{2}.J_1+\frac{1}{2}.J_2\right)-2F_2\left(J+\frac{3}{2}.J_1-\frac{1}{2}.J_2\right)-2F_2\left(J+\frac{1}{2}.J_1+\frac{1}{2}.J_2\right)-2F_2\left(J+\frac{1}{2}.J_1-\frac{1}{2}.J_2\right)+2F_2\left(-J+\frac{3}{2}.J_1+\frac{1}{2}.J_2\right)+2F_2\left(-J+\frac{3}{2}.J_1-\frac{1}{2}.J_2\right)+2F_2\left(-J+\frac{1}{2}.J_1+\frac{1}{2}.J_2\right)+2F_2\left(-J+\frac{1}{2}.J_1-\frac{1}{2}.J_2\right)\Big]$$

$$E_{19}=\frac{1}{16}.\Big[F_2\left(2.J+\frac{3}{2}.J_1+\frac{1}{2}.J_2\right)-F_2\left(2.J+\frac{3}{2}.J_1-\frac{1}{2}.J_2\right)+3F_2\left(2.J+\frac{1}{2}.J_1+\frac{1}{2}.J_2\right)-3F_2\left(2.J+\frac{1}{2}.J_1-\frac{1}{2}.J_2\right)-F_2\left(-2.J+\frac{3}{2}.J_1+\frac{1}{2}.J_2\right)+F_2\left(-2.J+\frac{3}{2}.J_1-\frac{1}{2}.J_2\right)-3F_2\left(-2.J+\frac{1}{2}.J_1+\frac{1}{2}.J_2\right)+3F_2\left(-2.J+\frac{1}{2}.J_1-\frac{1}{2}.J_2\right)-2F_2\left(J+\frac{3}{2}.J_1+\frac{1}{2}.J_2\right)+2F_2\left(J+\frac{3}{2}.J_1-\frac{1}{2}.J_2\right)-6F_2\left(J+\frac{1}{2}.J_1+\frac{1}{2}.J_2\right)+6F_2\left(J+\frac{1}{2}.J_1-\frac{1}{2}.J_2\right)+2F_2\left(-J+\frac{3}{2}.J_1+\frac{1}{2}.J_2\right)-2F_2\left(-J+\frac{3}{2}.J_1-\frac{1}{2}.J_2\right)+6F_2\left(-J+\frac{1}{2}.J_1+\frac{1}{2}.J_2\right)-6F_2\left(-J+\frac{1}{2}.J_1-\frac{1}{2}.J_2\right)\Big]$$

$$E_{20}=\frac{1}{4}.\Big[F_2\left(2.J+\frac{3}{2}.J_1+\frac{1}{2}.J_2\right)+F_2\left(2.J+\frac{3}{2}.J_1-\frac{1}{2}.J_2\right)-3F_2\left(2.J+\frac{1}{2}.J_1+\frac{1}{2}.J_2\right)-3F_2\left(2.J+\frac{1}{2}.J_1-\frac{1}{2}.J_2\right)-F_2\left(-2.J+\frac{3}{2}.J_1+\frac{1}{2}.J_2\right)-F_2\left(-2.J+\frac{3}{2}.J_1-\frac{1}{2}.J_2\right)+3F_2\left(-2.J+\frac{1}{2}.J_1+\frac{1}{2}.J_2\right)+3F_2\left(-2.J+\frac{1}{2}.J_1-\frac{1}{2}.J_2\right)-2F_2\left(J+\frac{3}{2}.J_1+\frac{1}{2}.J_2\right)-2F_2\left(J+\frac{3}{2}.J_1-\frac{1}{2}.J_2\right)+6F_2\left(J+\frac{1}{2}.J_1+\frac{1}{2}.J_2\right)+6F_2\left(J+\frac{1}{2}.J_1-\frac{1}{2}.J_2\right)+2F_2\left(-J+\frac{3}{2}.J_1+\frac{1}{2}.J_2\right)+2F_2\left(-J+\frac{3}{2}.J_1-\frac{1}{2}.J_2\right)-6F_2\left(-J+\frac{1}{2}.J_1+\frac{1}{2}.J_2\right)-6F_2\left(-J+\frac{1}{2}.J_1-\frac{1}{2}.J_2\right)\Big]$$

$$E_{21}=\frac{1}{4}.\Big[F_2\left(2.J+\frac{3}{2}.J_1+\frac{1}{2}.J_2\right)-F_2\left(2.J+\frac{3}{2}.J_1-\frac{1}{2}.J_2\right)-F_2\left(2.J+\frac{1}{2}.J_1+\frac{1}{2}.J_2\right)+F_2\left(2.J+\frac{1}{2}.J_1-\frac{1}{2}.J_2\right)-F_2\left(-2.J+\frac{3}{2}.J_1+\frac{1}{2}.J_2\right)+F_2\left(-2.J+\frac{3}{2}.J_1-\frac{1}{2}.J_2\right)+F_2\left(-2.J+\frac{1}{2}.J_1+\frac{1}{2}.J_2\right)-F_2\left(-2.J+\frac{1}{2}.J_1-\frac{1}{2}.J_2\right)-2F_2\left(J+\frac{3}{2}.J_1+\frac{1}{2}.J_2\right)+2F_2\left(J+\frac{3}{2}.J_1-\frac{1}{2}.J_2\right)+2F_2\left(J+\frac{1}{2}.J_1+\frac{1}{2}.J_2\right)-2F_2\left(J+\frac{1}{2}.J_1-\frac{1}{2}.J_2\right)+2F_2\left(-J+\frac{3}{2}.J_1+\frac{1}{2}.J_2\right)-2F_2\left(-J+\frac{3}{2}.J_1-\frac{1}{2}.J_2\right)-2F_2\left(-J+\frac{1}{2}.J_1+\frac{1}{2}.J_2\right)+2F_2\left(-J+\frac{1}{2}.J_1-\frac{1}{2}.J_2\right)\Big]$$

$$E_{22} = \frac{1}{8}.\Big[F_2\Big(2.J + \frac{3}{2}.J_1 + \frac{1}{2}.J_2\Big) + F_2\Big(2.J + \frac{3}{2}.J_1 - \frac{1}{2}.J_2\Big) - F_2\Big(2.J + \frac{1}{2}.J_1 + \frac{1}{2}.J_2\Big) - F_2\Big(2.J + \frac{1}{2}.J_1 - \frac{1}{2}.J_2\Big) + F_2\Big(-2.J + \frac{3}{2}.J_1 + \frac{1}{2}.J_2\Big) + F_2\Big(-2.J + \frac{3}{2}.J_1 - \frac{1}{2}.J_2\Big) - F_2\Big(-2.J + \frac{1}{2}.J_1 + \frac{1}{2}.J_2\Big) - F_2\Big(-2.J + \frac{1}{2}.J_1 - \frac{1}{2}.J_2\Big) - 4F_2\Big(J + \frac{3}{2}.J_1 + \frac{1}{2}.J_2\Big) - 4F_2\Big(J + \frac{3}{2}.J_1 - \frac{1}{2}.J_2\Big) + 4F_2\Big(J + \frac{1}{2}.J_1 + \frac{1}{2}.J_2\Big) + 4F_2\Big(J + \frac{1}{2}.J_1 - \frac{1}{2}.J_2\Big) - 4F_2\Big(-J + \frac{3}{2}.J_1 + \frac{1}{2}.J_2\Big) - 4F_2\Big(-J + \frac{3}{2}.J_1 - \frac{1}{2}.J_2\Big) + 4F_2\Big(-J + \frac{1}{2}.J_1 + \frac{1}{2}.J_2\Big) + 4F_2\Big(-J + \frac{1}{2}.J_1 - \frac{1}{2}.J_2\Big) + 6F_2\Big(\frac{3}{2}.J_1 + \frac{1}{2}.J_2\Big) + 6F_2\Big(\frac{3}{2}.J_1 - \frac{1}{2}.J_2\Big) - 6F_2\Big(\frac{1}{2}.J_1 + \frac{1}{2}.J_2\Big) - 6F_2\Big(\frac{1}{2}.J_1 - \frac{1}{2}.J_2\Big)\Big]$$

$$E_{23} = \frac{1}{8}.\Big[F_2\Big(2.J + \frac{3}{2}.J_1 + \frac{1}{2}.J_2\Big) - F_2\Big(2.J + \frac{3}{2}.J_1 - \frac{1}{2}.J_2\Big) + F_2\Big(2.J + \frac{1}{2}.J_1 + \frac{1}{2}.J_2\Big) - F_2\Big(2.J + \frac{1}{2}.J_1 - \frac{1}{2}.J_2\Big) + F_2\Big(-2.J + \frac{3}{2}.J_1 + \frac{1}{2}.J_2\Big) - F_2\Big(-2.J + \frac{3}{2}.J_1 - \frac{1}{2}.J_2\Big) + F_2\Big(-2.J + \frac{1}{2}.J_1 + \frac{1}{2}.J_2\Big) - F_2\Big(-2.J + \frac{1}{2}.J_1 - \frac{1}{2}.J_2\Big) - 4F_2\Big(J + \frac{3}{2}.J_1 + \frac{1}{2}.J_2\Big) + 4F_2\Big(J + \frac{3}{2}.J_1 - \frac{1}{2}.J_2\Big) - 4F_2\Big(J + \frac{1}{2}.J_1 + \frac{1}{2}.J_2\Big) + 4F_2\Big(J + \frac{1}{2}.J_1 - \frac{1}{2}.J_2\Big) - 4F_2\Big(-J + \frac{3}{2}.J_1 + \frac{1}{2}.J_2\Big) + 4F_2\Big(-J + \frac{3}{2}.J_1 - \frac{1}{2}.J_2\Big) - 4F_2\Big(-J + \frac{1}{2}.J_1 + \frac{1}{2}.J_2\Big) + 4F_2\Big(-J + \frac{1}{2}.J_1 - \frac{1}{2}.J_2\Big) + 6F_2\Big(\frac{3}{2}.J_1 + \frac{1}{2}.J_2\Big) - 6F_2\Big(\frac{3}{2}.J_1 - \frac{1}{2}.J_2\Big) + 6F_2\Big(\frac{1}{2}.J_1 + \frac{1}{2}.J_2\Big) - 6F_2\Big(\frac{1}{2}.J_1 - \frac{1}{2}.J_2\Big)\Big]$$

$$E_{24} = \frac{1}{2}.\Big[F_2\Big(2.J + \frac{3}{2}.J_1 + \frac{1}{2}.J_2\Big) - F_2\Big(2.J + \frac{3}{2}.J_1 - \frac{1}{2}.J_2\Big) - 3F_2\Big(2.J + \frac{1}{2}.J_1 + \frac{1}{2}.J_2\Big) + 3F_2\Big(2.J + \frac{1}{2}.J_1 - \frac{1}{2}.J_2\Big) + F_2\Big(-2.J + \frac{3}{2}.J_1 + \frac{1}{2}.J_2\Big) - F_2\Big(-2.J + \frac{3}{2}.J_1 - \frac{1}{2}.J_2\Big) - 3F_2\Big(-2.J + \frac{1}{2}.J_1 + \frac{1}{2}.J_2\Big) + 3F_2\Big(-2.J + \frac{1}{2}.J_1 - \frac{1}{2}.J_2\Big) - 4F_2\Big(J + \frac{3}{2}.J_1 + \frac{1}{2}.J_2\Big) + 4F_2\Big(J + \frac{3}{2}.J_1 - \frac{1}{2}.J_2\Big) + 12F_2\Big(J + \frac{1}{2}.J_1 + \frac{1}{2}.J_2\Big) - 12F_2\Big(J + \frac{1}{2}.J_1 - \frac{1}{2}.J_2\Big) - 4F_2\Big(-J + \frac{3}{2}.J_1 + \frac{1}{2}.J_2\Big) + 4F_2\Big(-J + \frac{3}{2}.J_1 - \frac{1}{2}.J_2\Big) + 12F_2\Big(-J + \frac{1}{2}.J_1 + \frac{1}{2}.J_2\Big) - 12F_2\Big(-J + \frac{1}{2}.J_1 - \frac{1}{2}.J_2\Big) + 6F_2\Big(\frac{3}{2}.J_1 + \frac{1}{2}.J_2\Big) - 6F_2\Big(\frac{3}{2}.J_1 - \frac{1}{2}.J_2\Big) - 18F_2\Big(\frac{1}{2}.J_1 + \frac{1}{2}.J_2\Big) + 18F_2\Big(\frac{1}{2}.J_1 - \frac{1}{2}.J_2\Big)\Big]$$

$$A = [A_2 + (2A_1 + 2A_3.q_0).(2B_2 + 2B_6.q_0)].[1 - (2A_1 + 2A_3.q_0).(2B_1 + 4B_3.q_0 + 2B_7.q_0^2)]^{-1}$$

$$B = [A_2.(2B_1 + 4B_3.q_0 + 2B_7.q_0^2) + 2B_2 + 2B_6.q_0].[1 - (2A_1 + 2A_3.q_0).(2B_1 + 4B_3.q_0 + 2B_7.q_0^2)]^{-1}$$

$$C = 6C_1 + 30C_2.q_0 + 60C_3.q_0^2 + 60C_4.q_0^3 + 30C_5.q_0^4 + 6C_6.q_0^5$$

$$D = E.(2A_1 + 2A_3.q_0) + 2A_3.q_9.B + A_4.B^2$$

$$E = \{2B_6.q_9 + A_2.(4B_3.q_9 + 4B_7.q_0.q_9) + 2A_2.B_5 + A_2^2.(2B_4 + 2B_8.q_0) + B.(2A_1 + 2A_3.q_0).(2B_5 + 4B_3.q_9 + 4B_7.q_0.q_9 + 4A_2.B_4 + 4A_2.B_8.q_0) + (2B_1 + 4B_3.q_0 + 2B_7.q_0^2).(2A_3.q_9.B + A_4.B^2) + B^2.(2B_4 + 2B_8.q_0).(2A_1 + 2A_3.q_0)^2\}.\{1 - (2A_1 + 2A_3.q_0).(2B_1 + 4B_3.q_0 + 2B_7.q_0^2)\}^{-1}$$

$$F = 20C_7 + 60C_8.q_0 + 60C_9.q_0^2 + 20C_{10}.q_0^3 + 30C_2.q_9 + 120C_3.q_9.q_0 + 180C_4.q_9.q_0^2 + 120C_5.q_9.q_0^3 + 30C_6.q_9.q_0^4$$

$$q_1 = \frac{E_2+2E_{12}.q_0+E_{22}.q_0^2}{1-2E_5-2E_7.q_0} \quad , \quad q_2 = \frac{E_6}{1-2E_5-2E_7.q_0}$$

$$q_3 = \frac{2E_8+2E_{18}.q_0}{1-2E_5-2E_7.q_0} \quad , \quad q_4 = \frac{4E_8+4E_{18}.q_0}{1-2E_5-2E_7.q_0}$$

$$q_5 = \frac{2E_9+2E_{19}.q_0}{1-2E_5-2E_7.q_0} \quad , \quad q_6 = \frac{2E_2+4E_{12}.q_0+2E_{22}.q_0^2}{1-2E_5-2E_7.q_0}$$

$$q_7 = \frac{E_3+2E_{13}.q_0+E_{23}.q_0^2}{1-2E_5-2E_7.q_0} \quad , \quad q_8 = \frac{2E_3+4E_{13}.q_0+2E_{23}.q_0^2}{1-2E_5-2E_7.q_0}$$

$$q_9 = q_1.B^2 + q_2 + q_3.A + q_4.B + q_5.C + q_6.A.B + q_7.A.C + q_8.B.C$$

Journal of Physics: Conference Series **758** (2016) 012006 doi:10.1088/1742-6596/758/1/012006

Dynamic model of heat and mass transfer in rectangular adsorber of a solar adsorption machine

W Chekirou[1*], N Boukheit[1] et A Karaali[1]
[1]Laboratoire de thermodynamique et traitement de surfaces de matériaux, Université des frères Mentouri Constantine, Constantine 25000, Algérie

E-mail : chekirouw@yahoo.fr

Abstract. This paper presents the study of a rectangular adsorber of solar adsorption cooling machine. The modeling and the analysis of the adsorber are the key point of such studies; because of the complex coupled heat and mass transfer phenomena that occur during the working cycle. The adsorber is heated by solar energy and contains a porous medium constituted of activated carbon AC-35 reacting by adsorption with methanol. To study the solar collector type effect on system's performances, the used model takes into account the variation of ambient temperature and solar intensity along a simulated day, corresponding to a total daily insolation of 26.12 MJ/m^2 with ambient temperature average of 27.7 °C, which is useful to know the daily thermal behavior of the rectangular adsorber.

1. Introduction

Since the Montreal protocol called for a ban on the use of CFCs, the research efforts have been increased over the last twenty years on the development of refrigeration technologies, which address the environmental concerns of Ozone layer depletion and global warming. Solar adsorption cooling machines constitute very attractive solutions. They are of significance to meet the needs for cooling requirements such as air conditioning, ice making, and medical or food preservation in remote areas far from electric grid. Compared with the electric driven vapor compression refrigerator systems, these machines are advantageous because: they are noiseless; they usually employ environmentally friendly substances as refrigerant, such as water, methanol and ammonia; they operate with no moving part; no compression work. They are also advantageous if they are compared with the absorption systems, mainly because: their cycle is intermittent, which is well adapted to solar energy; the adsorbent bed can be implanted directly in the solar collector; there is no need for circulation pump or rectifier; They can drive by low temperature heat sources, which can be provided by solar energy with single solar flat plate collector. Although, the adsorption solar cooling systems have all the advantageous discussed above, some drawbacks have become obstacles for its real applications and commercialization, such as: the discontinuous operation of the cycle; the large volume and weight relative to traditional refrigeration systems; the low coefficient of performance [1-3] and the long cycle time.

The adsorber is the most important component in an adsorption cooling systems and the enhancement of heat and mass transfer inside it is the most important factor to improve the performance of such

[1] To whom any correspondence should be addressed.

systems. Thus, great efforts have been made to investigate the transfer occurring in the adsorption and desorption process.
The model presented in this study simulates a transitory behaviour of the solar reactor (adsorber). For this purpose a mathematical model based on uniform pressure and non-uniform temperature distribution inside the adsorbent bed has been developed, using the meteorological data of Constantine. The influence of the type of solar collector on the system's performance is also discussed using COP_s (solar performance coefficient) as an optimization criteria. The effect of the adsorbent mass, solar collector area and total solar radiation is also discussed.

2. Solar intensity and ambient temperature

In order to simulate the system in more realistic manner, the solar intensity and ambient temperature are modelled in this study as varying along the day (hypothetical clear day). These meteorological conditions are taken for Constantine region, which is situated at the north- east of Algeria, at 6°,37' East (longitude) and 37°,17' North (latitude), with an average altitude of about 625 m [4].
The solar intensity is assumed to vary sinusoidally from sunrise to sunset according to:

$$G(t) = G_{\max} \sin(\frac{\pi t}{D_j}) \tag{1}$$

Where $G_{\max}$ is the maximum solar intensity occurring at solar noon; D_j is the length of the day; t in equation (1) is the difference between the time of the day (at a given instant) and the sunrise time in hours.
The total solar intensity G_{tot} per 1 m^2 of collector area is obtained by:

$$G_{tot} = \int_{Sunser}^{Sunset} G(t)\, dt \tag{2}$$

The ambient temperature is calculated by the following equation [4]:

$$T_{amb} = \frac{(T_{\max} + T_{\min})}{2} + \frac{(T_{\max} - T_{\min})}{2} \sin\left(\frac{\pi\,(t-8)}{D_j}\right) \tag{3}$$

Where, $T_{\max}$ and $T_{\min}$ are the maximal and minimal daily temperatures.
The distribution of the solar intensity and ambient temperature during a whole day is shown in figure 1. This simulated day corresponds to a total daily insolation of 26,12 MJ/m^2 and an average ambient temperature of 27,7 °C.

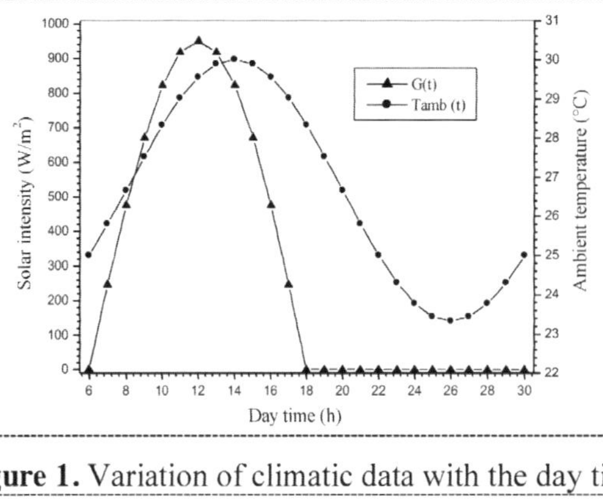

Figure 1. Variation of climatic data with the day time.

3. Performance analysis

The performance of the adsorption cooling system considered in this study is evaluated by the solar performance coefficient COP_s, which is calculated as the ratio between the cooling production and the total daily solar energy absorbed by the collector during the whole day.
It is given by:

$$COP_s = \frac{Q_f}{A \int_{sunrise}^{sunset} G(t)\, dt} \quad (4)$$

Where, A is the total solar collector area. Q_f is the cooling production, it is defined by the adsorbate latent heat of evaporation minus the sensible heat to cool the adsorbate from the condensation temperature to the evaporation temperature, between the period of time when isosteric and isobaric cooling is completed, respectively.

4. Results

A computer program was done on the basis of the numerical model, which is well discussed in our previous work by Chekirou et al. [5]. Al the basic parameters used in the model are also cited in reference [5].
The estimates of the saturated vapour pressure, the adsorbed phase density and the latent heat of methanol are given by Weast [6] and Bejan and Kraus [7].
The figure 2 represents a schema of rectangular adsorber used in this study. The rear and the lateral insulation are used to limit the thermal losses. The activated carbon AC-35 is packed inside a rectangular adsorber.

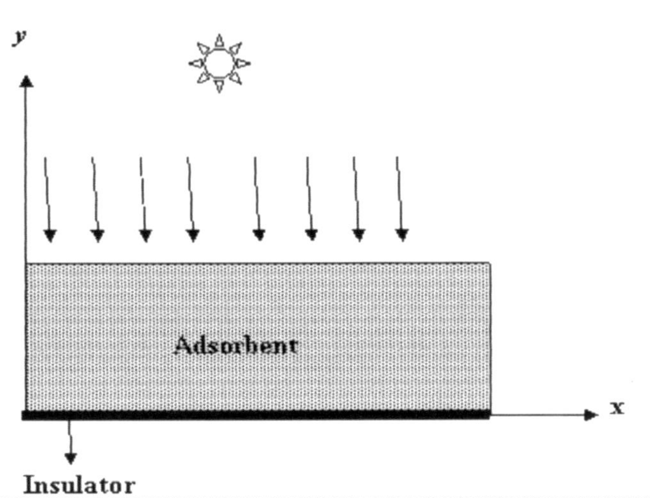

Figure 2. Schema of rectangular adsorber.

The activated carbon/methanol pair has proved to be the best pair among those studies so far, because it is reasonably stable chemically, has a high performance coefficient and is less expensive than other pairs. Methanol has been selected as an adsorbate, because it can evaporate at a temperature largely below 0°C, its enthalpy of evaporation is high (~ 1200 KJ/kg at -5 °C), its molecule is small enough (4. 10^{-4} μm) to be easily adsorbed into micropores, with a diameter smaller than 2. 10^{-3} μm. Its pressure is always lower than the atmospheric one, which means a safety factor in case of leakage. The activated carbon AC-35 has been selected as an adsorbent material, because of its affinity with methanol. It has high adsorptive capacity at ambient temperature and low pressures, maintaining small capacity of adsorption at high temperatures and pressures. It has a significant volume of micropores of convenient size for adsorption. The thermo-physical properties of the AC-35, is produced by CECA (France), its specific surface BET is 1150 m^3/g. The void space in the AC-35 adsorbent bed represents 78 % of its total volume (39 % of the volume is constituted by the interparticles space, and 39 % by the pores within the grains).

The average of the temperature inside the adsorber during one day is presented in figure 3.

The temperature is considered equal to the ambient temperature in the morning before sunrise and under a low pressure of methanol (equal to the evaporation pressure $P_e = 28.72\ hPa$). When the adsorber is heated by solar radiation, the adsorbent bed temperature increases rapidly with time. The heating continues until the condensation pressure is reached (equals to $P_c = 216.76\ hPa$). The reached temperature is known as limit temperature of desorption . The heating continues also as much as the solar radiation is sufficient to increase the temperature of the adsorbent bed and until the maximum regenerating temperature is reached. This last is assumed as the temperature where the adsorbent bed and the adsorber wall temperature are the same. Consequently, there are no heat exchanges between the adsorber wall and the adsorbent bed at this temperature. We can observe also that the slope of the curve, during the first period of isosteric heating is larger than that during the isobaric heating period. This is explained by the fact that the heat absorbed from the heating of the adsorber by the solar flux goes mainly to increasing the temperature of the adsorbent bed and adsorber wall. While, in the isobaric heating, the absorbed heat not only increases the temperature of the adsorbent bed and adsorber wall, but also contributes to the heat of desorption.

Solar flux decreases and the cooling starts down after it reaches the maximum regenerating temperature. The temperature decreases until the pressure becomes equal to that which reigns in the evaporator ($P_e = 28.72\ hPa$).The reached temperature is known as limit temperature of adsorption. Cooling continues until the temperature 25 °C, which is determined by the temperature in which there is no heat exchange between the adsorber wall and the adsorbent bed at this temperature. This value corresponds to the adsorption temperature given at the beginning of the cycle ($T_a = 25\ °C$).

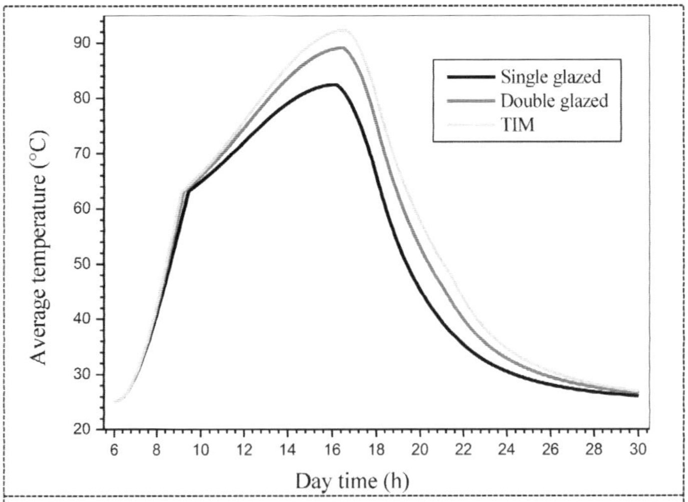

Figure 3. Effect of collector configuration on average temperature inside the adsorber during one cycle.

The knowledge of the behaviour of the adsorber allows study the effect of some parameters on the solar performance coefficient and cooling production such as the type of the solar collector (collector configuration). In this study, we choose three types of collector (simple, double glassed and TIM cover (Transparent Insulation Material)).
Usually, the increase of glazing cover number is limited by collector structure and dimension; in common practice, there is no more than two or three glazing. Figure 3 shows also the effect of the three types of collector on the average temperature inside the adsorber.
We can see that the average temperature inside the adsorber increases when the double glass cover or TIM cover are used, this result is due to the fact that the heat losses with the TIM cover is lower than those related to a single glass cover and double glass cover.
We represent in Figure 4 and 5, the effect of the solar collector surface area on the solar performance coefficient COP_s and cooling production Q_f. It can be observed from Figure 4 that the COP_s decreases with an increase in the solar collector surface area.
This can be justified by the fact that the increase in solar collector surface area induces an increase in the cooling production Q_f (Figure 5) and once the solar collector surface area reaches an optimal value, the cooling production decreases (Figure 5). Moreover, the total solar energy absorbed by the collector throughout the whole day, also strongly increases. Consequently, the variation of cooling production Q_f relative to the variation of the total solar energy absorbed by the collector, will be lower contributing to a decrease in the solar performance coefficient COP_s.

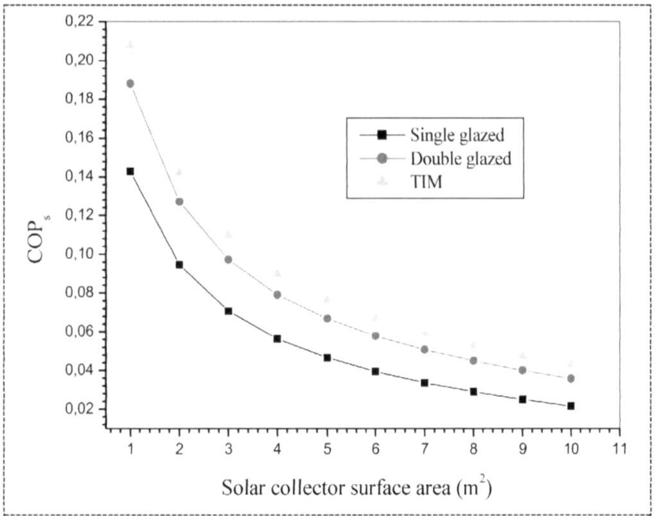

Figure 4. Effect of solar collector area and its configuration on system's performance.

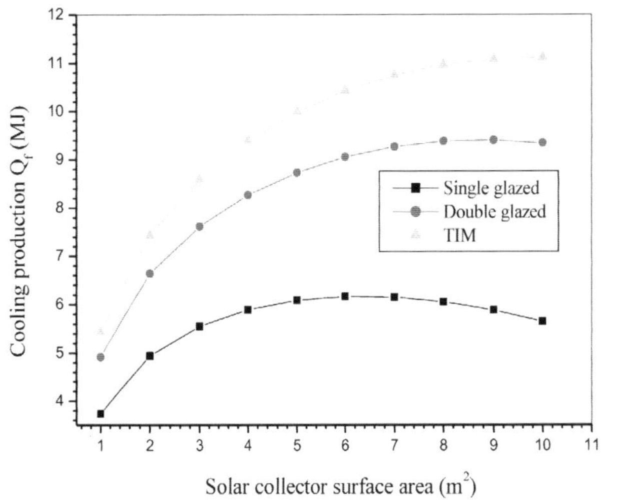

Figure 5. Effect of solar collector area and its configuration on cooling production.

The adsorbent mass is an important parameter for system optimization. Its influence on the system's performances is presented in Figure 6. It can be seen from this figure that the solar performance coefficient COP_s increases as long as the adsorbent mass is less than an optimal value. Moreover, when the adsorbent mass is increased to higher values, the COP_s decreases. This trend reflects the fact that the required amount of methanol is desorbed at this optimal value of adsorbent mass and beyond this value the energy consumed by the adsorber increases only the sensible heat of its components (adsorbent, adsorbate and wall adsorber) and doesn't contribute to desorption. And the same justification holds for the cooling production (Figure7).

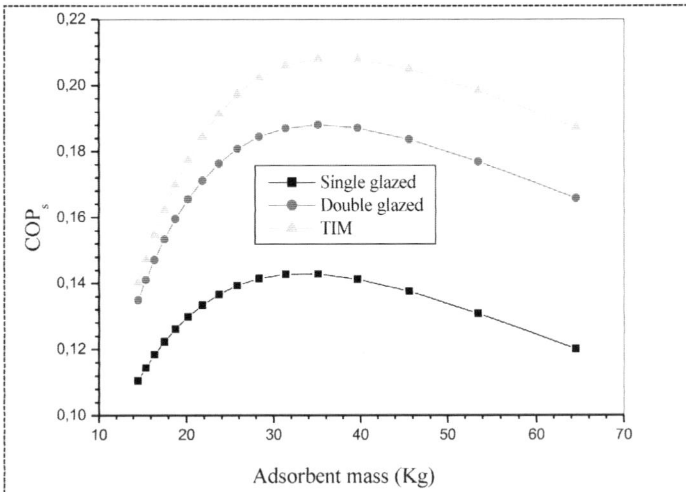

Figure 6. Effect of adsorbent mass and collector configuration on system's performance.

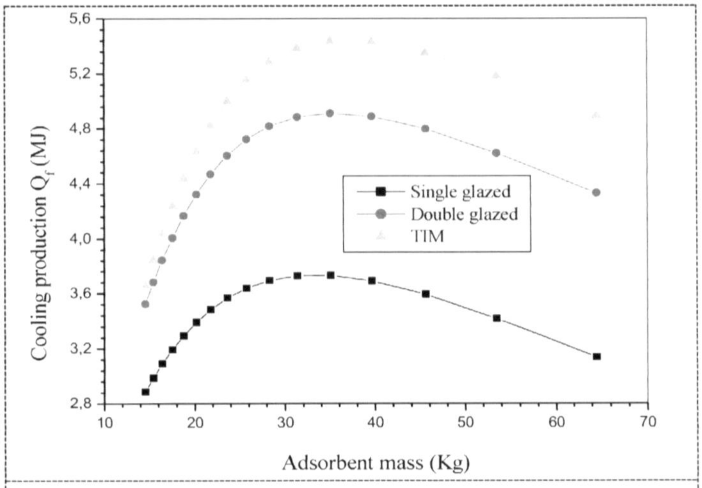

Figure 7. Effect of adsorbent mass and collector configuration on cooling production.

The solar refrigerator is powered by solar radiation energy, therefore the solar energy intensity determines the cooling power as well as the solar performance coefficient COP_s value, and these effects are shown in Figure 8. We can see that both the cooling power Q_f (Figure 9) and the solar performance coefficient COP_s increase with the increase of the total solar energy absorbed by the collector.

However, there is a minimum value of solar radiation intensity under which no ice can be produced in practical application, because the amount of desorbed methanol, when evaporated, is just enough to provide cooling to evaporator. This minimum value of solar intensity depends on the variations of the atmospheric temperature during the day and characteristics of solar refrigerator device, this minimum value is about 11 MJ/m^2 [8].

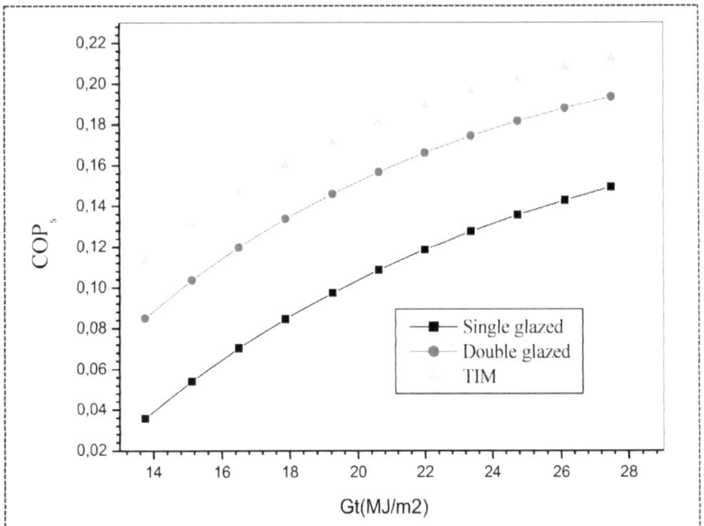

Figure 8. Effect of total solar radiation and collector configuration on system's performance.

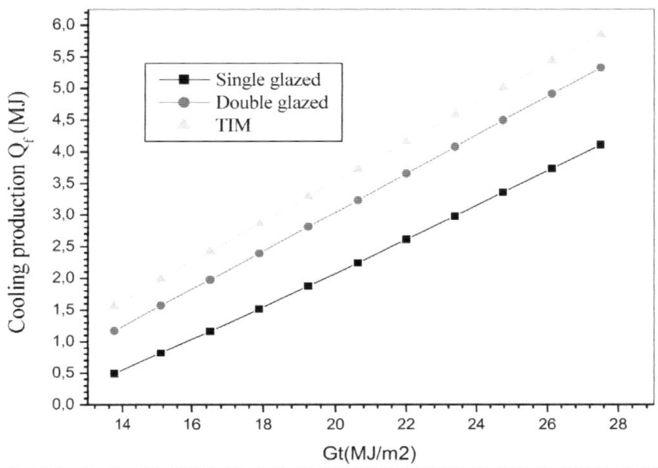

Figure 9. Effect of total solar radiation and collector configuration on cooling production.

5. Conclusion

A theoretical model and a numerical program have been developed in order to evaluate the performance of the adsorption cooling system powered by solar energy. The system uses the activated carbon AC-35/methanol as a working pair. The ambient and solar radiation time variations are taken into account. A uniform pressure and non uniform temperature model is proposed, in order to describe the behaviour of heat and mass transfer in the adsorbent bed, using a meteorological data of Constantine for a hypothetical clear day. The performance of the system considered in this study is evaluated by solar performance coefficient, COP_s. The obtained results show that the solar performance machine increase when a TIM or double glass cover is used. For fixed operating conditions, the solar performance coefficient presents a maximum for an optimal adsorbent mass per square meter of solar collector surface area. The solar collector surface area plays an important role in determining the performance of the system. COP_s decreases with an increase in the solar collector surface area. Solar performance coefficient COP_s increases with the increase of the total solar energy absorbed by the collector.

6. References

[1] Anyanwu E E 2003 *Energ. Conv.Mgn.* **44** 301
[2] Passos E, Meunier F, Gianola G C 1986 *J. Heat recovery systems* **6** 259
[3] Critoph R E 1989 *Carbon* **27** 63
[4] Capderou M 1986 Solar atlas of Algeria, (in French), University Publications Office, Alger
[5] Chekirou et al. 2014 Int. J. Refrig. **39** 137
[6] Weast R C 1995 *Handbook of chemistry and physics* 75 th Edition C RC press
[7] Bejan A and Kraus A D 2003 *Heat transfer handbook* John wiley and Sons New York
[8] Pons M and Grenier Ph 1987 *Transaction of the ASME.* 109/303

Concentration influence on structural and optical properties of SnO_2 thin films synthesized by the spin coating technique.

Soumia Belhamri, Nasr-Eddine Hamdadou
Laboratoire de Micro et de Nanophysique (LaMiN), Ecole Nationale Polytechnique d'Oran (ENPOran), BP 1523 El Mnaouer, Oran 31000, ALGERIE
Email: somimira@live.fr

Abstract. Tin dioxide is an n-type semiconductor, with wide band gap 3.6 eV and special properties such as high optical transmission in the visible range, the infrared reflection and chemical stability. The objective of our work is to study the effect of solution concentration on the properties of SnO_2 thin films, which were deposited on glass substrate by sol-gel spin coating technique and characterized by X-ray diffraction, UV–visible spectroscopy after annealing for one hour at 500°C.

X ray diffraction spectra (XRD) showed that the films deposited at different concentrations (0.7 mol/l, 1 mol/l, 1.5 mol/l) are polycrystalline with a rutile type tetragonal. The grains have two preferred orientations along the directions (110) and (101) corresponding to $2\theta = 26{,}744°$ and 34,113° respectively. We have also noted that the grain size change between 20 and 40 nm. The peak of diffraction becomes less intense when the solution concentration is more than 0.7 mol / l. The opticall transmittance of the films in the visible spectrum was in the range of 59 - 44%.

Keywords: Thin film, SnO_2, spin coating, structural and optical properties.

1. Introduction

Environmental protection has become the objective of many the current researches and the most important of these inventions use the technology photovoltaic. The best known are probably the modules based on polycrystalline silicon, which still occupy the greater part of the PV market. Less known are the thin-film modules, which are often described as second-generation photovoltaic modules. These thin-film solar modules offer several advantages like the absence of raw materials and use the low-cost materials such as tin dioxide, which is the subject of our study. The tin dioxide is one of the most used oxides in the photovoltaic cells [1] because of its specific properties such as an n-type semiconductor with a large gap (3.6-4.0 eV) [2,3] , a high optical transmission in the visible range [4], with a rutile structure (tetragonal), the lattice parameters a = b = 4.737 Å and c = 3.816 Å, [5]. The grain size is typically between 200- 400 Å which are highly dependent on the deposition technique, temperature, doping level etc. [6-7].

2. Experimental

2.1. Preparation of thin films:

In our work we used the spin coating technique or centrifugation. This method is simple to implement, low energy costs and lead to very pure and homogeneous materials with a thin thickness.

SnO_2 thin films are grown on glass substrates rectangular (25 x 10 x 1) mm^3 cut by a rod made of alumina. The cleaning of the substrates is very important to obtain films with good adhesion to the substrate and good structural and morphological properties. The substrates were first cleaned with acetone for 15 minutes at room temperature, and then dipped in distilled water for 10 minutes. Afterwards, the substrates were cleaned with Betadine (medical soap) to remove the last traces. Finally, they were rinsed in distilled water.

The solution was prepared by dissolving the precursor of tin chloride ($SnCl_2.2H_2O$) in ethanol (C_2H_5OH) as solvent at room temperature; the solution was stirred at 80°C for 1h until obtaining a yellow color. Then put it in a flask under room temperature for 24h. After cleaning the substrate, we are putting in the spin coater for depositing a small puddle of solution onto the center of them and then spinning the substrate at high speed. Centripetal acceleration will cause the solution to spread off the edge of the substrate. The following step is the drying: When we evaporating the volatile liquids on the heating plate, we get a thin film. We have repeated these steps 10 times to obtain a considerable thickness.

Table 1. Deposition parameters for SnO_2 thin film.

Substrate temperature	120°C for 10 min
Concentration of solution	0.7 - 1 - 1.5 mol l^{-1}
Rotation speed	3000 rpm
Number of cycles	10
Deposit time	30 s
Solutions	tin chloride [$SnCl_2$] in ethanol
Acceleration	600 rpm / min.
PH	2

After the deposition, thin films of SnO_2 are annealed in an oven for 1 hour at 500°C. This step improves the crystallization of the film and the thermally stabilize.

2.2. Characterization of SnO_2 thin films:
Structural analysis of the films were carried out by BRUKER diffractometer using Cu $K\alpha_1$ radiation (λ=1, 54056 Å). Diffractograms were recorded from 10° to 80° with a step of 0.01°.The optical absorption and transmittance were studied with UV–vis spectrophotometer, 210 SPECORD Plus® in the wavelength range of 100–1200 nm at room temperature.

3. Results and discussion

3.1. X-ray diffraction studies:
Figs. (1), (2) and (3) presents the XRD patterns of the SnO_2 thin films prepared with different concentration. As can be observed, the SnO_2 thin film appears to have a polycrystalline structure with a cassiterite tetragonal crystalline phase according to reference pattern of JCPDS (41-1445). The value of lattice parameters were found to be a = b = 4.69 Å and c =3.16 Å, which are in good agreement with the reported bulk values (a = 4.7382 Å, c = 3.1871 Å). [8]. SnO_2 film showed a random orientation along (110) and (101) planes.
The average grain size was calculated using Scherrer formula:

$$D = K\frac{\lambda}{\beta \cos\theta} \quad (1)$$

Where D is the mean grain size of the nanoparticles, K = 0.9 and λ= 1.5406 Å and β is the full width half maximum (FWHM) of the diffraction peaks.

Crystallite sizes of the samples were found between 17 and 39.8 nm, see Tables (2), (3) and (4). The grain size is small which leads to a very intense grain boundary scattering, and then a low mobility, that means the films are resistive. This small grain size causes high surface energy.

It is manifested that as the concentration solution decreases, the intensity of the peaks particular (110) plane gets enhanced, which shows that the films deposited at low concentration have better crystallinity.

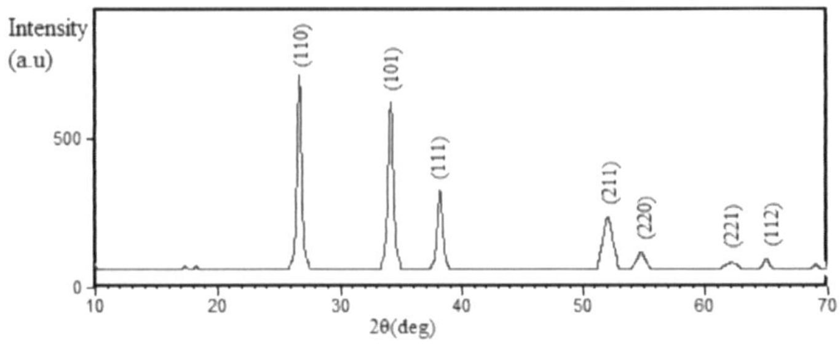

Figure 1. XRD patterns of SnO_2 thin films with a concentration 0.7 mol / l after annealing at 500°C.

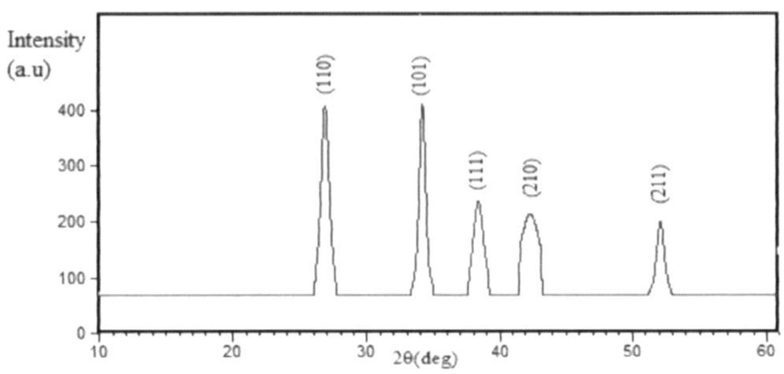

Figure 2. XRD patterns of SnO_2 thin films with a concentration 1 mol / l after annealing at 500°C.

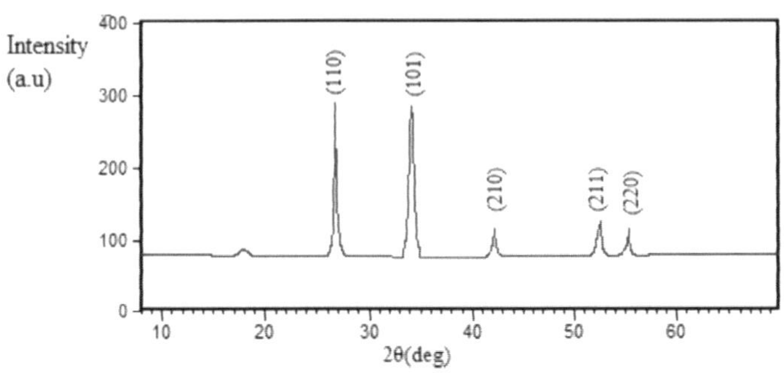

Figure 3. XRD patterns of SnO_2 thin films with a concentration 1.5 mol / l after annealing at 500°C.

Table 2. The crystalline parameter values of SnO_2 thin film with a concentration of 0.7 mol/l after annealing at 500°C.

2θ (deg)	hkl	D (nm)	I (u.a)	I_0 (%)	d_{hkl} (Å)	FWHM (deg)
26.744	(110)	36.808	645	100	3.335	0.387
34.113	(101)	31.177	833	75	2.6262	0.465
38.117	(111)	31.535	580	21	2.359	0.465
51.987	(211)	16.598	198	57	1.757	0.929
54.714	(220)	20.136	446	14	1.6763	0.775
64.917	(112)	28.970	78	12	1.435	0.567

Table 3. The crystalline parameter values of SnO_2 thin film with a concentration of 1 mol/l after annealing at 500°C.

2θ (deg)	hkl	D (nm)	I (u.a)	I_0 (%)	d_{hkl} (Å)	FWHM (deg)
26.796	(110)	20.441	372	100	3.324	0.697
34.066	(101)	26.745	396	75	2.629	0.542
38.326	(111)	31.554	207	21	2.346	0.465
42.122	(210)	24.044	284	1	2.143	1.859
52.200	(211)	39.880	161	57	1.750	0.387

Table 4. The crystalline parameter values of SnO_2 thin film with a concentration of 1.5 mol/l after annealing at 500°C.

2θ (deg)	hkl	D (nm)	I (u.a)	I_0 (%)	d_{hkl} (Å)	FWHM (deg)
26.872	(110)	30.59	278	100	3.315	0.465
34.090	(101)	23.38	279	75	2.627	0.620
42.036	(210)	31.930	98	1	2.147	0.465
52.302	(211)	33.1	124	57	1.747	0.465
55.051	(220)	25.208	72	14	1.666	0.620

3.2. UV–vis characterization:

The UV–vis spectra of films SnO_2 deposited at different concentration are given in Figs. (4) and (5). The absorption spectra UV-visible of SnO_2 thin film developed at different concentration (0.7-1-1.5 mol / l) after annealing at 500°C for 1 hour (Fig.4) showed that there is absorption over the range (300-350nm) due to the formation of SnO_2 crystallites. That means this concentration cause the formation of tin dioxide crystals.

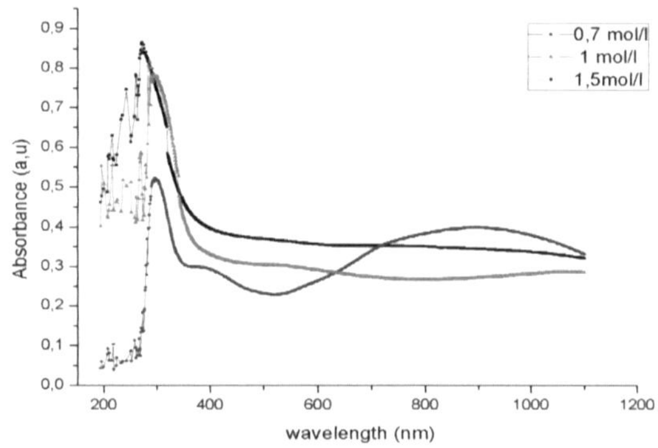

Figure 4. UV-visible absorption spectra of the SnO_2 thin film prepared with different concentrations.

The optical transmission spectra of thin films of SnO_2 annealing at 500°C for 1 hour are shown in the Fig (5). The transmittance in the visible region was found to be in the range 59 - 44%. In general, in this region of the spectrum (400 and 800 nm), the transmission is high which is due to the fact that there is no absorption and the reflectivity is low.
The onset absorption edge at around 350 nm was observed in the transmission spectra of the SnO_2 thin films. The decrease of transmission SnO_2 with increasing concentration of solution can be attributed to the increase in surface roughness of the SnO_2.

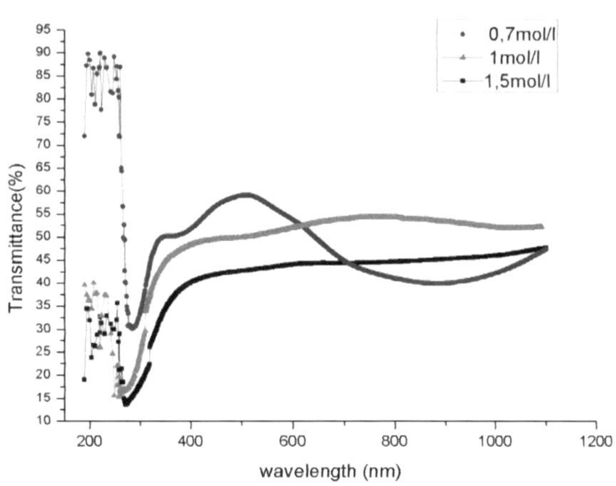

Figure 5. Transmittance spectra of the SnO_2 film prepared with different concentrations.

4. Conclusion

In summary, we present our work on tin dioxide thin films prepared by spin coating technique, regarding its structural and optical properties.

The X-ray diffraction helped to identify the tetragonal structure (rutile) of SnO_2 with the maximum intensity peaks from (110) and (101) planes. However, the relative intensity of the peaks decreases with an increase in the concentration of solution

Optical transmission spectra show high transparency between 400 and 800 nm with a value from 59 to 44% in the visible.

The experimental results suggest that the concentration used has an effect on the structural and optical properties

References

[1] S.Rani, S.C.Roy, M.C. Bhatnagar, Effect of Fe doping on the gas sensing properties of nano-crystalline SnO_2 thin films, Sens. Actuators B 122 (2007) 204–210.

[2] K. G. Godinho, A. Walsh, G. W. Watson, J. Phys. Chem. C 113 (2009) 439.

[3] T. J. Coutts, D. L. Young, and X. Li, MRS. Bull., 25, (2000), 58.

[4] M. Batzill and U. Diebold, "The surface and materials science of tin oxide," Progress in Surface Science, vol. 79, no. 2–4, pp. 47–154, 2005. View at Publisher · View at Google Scholar · View at Scopus.

[5] A. L. Dawar and J. C. Joshi, J. Mater. Sci., 19, (1984), 1.

[6] K. L. Chopra, S. Major, and D. K. Pandya, Thin Solid Films, 102, (1983), 1.

[7] D.S. Ginley, C. Bright, Transparent conducting oxides, MRS Bull. 25 (2000) 15.

[8] JCPDS (Joint Committee for Powder diffraction Standards) Data Cards. Swarthmore, PA, 41–1445, 1997.

Journal of Physics: Conference Series **758** (2016) 012008 doi:10.1088/1742-6596/758/1/012008

The effects of synthesis conditions on the magnetic properties of zinc ferrite spinel nanoparticles

K El Maalam[1,2,*], L Fkhar[1,2,3], Z Mahhouti[1,2], O Mounkachi[1], M AitAli[3], M Hamedoun[1] and A Benyoussef[1,2,4]

[1]Materials_Nanomaterials Center, MAScIR Foundation, B.P. 10100-Rabat, Morocco.
[2]Magnetism and Physics of the high Energies Laboratory, URAC 12, B.P. 1014, Faculty of science, Mohammed V University, Rabat, Morocco,
[3]Coordination Chemistry Laboratory, Cadi Ayyad University, Faculty of sciences Semlalia (UCA-FSSM), B.P. 2390 - 40000 Marrakech, Maroc
[4]Hassan II Academy of Science and Technology, Rabat, Morocco,

E-mail : khadija.elmaalam@gmail.com

Abstract. Zinc ferrite nanocrystals were synthesized from metal chloride precursors via chemical co-precipitation method, using different synthesis conditions. Characterization measurements including X-ray diffraction (XRD), transmission electron microscopy (TEM) and super conductiong quantum device (SQUID) were used to study the influence of precursor's concentration and reaction time on the crystalline structure, average sizes and magnetic properties of zinc ferrite nanoparticles. The transmission images show spherical, homogenous shape and particle size ranging from 16 to 22 nm. DC magnetization (2-300 K) measurements reveal a superparamagnetic behavior for the $ZnFe_2O_4$ samples with a blocking temperature in the range of 18-24 K. Our results demonstrate that magnetic properties of magnetic particles can be largely modified by just changing the reaction condition such as concentration and reaction time, which might be a useful way to design novel magnetic materials.

1. Introduction

Spinel ferrites with general formula AB_2O_4 are very important magnetic materials because of their interesting magnetic properties combined with chemical and thermal stability. Spinel oxides have long played an important role in various applications. In normal spinel AB_2O_4, A is generally a divalent cation occupying tetrahedral sites, while B is a trivalent cation occupying octahedral sites. In inverse spinels, half of B cations occupy the tetrahedral sites, and the formula is rewritten as B [AB] O_4[1].

The zinc ferrite $ZnFe_2O_4$ is one of the iron based cubic spinel series (S.G. = Fd3m), show striking changes in its magnetic properties by reducing the grain size to the nanometer-sized range. Bulk zinc ferrite is a completely normal spinel structure with Zn ions in the tetrahedral or A sites and Fe ions in the octahedral or B sites. Due to antiferomagnetic (AFM) super-exchange interaction between B-B ions, bulk zinc ferrite is antiferomagnetic at T_N = 10 K[2]. However, when it is prepared in the nanoscale, the energy associated to the low particle size favors a mixed cation distribution in wich the Zn^{2+} and Fe^{3+} ions are distributed along the A and B sites giving rise to the mixed spinel structure $(Zn^{2+}_{1-x}Fe^{3+}_x)\,[Fe^{3+}_{2-x}]O^{2-}_4$,where parentheses and square brackets denote the A and B sites, respectively and the x is the inversion degree[3].

Zinc ferrites spinels have been investigated extensively due to their potential and technological applications such as magnetic and optical materials, electronic, semiconductors, pigments, biomedicine, magnetic date storage, medical imaging, drug targeting, and catalysis[4-13].Various synthesis methods, such as co-precipitation, sol-gel, micro-emulsions, thermal decomposition, and ultrasonic cavitation approach, have been reported to prepare nanometer-sized $(Zn_{1-x}^{2+}Fe_{x}^{3+})\,[Fe_{2-x}^{3+}]O_4^{2-}$ particles[14-18]. Here we report a simple method to prepare zinc ferrite nanoparticle by co-precipitation reaction that is a useful and attractive technique for the preparation of nanosized particles because of its advantages like good stoichiometric control and the production of ultrafine particles with low sintering temperature and smaller duration.

Concentration and reaction time of reaction mostly result in the appearance of the disorder of Zn and Fe ions over the tetrahedral and octahedral sites, which in turn lead to a significant variation of the magnetic properties. In this work, we have studied the influence of synthesis conditions (concentration and reaction time) in the structural and magnetic properties of zinc ferrite spinel nanoparticles. The phase structure, morphology and magnetic properties of as-prepared products were investigated by X-ray diffraction (XRD), transmission electron microscopy (TEM) and superconducting quantum interference device (SQUID).

2. Experimental details:

The co-precipitation processing for synthesis zinc ferrite nanoparticles is based on the following reaction:

$$ZnCl_2 + 2\ FeCl_3 + 8\ NaOH \rightarrow ZnFe_2O_4 + 4\ H_2O + 8\ NaCl$$

In this work, $ZnFe_2O_4$ nanoparticles have been obtained using different conditions (Table.1). The required quantities of zinc chloride ($ZnCl_2$, 99%) and iron chloride ($FeCl_3$, 99%) solution are mixed in a stoichiometric proportion. The precursors were added in H_2O and mixed homogeneously to yield a uniform mixture of precursor at 80 °C for 40 min. Then, aqueous hydroxide sodium solution (NaOH) is added drip (pH of the medium 11). The reaction temperature is controlled at 100 °C. All samples were washing at several times with water and ethanol to remove the soluble salt, and finally drying at 70 °C overnight. The dried samples were subsequently annealed at 500 °C for 8h to get the desired nanoparticles.

Table 1. Sample symbol and conditions reaction for as-synthesized $ZnFe_2O_4$ nanoparticles

Sample Symbol	Reaction conditions		
	reaction time (min)	$[ZnCl_2]$ (mol/l)	$[FeCl_3]$ (mol/l)
ZF1	15	0.308	0.616
ZF2	120	0.308	0.616
ZF3	15	0.462	0.924
ZF4	120	0.462	0.924

Thermogravimetric/Differential Thermal Analysis (TG/DTA) equipment was used to determine the thermal decomposition and weight loss of the precursor powder, by varying the temperature from 25 °C up to 1000 °C with a heating rate of 10 °C/min. The structure and particle sizes of all samples were performed by X-ray diffraction (Model: D8 Discover Bruker AXS Detector 2D Hi-Star Turbo Source Rotating Anode X-Ray Cu). The morphology and particle size were observed by transmission electron microscopy (TEM) with an accelerating voltage of 200 kV, while the magnetic characterization was done by Magnetic Properties Measurement System (MPMS-7XL, Quantum Design).

The Thermal Characterization

The thermal characterization of all sample synthesized $ZnFe_2O_4$ powder has been done by TGA (Thermogravimetric Analyzer) analysis. The results of TGA measurements of $ZnFe_2O_4$ are shown in Fig. 1. It shows exothermic peaks at around 220 °C which may correspond to the removal of water from precursors. In Figure.1, there is no more weight loss above 500 °C, which indicates the formation of crystalline $ZnFe_2O_4$ ferrite.

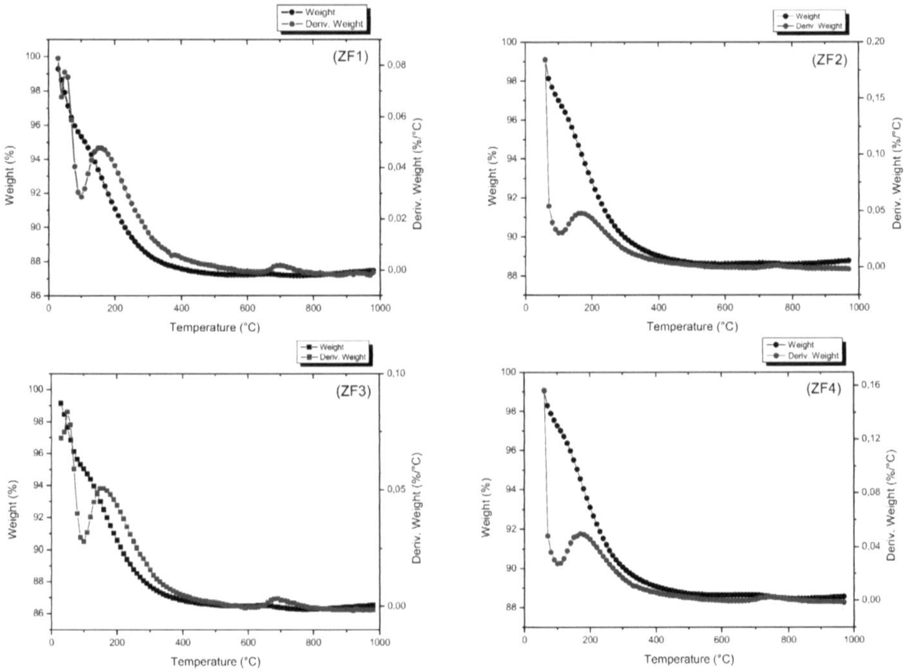

Figure 1.TG–DTG curves ofas-prepared $ZnFe_2O_4$ precursorpowders (ZF1), (ZF2), (ZF3) and (ZF4).

3. Results and discussion

The X-ray diffraction of (ZF1), (ZF2), (ZF3) and (ZF4) samples annealed at 500 °C for 8 h reveal that all the samples crystallized with the cubic spinel type structure, space group Fd3m, and permit us to exclude the presence of any impurities or secondary phases (Figure.2).

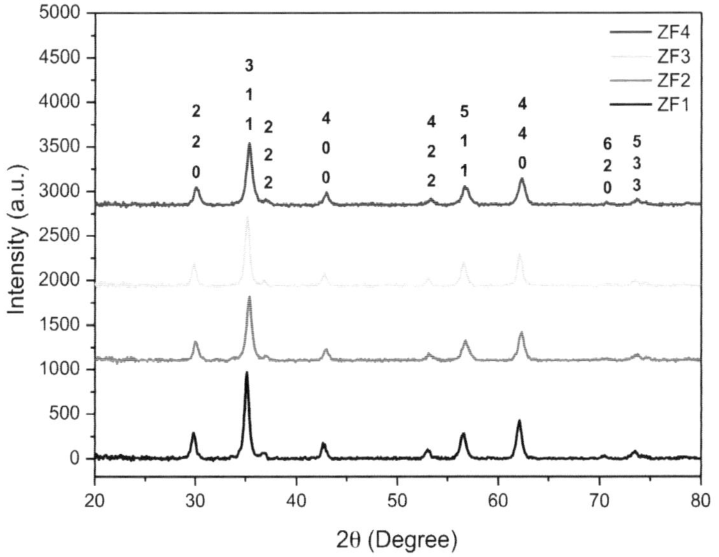

Figure 2. XRD patterns of the annealed samples (ZF1), (ZF2), (ZF3) and (ZF4).

The lattice constant (*a*) was calculated from the *XRD* spectra using the following relation:

$$a = [d^2(h^2 + k^2 + l^2)]^{1/2}$$

From Table.2, it can observed that ZF1 and ZF3 nanoparticles samples show a higher unit cell parameter comparing with ZF2 and ZF4 samples (a_{ZF1}= 8.47 Å, a_{ZF3}= 8.45 Å). This variation could be related to disorder of cation distribution. So the reaction time affects directly the unit cell parameter of zinc ferrite sample.

The broadening of the reflections indicates the nanometric size of the particles for all samples. The average particle size (D_{RX}) has been calculated using the well-known Scherrer's formula[19],

$$D_{RX} = \frac{0.9\,\lambda}{\beta \cos\theta}$$

Where λ is the wavelength (Cu K_α), β is the full width to half maximum (FWHM) of line broadening and θ is the Bragg angle of diffraction. The most intense peaks (311) are used to estimate the size of nanocrystallites and are found in the range of 12-16 nm (Table.2).

The average grain sizes of various $ZnFe_2O_4$ ferrites nanoparticles prepared with different synthesis conditions are listed in Table.2. It is observed that the average grain size decreases with increasing of reaction time, these results can be explained by the cations distribution between A and B sites.

The TEM micrographs for (ZF1), (ZF2), (ZF3) and (ZF4) samples are shown in Figure. 2. The particles are regular and uniform in size. The average particles size is 21.6, 20.2, 21.2 and 16.4 for (ZF1), (ZF2), (ZF3) and (ZF4), respectively, which is in good agreement with the particle sizes estimated by Scherer's formula (the same scale of variation).

Table 2. Lattice parameter (a), crystallites sizes (D of samples (ZF1), (ZF2), (ZF3) and *(ZF4)*.

Sample	ZF1	ZF2	ZF3	ZF4
a (A°)	8.47	8.40	8.45	8.40
D_{XRD} (nm)	15.8	14.4	15.5	12.7
D_{TEM} (nm)	21.6	20.2	21.2	16.4

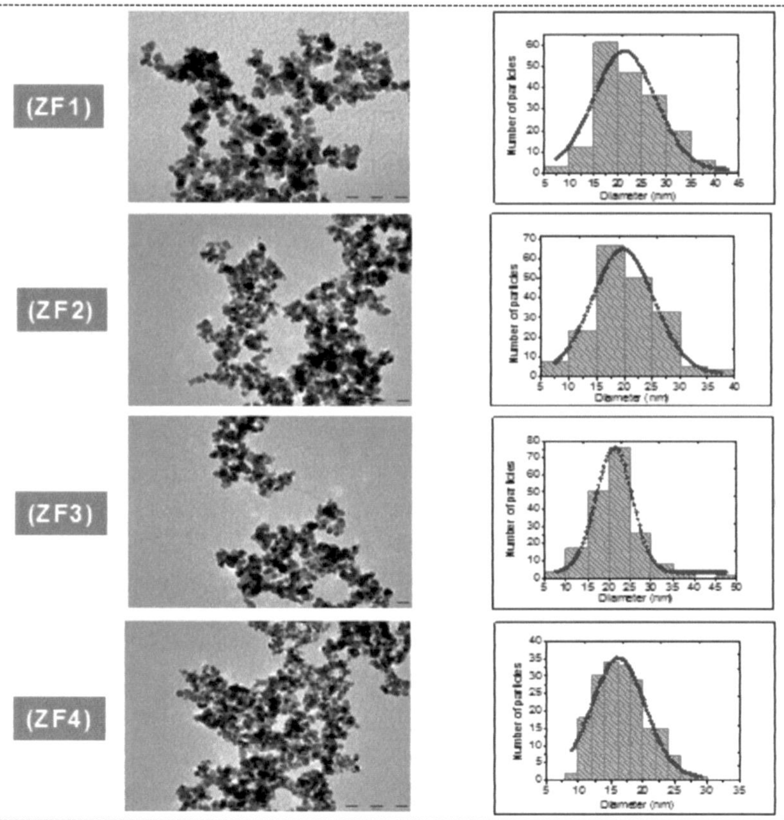

Figure 3. The TEM micrograph of samples (ZF1), (ZF2), (ZF3) and (ZF4).

The magnetic properties of the materials which originate from the quantum couplings at the atomic level, including the coupling between electron spins (S–S coupling) and the coupling between the electron spin and the angular momentum of the electron orbital (L–S coupling). Each of the magnetic nanoparticles tends to possess a single magnetic domain [20]. Therefore, the nanoparticles provide excellent opportunities for the fundamental studies on the relationship between magnetic behavior and the magnetic couplings at the atomic level. Figure.4 show the zero-field-cooled *(ZFC)* and field-cooled *(FC)*magnetization curves for (ZF1), (ZF2), (ZF3) and (ZF4) zinc ferritenanoparticles measured in the temperature range from 5 to 300 K. As expected, the *ZFC* magnetization increases before reaching a maximum value, that is, the blocking temperature (T_B) [21].It is interesting to note that the tow curves diverge below this temperature. At temperatures above T_B, the thermal energy, characterized by k_BT, is larger than the magnetic energy barrier and thus the materials become superparamagnetic following the Curie-Weiss law.

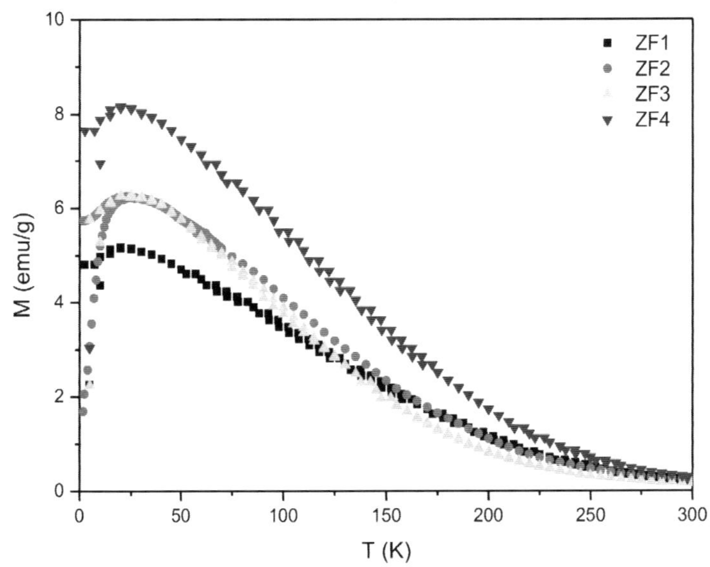

Figure 4. ZFC-FC curves for the obtained samples $ZnFe_2O_4$ nanoparticles (ZF1), (ZF2), (ZF3) and (ZF4).

The blocking temperatures T_B of zinc ferrite nanoparticles are found to be approximately 20, 24, 18 and 22 for (ZF1), (ZF2), (ZF3) and (ZF4), respectively (Table.3). The magnetic anisotropy constant K of the samples as-prepared can be calculated by the following formula [22]:

$$K = \frac{25K_BT_B}{V}$$

Where K_B is the Boltzmann constant ($K_B = 1.381\times10^{-23}$), T_B is the blocking temperature of the samples, and V is the volume of a single particle. The calculated magnetic anisotropy constant of our samples are given in Table.3. The T_B is controlled by the anisotropy energy that is the result of contributions from both anisotropy constant and the particle size. The anisotropy constant is influenced by the inversion parameter that, as mentioned before, increases when the particle size decreases but also affected by the synthesis conditions (concentration and reaction time).

The field dependence of the magnetization of the as-prepared zinc ferrite nanoparticles was measured at 5K with an applied field 0kOe$\leq H \leq$ 6kOe. Figure.5 presents the magnetization curves of (ZF1), (ZF2), (ZF3) and (ZF4) samples. As shown in Figure.5 and Table.3, the values of M_s are 45.4, 49.8, 46.6 and 49.7 emu/g corresponding to the (ZF1), (ZF2), (ZF3) and (ZF4) samples, respectively. The ZF2, ZF 4 samples have highest M_S values than the ZF1 and ZF3 samples, which can explained through the highest inversion degree. So, it is possible to increase the cations distribution between A and B sites if the reaction is maintained for a long time.

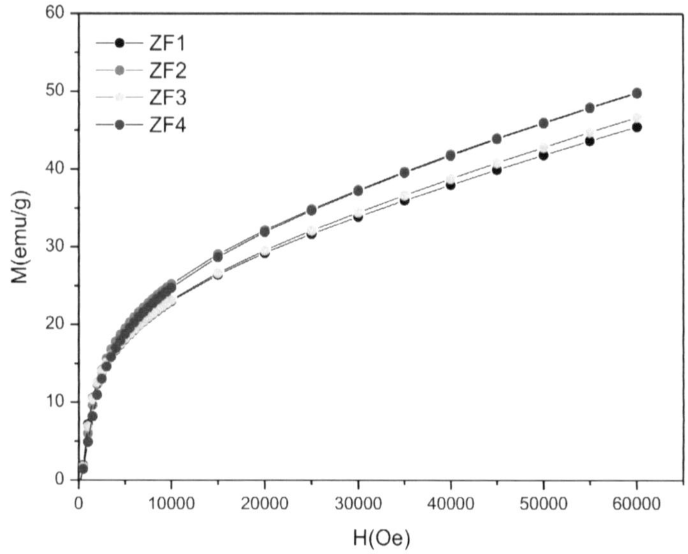

Figure 5. The magnetization M (emu/g) versus the applied magnetic field H (Oe) at 5K for all the obtained samples $ZnFe_2O_4$ nanoparticles *(ZF1), (ZF2), (ZF3) and (ZF4).*

Table 3. Blocking temperature T_B, magnetic anisotropy constant K, Saturation magnetization M_S, and Curie temperature T_C of *(ZF1), (ZF2), (ZF3) and (ZF4)* samples.

Sample	ZF1	ZF2	ZF3	ZF4
T_B (K)	20	24	18	22
$K*10^5$(ergs/cm^3)	0.33	0.53	0.31	0.7
M_s (emu/g)	45.4	49.8	46.6	49.7

4. Conclusion

The zinc ferrites nanoparticles were prepared via co-precipitation method.XRD studies of products have proved the formation of cubic spinel structure, single phase, for all samples. Particles sizes as estimated using transmission electronic microscopy were found in the range 16–22 nm. The magnetic measurements show that all the samples exhibit a superparamagnetic behavior with a blocking temperature between 18 and 24 K. The results obtained reveal that the magnetic properties of zinc ferrite nanoparticles can be largely modified by the variation of synthesis condition such as concentration and reaction time, which might be a useful way to design novel magnetic materials for potential applications.

References

[1] El Moussaoui H, Hamedoun M, Mounkachi O, Benyoussef A, Masrour R, Hlil E.K 2012 J. Supercond Nov Magn 25:1995–200
[2] Changwa Y, Qiaoshi Zeng, Goya G. F, Torres T, Jinfang L, Haiping W, Mingyuan G, Yuewu Zeng,Youwen W and Jiang J. Z 2007 J. Phys. Chem 111 12274-12278
[3] Blanco-Gutierrez V, Climent-Pascual E, Torralvo-Fernandez M.J, Saez-Puche R, Fernandez-Diaz M.T 2011 J. Solid State Chemistry 184 1608–1613
[4] Gomes J. A, Azevedo G. M, Depeyrot J, Mestnik-Filho J, Paula F. L. O, Tourinho F. A and Perzynski R 2012 J. Phys. Chem 116 24281–24291
[5] Zhongwu Wang, David Schiferl, Yusheng Zhao, H.St. C. O'Neill 2003 Journal of Physics and Chemistry of Solids 64 2517–2523
[6] Yongbo L, Ran Y, Aiguo Y, Lianwen D, Kechao Z, Xiaohe L 2009 Solid State Sciences 11 1319–1324.
[7] Yao C, Zeng Q, Goya G. F, Torres T, Liu, J.; Wu, H.; Ge, M.; Zeng, Y.; Wang, Y Jiang J. Z , C 2007 J. Phys. Chem., 111 12274
[8] Mathew D. S, Juang R. S 2007 J.Chem. Eng 129 51
[9] Sultan M, Singh R 2009 J. Phys. D: Appl. Phys 42 115306.
[10] ChenY. F, Spoddig D, Ziese M. 2008 J. Phys. D: Appl. Phys 41 205004.
[11] Xavier C. S, Candeia, R. A, Bernardi M. I. B, Lima S. J.G, Longo E, Paskocimas C. A, Soledade L. E. B, Souza A. G, Santos I. M. G 2007 J. Therm. Anal. Calorim 87 709.
[12] Lee H, Jung J. Ch, Kim H, Chun Y. M, Kim T. J, Lee S. J, Oh S. H, Kim Y. S, Song I. K 2008 Catal. Lett 122 28
[13] Toledo-Antonio J. A, Nava N, Martinez M, Bokhimi X A 2002 Appl.Catal 234 137
[14] AnimeshKundu, C. Upadhyay, H.C. Verma, 2003 PhysicsLetters A 311 410–415
[15] Raghavender A.T, Biliškov N, Skoko Ž 2011 Materials Letters 65 677–680
[16] Aiguo Y, Xiaohe L, Ran Y, Rongrong S, Ning Z and Guanzhou Q C 2008 J. Phys. Chem. 112 8558–8563
[17] Lu Z.L, Lv L.Y, Zhu J.M, Li S.D, Liu X.C, Zou W.Q, Zhang F.M, Du Y.W 2006 Solid State Communications 137 528-532
[18] Li Y, Yi R, Yan A, Deng L, Zhou K, Liu X 2009 Solid State Sciences 11 1319-1324
[19] Manish S, Chaubey S, Animesh K. Ojha M 2009 Chem. Phys 118 174–180
[20] Sharifi I, Shokrollahi H, Mahdi Doroodmand M, Safi R, Magn J 2012 Magn.Mater. 324 1854–1861
[21] Sanjeev K, Vaishali Sin, Saroj A, Uttam Kumar M, Ravinder Kumar K 2010 J. Phys. Chem. C 114 6272–6280.
[22] El Moussaoui H, Mounkachi O, Masrour R, Hamedoun M, Hlil E.K, Benyoussef A 2013 J. Alloys Comp. 581 776–781.

First-principles study of electronic, electrical and optical properties of $HoMn_2O_5$

M Ait Tamerd[2], B Abraime[1,2], A Abbassi[2], K El Maalam[1,2] , A El Kenz[2]

A Benyoussef[1,2,3], M Balli[4], M Hamedoun[1]and O Mounkachi[1,*]

[1]Institute of Nanomaterials and Nanotechnology, MAScIR (Moroccan Foundation for Advanced Science, Innovation and Research), Rabat, Morocco

[2]LMPHE (URAC 12), Physics department, Faculty of Sciences, Mohammed V University, Rabat, Morocco

[3]Hassan II Academy of Sciences and Technology, Rabat, Morocco

[4]Department of Physics, University of Sherbrooke, 2500 Blvd de l'Universite, Quebec J1K2R1, Canada

E-mail: o.mounkachi@mascir.com, kachiomar@gmail.com

Abstract. In this work, we have studied the electronic, electrical and optical properties of $HoMn_2O_5$ using first-principles density functional theory within the generalized gradient approximation (GGA). The electrical conductivity decreases with increasing temperature and it exhibits metal-like behavior. The maximum conductivity values as a function of relaxation time reach $11{,}44.10^{22}$ ($\Omega^{-1}m^{-1}s^{-1}$) at 300 K. The optical calculation presents this material a good absorbing light in the visible region which is good for many optoelectronic and photovoltaic applications.

1. Introduction

Multiferroics materials, which exhibit simultaneous magnetic and ferroelectric order [1-3], have attracted much attention recently because of the interesting physics of systems with coupled multiple order parameters and because of their potential for cross electric and magnetic functionality: magnetic field induced polarization and depolarization [4], colossal magnetodielectric effects in DyMn2O5 [5], Colossal magnetostriction effect in HoMn2O5 [6], magnetocaloric effect [7,8], spintronic devices[9-12]. Thus, electrically accessible magnetic memory and processing and vice versa are important possibilities. Fundamental interest in multiferrocity also derives from the strong interplay between magnetic frustration, ferroelectric order and fundamental symmetry issues in phase transformations that characterize these materials [13-16]. Multiferroics with magnetic and electric ordering united in a single phase were thought to be rare [17,18]. RMn_2O_5 (R = Tb, Dy, Ho, Y, etc.) belongs to a very special class of multiferroics because the ferroelectricity is driven by the magnetic ordering [19–21], the multiferroic compounds in this study is improper ferroelectric [22]. One of the most important features in the magnetocaloric effect is multiferroics, where a ferroelectric and an (anti)ferromagnetic order coexist and both the ordered temperatures strongly associate with each other [23,24]. The presence of magnetic ions Ho3+ with high magnetic moment could enhance the magnetocaloric effect in HoMn2O5 [7]. $HoMn_2O_5$ has an orthorhombic structure with Pbam symmetry at room temperature, where edge-sharing $Mn^{4+}O_6$ octahedra align along the c-axis and pairs of $Mn^{3+}O_5$ pyramids link the $Mn^{4+}O_6$ chains in the ab-plane[25] shown in **Fig.1.**

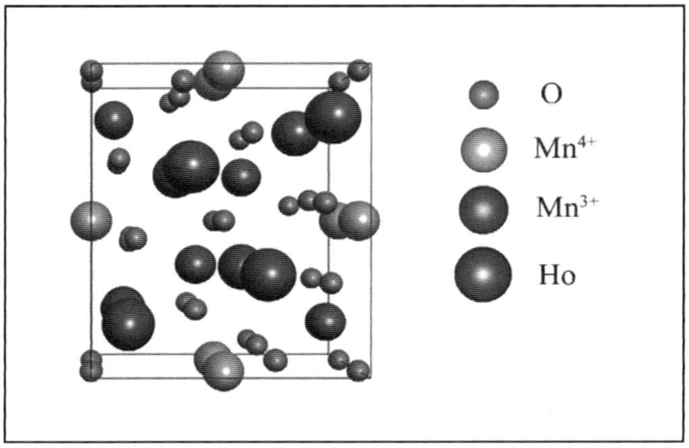

Figure1: Orthorhombic structure of $HoMn_2O_5$ bulk

2. Computational Details

Electronic structure calculations are performed through full potential linearized augmented plane waves (FP-LAPW) method based on the density functional theory implemented in the WIEN2k code [26]. For the exchange–correlation functionals, we have utilized the generalized gradient approximation (GGA) [27]. Total energies were converged within 10^{-5} (eV) precision in each self-consistency cycle. In addition, we have employed the BoltzTraP code to calculate the electrical properties [28,29], which is based on the analytical expressions of the electronic bands. We have calculated the electrical conductivity and mobility verses the temperature variation. $HoMn_2O_5$ crystallize in orthorhombic structure with space group Pbam (55). The lattice parameters a=7.272 Å, b = 8.493 Å, and c=5.679 Å. The atoms and Wyckoff positions are described in Table.1

Atom	**Wy.**	**x**	**y**	**Z**
Ho	4g	0.1367	0.1714	0
Mn^{4+}	4f	0	0.5	0.2611
Mn^{3+}	4h	0.4126	0.3496	0.5
O_1	4e	0	0	0.2702
O_2	4g	0.1637	0.4438	0
O_3	4h	0.1503	0.43	0.5
O_4	8i	0.3948	0.2063	0.2457

Table 1: Wyckoff and atomic positions

3. Results and discussion

3.1. Electronic properties

In the case of ferromagnetic phase, the density of states of $HoMn_2O_5$ is plotted versus energy with spin polarized calculation illustrated in Figure 2. The highest valence band is located at -18 (eV) formed completely by O(2s), and the top of the valence band between -7 (eV) and Fermi level, formed states of hybridized Ho(4f), Mn^{3+}(3d) and Mn^{4+}(3d) atoms with a small contribution of 2p states of O.Near of the Fermi level the density of state is mainly formed by states of Ho(f). Between 0.5 and 5 (eV) the conduction band is constituted by strong hybridition of Mn^{3+}(3d) and Mn^{4+}(3d) orbitals. The top of the conduction band occupied by the Ho (4d) orbitals.

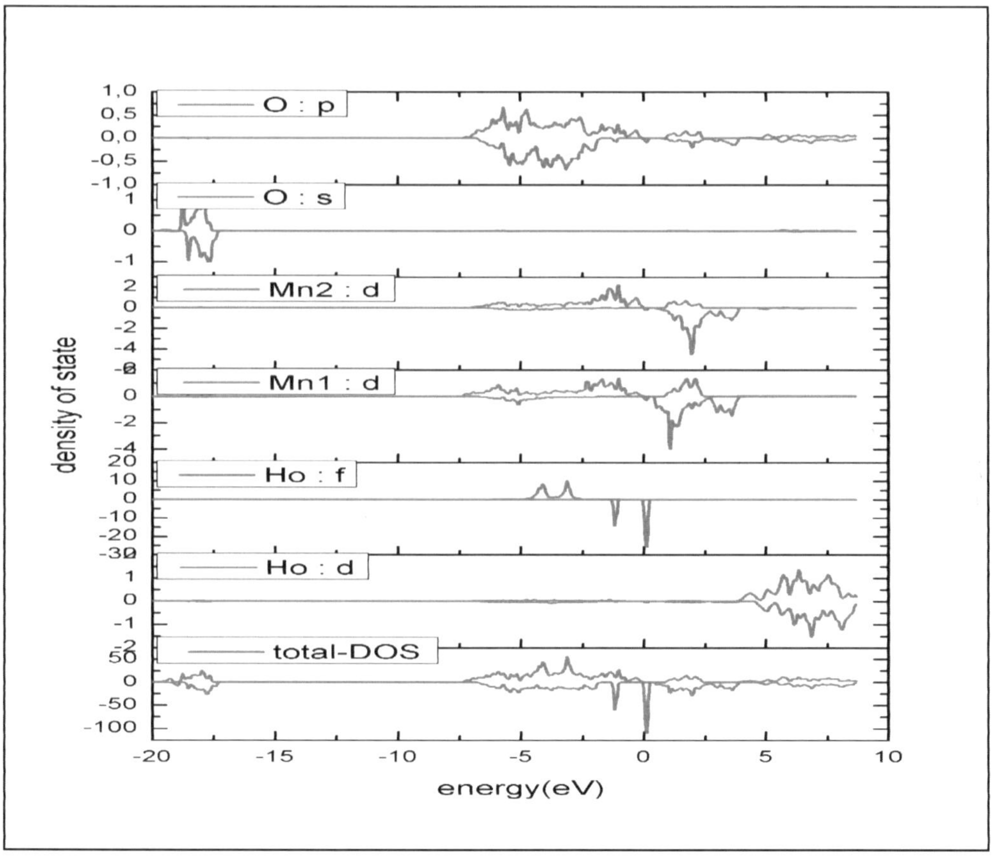

Figure 2.Density of states of $HoMn_2O_5$

3.2. Electrical properties

For the conductivity coefficient calculation, the relaxation time term τ should be treated as a constant parameter [30]. Figure3 presents the calculated temperature dependent electrical conductivities for orthorhombic $HoMn_2O_5$ compound as a function of relaxation time σ (τ) within the temperature interval from 300 K to 800 K .The maximum conductivity values as a function of relaxation time reach $11,44.10^{22}(\Omega^{-1}m^{-1}s^{-1})$ at 300 K.

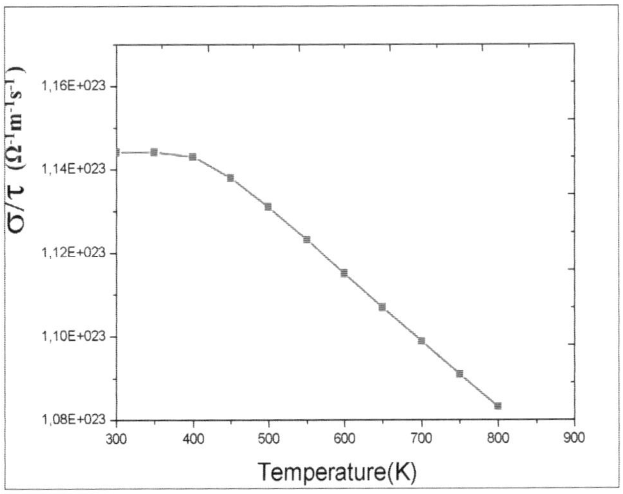

Figure 3.the electrical conductivity as a function of the temperature

Within the framework of the Boltzmann transport equation in constant relaxation time (τ) approximation, the electrical conductivity (σ) is expressed in the form of the ratio σ /τ. To calculate the electrical conductivity σ, we must determine the relaxation time τ. We assume that the relaxation time τ is independent to the direction, and treat relaxation time as a constant at a certain specific temperature and carrier concentration. The results follow the trend that the electrical conductivity decreases with increasing temperature and it exhibits metal-like behavior. From the σ/τ values, we obtain a set of the temperature dependent relaxation times as shown in Figure 3. The relaxation time shows a decreasing trend with increasing the temperature due to the reduction of mobility at the higher temperature. As we see that by increasing the temperature the conductivity of the investigated material decreases, which confirm the metal-like behavior.

3.3. Optical properties

In order to study the optical properties, we have investigated the optical absorption using the GGA approximation .The optical properties of $HoMn_2O_5$ were investigated on the basis of electronic structure, the absorption phenomena depends directly on the number of light-absorbing particles.

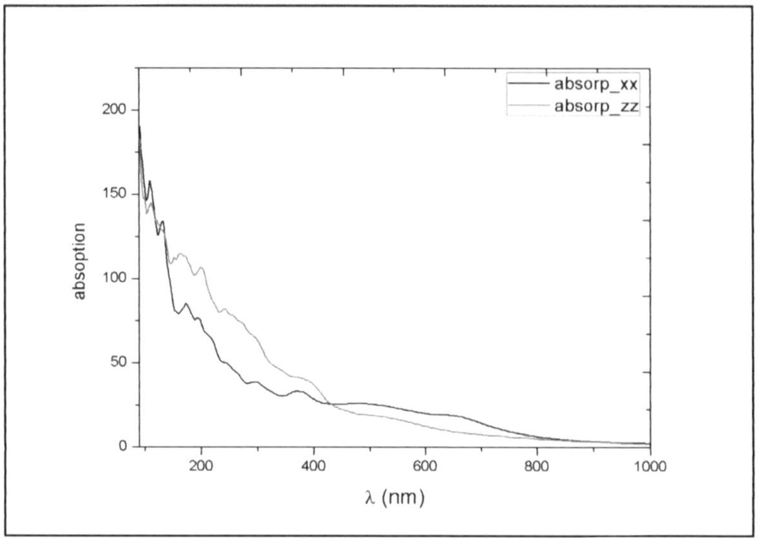

Figure 4. The absorption as a function of wavelength(nm)

According to the Figure 4, the calculation made shows the variation of optical absorption as a function of wavelength (nm). Based on the calculated optical absorption in different directions xx and zz, we noted that HoMn2O5 is an anisotropic system. The radiation hν which penetrates the material in E//xx direction present a different behavior to E//zz,which can be explained by the structural non-symmetry of HoMn2O5 matrix, in particular in the UV region. The absorption is high in the range of UV and decreases slightly in the region of visible light. From 300 nm the absorption becomes low, in this range the material did not absorb much radiation hν. The large peaks which appears in the optical absorption spectrum correspond to the inter band transitions between $HoMn_2O_5$ states, and the peaks are shifted to high energies, this can explain the reason in which this material is an absorbent in the UV region.

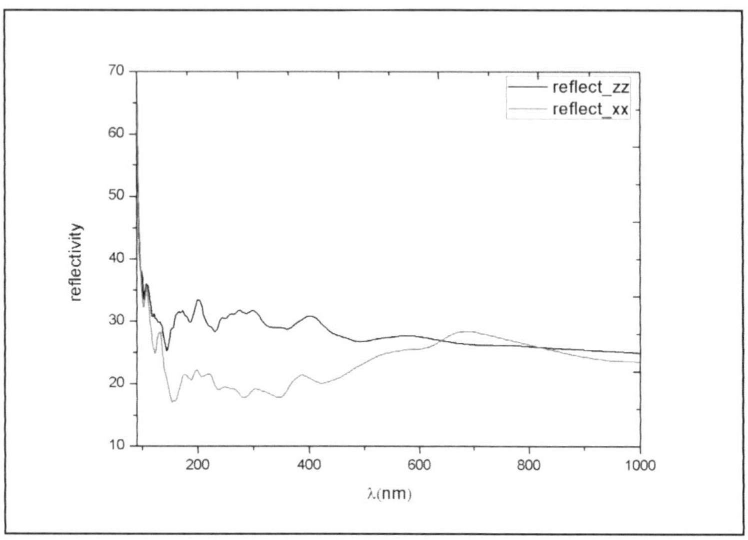

Figure 5. The reflectivity as a function of wavelength (nm)

According to the Figure 5, we show the variation of the reflectivity as a function of wavelength (nm), the average of reflectivity is 30% in the visible light region, it means that $HoMn_2O_5$is not a good transparent in this region, this is mainly due to the absorbance of energy.

4. Conclusion

The electronic and optical calculations are performed within the density functional theory using the full potential linearized augmented plane waves (FP-LAPW) method employing GGA as embodied in the WIEN2k code, while the electrical properties calculation was carried out applying the semiclassical Boltzmann transport theory utilizing the BoltzTraP package. The electrical properties were obtained from the calculated band structures by performing additional postprocessing calculations applying the BoltzTraP package. Electrical conductivity for HoMn2O5has maximum values of $(\sigma/\tau)=11{,}44.10^{22}$ $(\Omega^{-1}m^{-1}s^{-1})$ at 300 K. Our calculation showed that this material is absorbent at UV region so it could be used as candidate in several optoelectronic and photovoltaic applications.

References

[1] N. A. Hill, *J. Phys. Chem.* B **104**, 6694 (2000).
[2] D. I. Khomskii, J. *Magn. Magn. Mater*. **306**, 1 (2006).
[3] S. W. Cheong and M. Mostovoy, *Nat. Mater.* **6**,13(2007).
[4] D. Higashiyama et al, *Phys. Rev*. B **72**, 064421 (2005).
[5] N. Hur et al., Phys. *Rev. Letters* **93**, 107207(2004)
[6] I. Radulov et al., Eur. *Phys. J. B* **52**, 361-364 (2006).
[7] M. Balli, S. Jandl, P. Fournier, and M. M. Gospodinov. *APPLIED PHYSICS LETTERS* **104**, 232402 (2014)
[8] M. Balli, S. Jandl, P. Fournier and D. Z. Dimitrov. *Appl. Phys. Lett.* **108**, 102401 (2016)
[9] S. W. Cheong and M. Mostovoy, Nat. Mater. **6**, 13 (2007).
[10] N. Hur, S. Park, P. A. Sharma, S. Guha, and S.-W. Cheong, *Phys. Rev. Lett.* **93**, 107207 (2004).
[11] G. R. Blake, L. C. Chapon, P. G. Radaelli, S. Park, N. Hur, S.-W. Cheong,and J. Rodrıguez-Carvajal, *Phys. Rev. B* **71**, 214402 (2005).
[12] N. Hur, S. Park, P. A. Sharma, J. S. Ahn, S. Guha, and S.-W. Cheong,*Nature* **429**, 392 (2004).
[13] Kimura T, Goto T,Shintani H, Ishizaka K, Arima T and Tokura Y 2003 *Nature* **426** 55.
[14] Hur N, Park S, Sharma P, Ahn J, Guha S and Cheong S 2004 *Nature* **429** 392.
[15] Lawes G, Harris A B, Kimura T, Rogado N, Cava R J, Aharony A, Entin Wohlman O, Yildirim T, Kenzelmann M and Broholm C *2005 Phys. Rev. Lett.* **95** 087205.
[16] Mostovoy M 2006 *Phys. Rev. Lett.***96** 067601.
[17] Fiebig M 2005 *J. Phys. D: Appl. Phys*. **38**R123.
[18] Hill N A 2000 *J. Phys. Chem.* B**104** 6694.
[19] Eerenstein W et al 2006 *Nature* (London) **442** 759.
[20] Cheong S W et al 2007 *Nat.Mater*. **6** 13.
[21] WangC, Guo G C and HeL 2007 *Phys. Rev. Lett.***99** 177202.
[22] C. R. dela Cruz, B. Lorenz, Y. Y. Sun, Y. Wang, S. Park, S-W. Cheong, M. M. Gospodinov, and C. W. Chu. *PHYSICAL REVIEW B* **76**, 174106 2007
[23] T. Kimura, T. Goto, H. Shintani, K. Ishizaka, T. Arima and Y. Tokura, *Nature* **426,** 55 (2003).
[24] N. Hur, S. Park, P. A. Sharma, J. Ahn, S. Guha and S.-W. Cheong, *Nature* **429**, 392 (2004).
[25] Abrahams S C and Bernstein J L 1967J. *Chem. Phys*. **46** 3776.
[26] Blaha P, Schwarz K, Madsen G, Kvasnicka D and Luitz J 2011 WIEN2k User's Guide, Vienna University of Technology, ISBN 3-9501031-1-2.
[27] Perdew P J, Burke K and Ernzerhof M 1996 *Phys. Rev. Lett.*77 3865.

[28] T. J. Scheidemantel, C. Ambrosch-Draxl, T. Thonhauser, J. V. Badding, J. O. Sofo, *Phys. Rev. B* **68** (2003) 125210.
[29] G. K.H. Madsena, D. J. Singh, *Comput. Phys. Commun.* **175** (2006) 67-71.
[30] K. P. Ong. D. J. Singh, P. Wu, *Phys. Rev. B* **83** (2014) 115110.

Elastic and electronic properties calculations of the filled skutterudite $CeOs_4P_{12}$

M Merabet[a,b]*, L Djoudi[a,b], S Benalia[a,b], F Dahmane[a,b], M Boucharef[a] , D Rached[a], H Rached[a]
[a]Laboratoire des Matériaux Magnétiques, Faculté des Sciences, Université Djillali Liabès de Sidi Bel-Abbès, Sidi Bel-Abbès (22000), Algeria
[b]University of Tissemsilt, Institute of Science and Technology 38000, Algeria

*E-mail address: merabetmostefa@yahoo.fr (M. Merabet).

Abstract. The full-potential linear muffin-tin orbital method (FP-LMTO) within the local density approximation (LDA) using the Perdew-Wang parameterization is used to calculate the structural, electronic and elastic properties of the filled skutterudite $CeOs_4P_{12}$. The results of the electronic properties show that this compound is an indirect band gap material (Γ-N). A special interest has been made to the determination of the elastic constants since there have been no available experimental and theoretical data. The energy band gaps and their volume and pressure dependence are also investigated.

Keywords: Skutterudite, FP-LMTO, electronic properties, elastic constants.

1. Introduction

The pursuit of new and interesting strongly correlated electron phenomena has led to a systematic investigation of the rare earth-based filled skutterudite compounds. This pursuit has been driven by the remarkable fact that one can observe metal insulator transitions [1], magnetic order [2], heavy fermion behaviour [3], superconductivity [4, 5, 6, 7], quadrupolar ordering [8, 9, 10], Kondo insulating behaviour [11], non-Fermi liquid behaviour [12], quantum critical points [13, 14], and heavy fermion superconductivity all in the same family of compounds [3]. The filled skutterudite compounds have the general formula MT_4X_{12}
(M = alkali metal, alkaline-earth, lanthanide, or actinide; T = Fe, Ru, or Os; and X = pnictogen such as P, As, or Sb) with the $LaFe_4P_{12}$-type structure ($Im\bar{3}$ space group) [15]. These compounds are interesting materials on the viewpoint not only of strongly correlated electron systems but also for industrial application [16]. Recently, these compounds attracted renewed attention due to their potential application as thermoelectric materials (TE) [17, 18]. TE performance is characterized by a dimensionless figure of merit : $ZT = \frac{\sigma S^2 T}{\kappa}$, where σ is the electric conductivity, S is the thermopower, and κ is the thermal conductivity; ZT up to 1.4 at 600 K has been measured in skutterudites.

The compound $CeOs_4P_{12}$ belongs to the family filled skutterudites, which have exhibit several physical characteristics, $CeOs_4P_{12}$ is a narrow gap semiconductor with a gap of $\approx$0.036 eV ($\approx$400 K) estimated from resistivity data [19]. The room temperature resistivity is $\approx 10^{-2}$ $\Omega - cm$ increasing to $\approx 10^4$ $\Omega - cm$ at 2 K. The magnetic susceptibility is similar to that of $CeRu_4P_{12}$ [19]. The as-grown samples are p-type with a room temperature Hall mobility of 73 $cm^2/V - \sec$ for a carrier concentration of $5.10^{19} cm^{-3}$. The room temperature values of the Seebeck coefficient and thermal conductivity are $147 \mu V/K$ and $105 mW/cmK$, respectively [20].

For this compound, there exist a few experimental data for the electrical, magnetic and thermoelectric properties [19, 20]. Point of view theoretical, there are no available calculations for this compound. We therefore think that it is worthwhile to perform these calculations using the full-potential linear muffin-tin orbital (FP-LMTO) method in order to complete the exciting experimental works for this compound. In this works, we present the results of a theoretical investigation of electronic and elastic properties for filled skutterudite compounds $CeOs_4P_{12}$. The organization of the paper is a follow: we explain the computational method in Sec. 2. The results are presented and discussed in Sec. 3 for structural, electronic and elastic properties and a brief conclusion is drawn in the last section.

2. Calculations:

The atomic positions in the ternary skutterudite $CeOs_4P_{12}$ are as follows: the Ce atoms are located at (0, 0, 0), Os atoms at (1/4, 1/ 4, 1/4) and the phosphide atoms at (0, *u*, *v*). The calculations reported in this work were carried out by means of the full-potential linear muffin-tin orbital (FP-LMTO) [21, 22] within the framework of density functional theory (DFT). In this method, the space is divided into an interstitial region (IR) and non-overlapping (MT) spheres centered at the atomic sites. In the IR region, Fourier series represent the basic functions. Inside the MT spheres, the basis sets is described by radial solutions of the one particle Schrödinger equation (at fixed energy) and their energy derivatives multiplied by spherical harmonics.

In order to achieve energy eigenvalues convergence, the wave functions in the interstitial region were expanded in plane waves with a 110.0345 Ry energy cut-off and a number of plane waves equal to 34412. The muffin tin spheres radius is taken to be 3.733, 2.317 and 2.145 a.u for Ce, Fe and P respectively. The valence wave functions in side the spheres are expanded up to $l_{max} = 6$. The exchange-correlation (XC) effects are treated by the local density approximation (LDA) [23].The integrals over the Brillouin zone (BZ) are performed up to (6, 6, 6) grid in the irreducible Brillouin zone (IBZ), using the tetrahedron method [24].

3. RESULTS AND DISCUSSIONS

3.1. The grounds states

The internal parameters u and v, which define the position of pnictogen atoms, are determined by minimizing the total energy while keeping the volume fixed at the experimentally observed value. The optimized *u* and *v* values are found to be equal 0.3589 and 0.1424, respectively. In order to calculate the ground state properties of $CeOs_4P_{12}$ the total energies are calculated in skutterudite structure for different volumes around the equilibrium cell volume V_0. The plot of calculated energies versus reduced volume for this compound in this structure is given in Figure. 1. Fitting of the Birch equation of state [25] to the total energies versus lattice parameters, yields the equilibrium lattice parameter a_{eq}, the bulk modulus B_0 and its pressure derivative of the bulk modulus $B_0^{'}$.

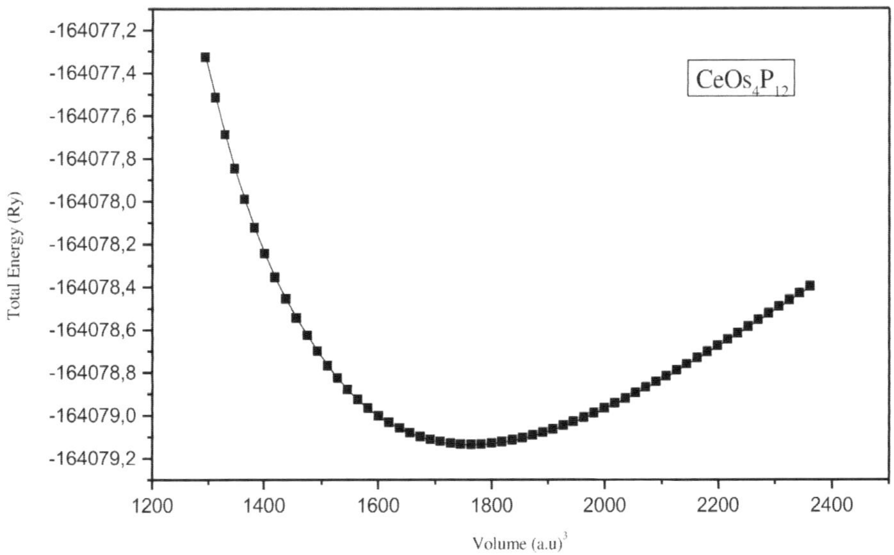

Figure 1: Calculated total energy as a function of volume

The optimized value of the lattice constant turns out to be 8.053 Å, as compared to the experimental value of 8.0626 Å [15], Our calculated values of a_0 is within 0.11% of the corresponding experimental value. This is due to the use of LDA which known to underestimate this quantities in comparison to experiment. The values of the bulk modulus and its pressure derivative (B_0 and B_0') evaluated at zero pressure are predicted to be (208.92 GPa and 5.87802). Our calculated value of the bulk modulus is overestimated compared to those obtained by the x-ray diffraction (150 ± 5) [26]. This overestimation is generally attributed to the use of LDA.

3.2. Elastic constants

The elastic properties define the properties of a material that undergo stress, deform and then recover and returns to its original shape after stress ceases. These properties play an important part in providing valuable information about the binding characteristic between adjacent atomic planes, anisotropic character of binding and structural stability. The elastic moduli require knowledge of the derivative of the energy as a function of the lattice strain. In the case of cubic system, this strain is chosen in such a way that the volume of the unit cell is preserved. Thus for the calculation of elastic constants C_{11}, C_{12} and C_{44} for this compound we have used the Mehl method [27, 28]. For cubic crystal, the bulk modulus is given by:

$$B=1/3\ (C_{11}+2C_{12}) \quad (1)$$

C_{11}, C_{12} and C_{44} comprise the complete set of elastic constants for a cubic system and the shear modulus G, Young's modulus E, and the Poisson's ration ν can be derived using the following standard relations:

$$E=9BG/(3B+G) \quad (2)$$

$$G=(C_{11}-C_{12}+3C_{44})/5 \quad (3)$$

$$\nu=(3B-E)/(6B) \quad (4)$$

$$A=2C_{44}/C_{11}-C_{12} \quad (5)$$

The calculated elastic constants C_{ij}, the shear modulus G, the Young's modulus E and the Poisson's ratio (ν) are summarized in Table 1. To date no experimental or theoretical data of the elastic constants of these compounds are given in the literature.

Table 1: Calculated elastic constant C_{11},C_{12}, C_{44} (in GPa), Young and shear modulus (in GPa), Poisson's ratios (v) and the anisotropic parameter for $CeOs_4P_{12}$

C_{11}	C_{12}	C_{44}	G	E	ν	A
503.04	43.86	304.58	274.58	572.81	0.043	1.32

Furthermore, this compound exhibit small value of C_{12}. From the elastic constants, we obtain the anisotropy parameter A=1.32, this indicates that the elastic properties of this compound are highly anisotropic. The requirement of mechanical stability in cubic crystal leads to the following restrictions on the elastic constants, $C_{44}>0$, $C_{11} \rangle |C_{12}|$, and $C_{11}+2C_{12}>0$ [29]. The elastic constant in Table 1, obey these stability conditions, including the fact that C_{12} must be smaller than C_{11}. With these results, we conclude that the skutterudite structure is the stable ground state.

The Debye temperature $(\theta_D)$is fundamental physical property, and is used to distinguish between high-and low temperature regions for a solid. If $T > \theta_D$ we expect all modes to have energy k_BT, and if $T< \theta_D$ one expects high-frequency modes to be frozen [30]. In the present case, Debye temperature $(\theta_D)$is estimated for $CeOs_4P_{12}$ by using the calculated elastic constant data, in terms of the following classical relations [31]:

$$\theta_D = \frac{h}{k}\left[\frac{3n}{4\pi}\left(\frac{N_a\rho}{M}\right)\right]^{1/3} v_m \quad (6)$$

Where v_m is the average wave velocity, h is Planck's constants, k is Boltzmann's constant, N_a is Avogadro's number, n is the number of atoms per formula unit, M is the molecular mass per formula unit, $\rho = M/V$ is the density, and v_m is obtained from [32]:

$$v_m = \left[\frac{1}{3}\left(\frac{2}{v_t^3}+\frac{1}{v_l^3}\right)\right]^{-1/3} \quad (7)$$

Where v_l and v_t, are the longitudinal and transverse elastic wave velocities, respectively, which are obtained from Navier's equations [33]:

$$v_l = \left(\frac{3B + 4G}{3\rho}\right)^{1/2} \qquad (8)$$

and

$$v_t = \left(\frac{G}{\rho}\right)^{1/2} \qquad (9)$$

The calculated average longitudinal and transverse elastic wave velocities and Debye temperature for $CeOs_4P_{12}$ are given in Table 2. Unfortunately, as far as we know, there are no data available to these properties in the literature for these compounds. Future experimental work will testify our calculated results.

Table 2: Calculated longitudinal, transverse and average sound velocity (v_l, v_t, v_m in m s^{-1}) calculated from polycrystalline elastic modulus and the Debye temperature (θ_D in K) for $CeOs_4P_{12}$.

v_t	v_l	v_m	θ_D
8239.07	11923.09	8963.32	329

3.3. Electronic properties

The calculated band structures of filled skutterudite $CeOs_4P_{12}$ at equilibrium lattice constant along the high symmetry lines of the BZ is presented in figure 2. This figure, show clearly the semiconducting behaviour of $CeOs_4P_{12}$, which are drawn along the symmetry directions in the first Brillouin zone.

Predicting the valence band maximum (VBM) at Γ point and the conduction band minimum (CBM) at N point resulting in an indirect band gap material $(\Gamma - \mathrm{N})$. The value of this gap turns out to be 0.218 eV.

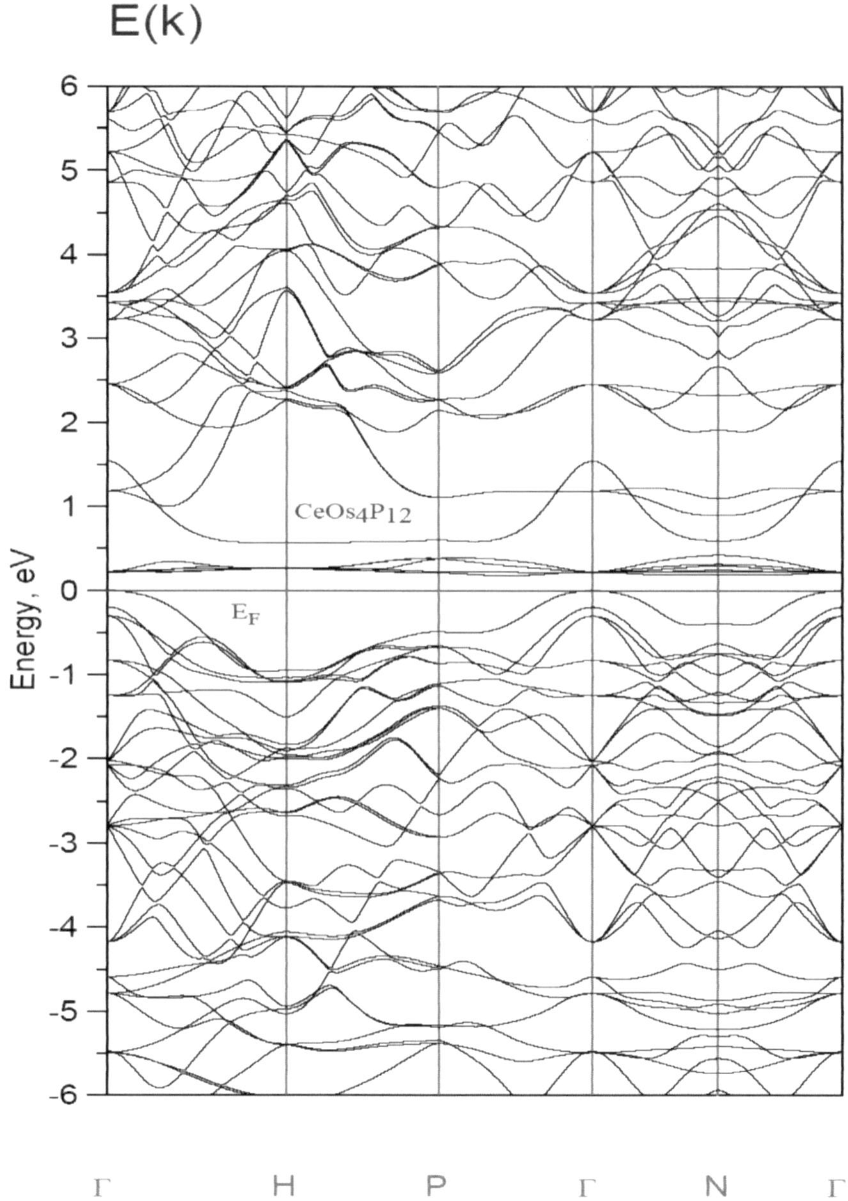

Figure 2: Band structure along the symmetry lines of the Brillouin zone at the equilibrium lattice constant. The position of the Fermi Level is shown by solid horizontal line.

From the figure the valence band is essentially dominated by the Os-5d below the Fermi energy with minor contributions of the others states of the elements forming the compound. The region including the Fermi energy arises predominately from the Ce-4f.

To further elucidate the nature of the electronic band structure, we have also calculated the total and the partial densities of states (DOS) of this compound these are displayed in Figure 3 and 4.

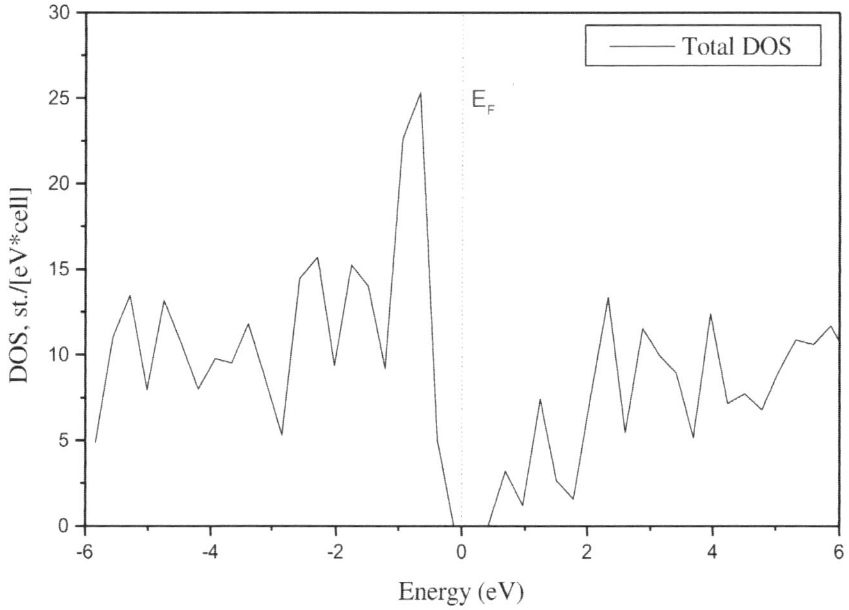

Figure 3: Total density of states (DOS) of $CeOs_4P_{12}$.

a)

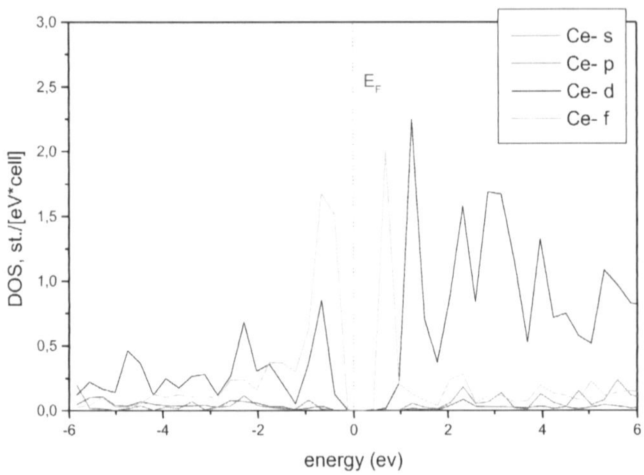
Ce- s
Ce- p
Ce- d
Ce- f
E_F
DOS, st./[eV*cell]
energy (ev)

b)

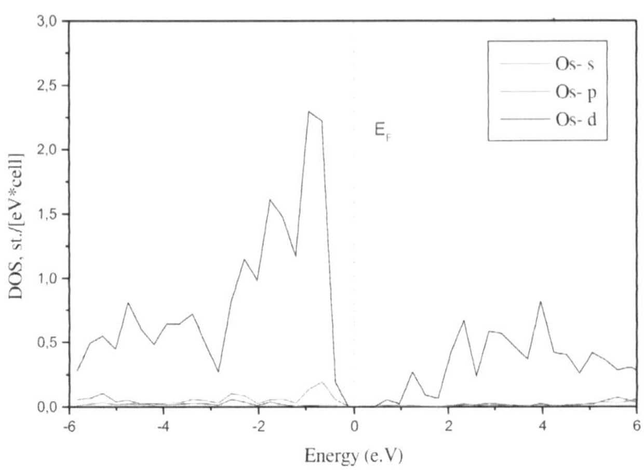
Os- s
Os- p
Os- d
E_F
DOS, st./[eV*cell]
Energy (e.V)

c)

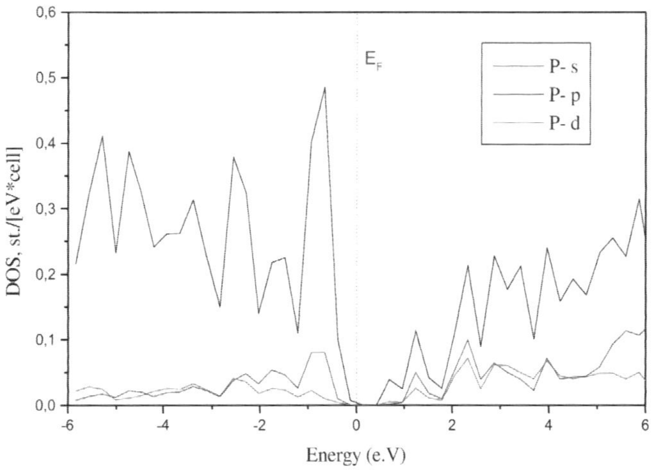

Figure 4: Partial DOS of $CeOs_4P_{12}$. (a) The states contributions of Ce,(b) the states contributions of Os and (c) the states contributions of P.

These results show that the top valence bands have characters of hybridized Os-5d and P-3p states. The low of the valence bands is mainly due to the P-3s states with a few contributions of Ce-5p. The remaining part until the Fermi energy is essentially dominated by Os-5d states with the contributions from P-3s, 3p and Ce-5d states. In conduction bands, from the Fermi energy up to strongly peak is dominated by Ce-4f states, the remaining part the conduction bands, there are mixture of the Ce-4f, Ce-5d and P-3p states.

In order to investigate the effects of the cell volume on the size of the energy gap of $CeOs_4P_{12}$ in the cubic skutterudite structure, the band energies at selected symmetry points are examined as a function of the cell volume. We calculated the pressure dependence of energy gap in this semiconductor using the following relation [34]:

$$\left(\frac{\partial E_P}{\partial P}\right)_T = \left(\frac{\partial E_g}{\partial V}\right)_T \left(\frac{\partial V}{\partial P}\right)_T = -\left(\frac{V_0}{B_0}\right)_T \left(\frac{\partial E_g}{\partial V}\right)_T \qquad (10)$$

The value of $\left(\frac{\partial E_g}{\partial V}\right)_T$ was obtained directly from the energy gap versus volume calculations and the value of $\left(\frac{\partial V}{\partial P}\right)_T$ was obtained from the compressibility relation ship:

$$K = \frac{1}{B_0} = -(\frac{1}{V_0})(\frac{\partial V}{\partial P})_T \qquad (11)$$

The results of our calculations for the nature of the band gap for CeOs4P12 versus the volume are shown in Figure 5.

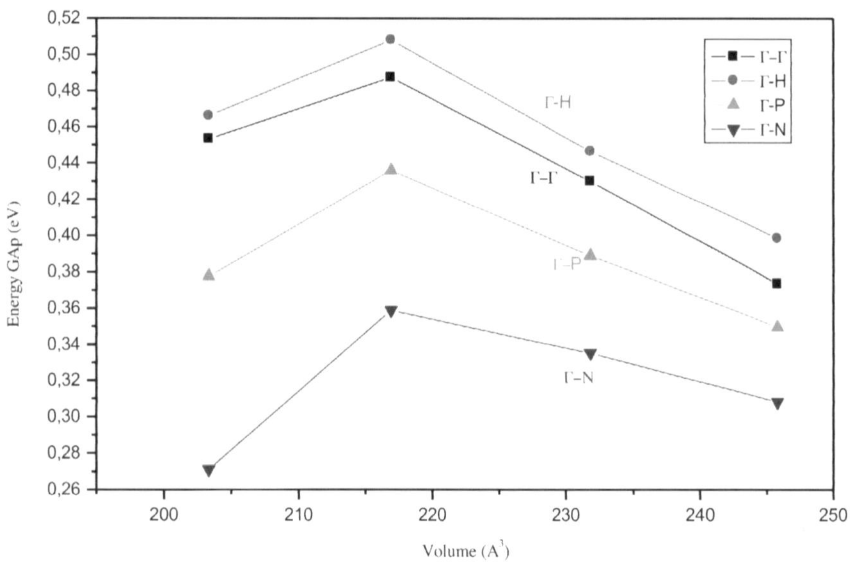

Figure 5: Calculated volume of energy gap in $CeOs_4P_{12}$

We notice that this compound is indirect gap (Γ-N) semiconductors at ambient conditions. Also we notice that the fundamental gaps present a nonlinear behaviour. The calculated linear pressure coefficients relative to different transitions (Γ-Γ-H-P-N) bands are listed in Table 3. The linear pressure increase with decrease of the pressure.

Table 3: Energy band gap (eV) for various points of the Brillouin zone and the calculated volume and pressure coefficients for $CeOs_4P_{12}$ skutterudite.

	$\Gamma-\Gamma$	$\Gamma-H$	$\Gamma-P$	$\Gamma-N$
$CeOs_4P_{12}$	0.218	0.261	0.215	0.198
$\frac{\partial E_g}{\partial V}$ (meV/A^3)	-3.95	-3.79	-2.99	-1.76
$\frac{\partial E_P}{\partial P}$ (meV/GPa)	4.93	4.73	3.73	2.19

4. Conclusion

We have performed first-principle FP-LMTO calculations of the elastic and electronic of filled skutterudite $CeOs_4P_{12}$. The ground states properties including the lattice parameter, bulk modulus and its pressure derivatives are compared with the available theoretical and experimental results. The elastic constants C_{ij}, shear modulus, Young's modulus and the Poisson's ratio for these compounds are derived. The Debye temperature is also estimated from the average sound velocity. Our results may be considered as reliable prediction of the elastic properties of this compound. Our results concerning the electronic properties show that this compound is an indirect gaps (Γ-N) semiconductor.

Reference:

[1] Sekine C, Uchiumi T, Shirotani I and Yagi T 1997 Phys. Rev. Lett. **79** 3218.

[2] Danebrock M E, Evers C B H and Jeitschko W 1996 J. Phys. Chem. Solids **57** 381.

[3] Bauer E D, Frederick N A, Ho P C, Zapf V S and Maple M B 2002 Phys. Rev. B **65** 100506(R).

[4] Meisner G P 1981 Physica B+C **108** 763.

[5] Shirotani I, Uchiumi T, Ohno K, Sekine C, Nakazawa Y, Kanoda K, Todo S, and Yagi T 1997 Phys. Rev. B **56** 7866.

[6] Maple M B, Bauer E D, Zapf V S, Freeman E J, and Frederick N A 2001 Acta Physica Polonica B **32** 3291.

[7] Maple M B, Ho P C, Zapf V S, Frederick N A, Bauer E D, Yuhasz W M, Woodward F M and Lynn J W 2002 J. Phys. Soc. Japan **71** Supple. 23.

[8] Hao L, Iwasa K, Nakajima M, Kawana D, Kuwahara K, Kohgi M, Sugawara H, Matsuda T D, Aoki Y, and Sato H 2003 Acta Phys. Pol. B **34** 1113.

[9] Aoki Y, Namiki T, Matsuda T D, Abe K, Sugawara H, and Sato H 2002 Phys. Rev. B **65** 064446.

[10] Kohgi M, Iwasa K, Nakajima M, Metoki N, Araki S, Bernhoeft N, Mignot J M, Gukasov A, Sato H, Aoki Y and Sugawara H 2003 J. Phys. Soc. Jpn. **72** 1002.

[11] Bauer E D, ´Slebarski A, Freeman E J, Sirvent C and Maple M B 2001 J. Phys.: Condens. Matter **13** 4495.

[12] Takeda N and Ishikawa M 1999 Physica B **259-261** 92.

[13] Maple M B, Ho P C, Frederick N A, Zapf V S, Yuhasz W M, Bauer E D, Christianson A D and Lacerda A H 2003 J. Phys.: Condens. Matter **15** S2071.

[14] Maple M B, Ho P C, Frederick N A, Zapf V S, Yuhasz W M, and Bauer E D 2003 Acta Physica Polonica B **34** 919.

[15] Jeitschko W and Braun D 1977 Acta. Cryst. **B33** 3401.

[16] Tsutsui S, Umemura J, Kobayashi H, Tazaki T, Nasu S, Kobayashi Y, Yoda Y, Onodera H, Sugawara H, Matsuda T D, Kikuchi D, Sato H, Sekine C, Shirotani I 2006 Hyperfine Interact **168** 1073.

[17] Mahan G, Sales B and Sharp J 1997 Phys. Today **50** 42.

[18] Nolas G S, Slack G A, Morelli D T, Tritt T M and Ehrlich A C 1996 J. Appl. Phys. **79** 4002.

[19] Shirotani I, Uchiumi T, Sekine C, Hori M, Kimura S and Hamaya N 1999 J. Solid State Chem. **142** 146.

[20] Sekine C, Akita K, Yanase N, Shirotani I, Inagawa I and Lee C H 2001 Jap. J. Appl.Phys. **40** 3326.

[21] Savrasov S Y 1996 Phys. Rev. B **54**, 16470.

[22] Savrasov S, Savrasov D 1992 Phys. Rev. B **46** 12181.

[23] Perdew J P and Wang Y 1992 Phys. Rev. B **46** 12947.

[24] Blochl P, Jepsen O and Andersen O K 1994 Phys. Rev. B **49** 16223.

[25] Birch F 1978 J. Geophys. Res. **83** 1257.

[26] Mehl M J, Osburn J E, Papaconstantopoulos D A and Klein B M 1990 Phys. Rev. B **41**, 10311.

[27] Mehl M J, Klein B M and Papaconstantopoulos D A, West- Brook J H, R L Fleisher (Eds) 1995 Principles Intermetallic Compounds,Vol. 1, Wiley, New York (Chapter 9).

[28] Wachter P, Filzmoser M and Rebizant J 2001 Physica B **293** 199.

[29] Anderson O L 1963 J. Phys. Chem. Solids **24**, 909.

[30] Schreiber E, Anderson O L, Soga N 1973 Elastic Constants and Their Measurements, McGraw-Hill, New York.

[31] Lach-hab M, Keegan M, Papaconstantopoulos D A, Mehl M J 2000 J. Phys. Chem. Solids **61** 1639.

Journal of Physics: Conference Series **758** (2016) 012011 doi:10.1088/1742-6596/758/1/012011

Effect of temperature on (IV) statics characteristics of GaAs Mesfet

Z. Fares [1], **Y. Saidi** [2], **W. Aliouat** [2].

[1] Department of Engineering, Faculty of Science, University of Constantine 1 Algeria.

[2] Department of Physics, Faculty of Exact Sciences University of Constantine 1 Algeria.

E-mail: saidiyas09@yahoo.fr

Abstract. The GaAs metal semiconductor field effect transistors called MESFETS are the most active components used in microwave applications. To better exploit the performance of these components circuits, it is necessary to develop techniques for sophisticated numerical computation based on physical mechanisms that govern the operation of these devices. The static properties of GaAs MESFET could be determined from an original analytical study based on the resolution of the semiconductor fundamental equations. Then we will study the equation of thermal resistance as a function of the physical parameters of MESFETs by analogy electric thermal resistance RTH will be determined as the ratio of the difference of temperature on the thermal dissipation. The model took into account the difference between the temperature of the component and the ambient temperature and the effect of temperature on the parameters of the component.

1-Introduction:

The GaAs MESFETS are attractive devices for the used in microwave applications because of their relatively simple processing and they high-speed and low noise performance. The principal object in the paper is to propose a physical and analytical model of the characteristics current voltage of these devices with different laws of temperature. In the first ,we calculate the potential in the depletion layer due to the electrical charge formed under the gate which can be obtained by resolving the Poisson's equation by the conventional approximation ,Then we determine the drain current Id, the characteristic I-V obtained by this model, using effect of temperature on IV characterisation.

2- Calculation of the potential and the current in the channel:

To calculate the potential and the electric field under the gate, the channel is divided into two principal regions [1] as shown in figure (1):

- The first region (1) below the gate directly, it is said a region controlled by the gate.
- The second region (2) outwards of the first region known as region not controlled by the gate.

Journal of Physics: Conference Series **758** (2016) 012011 doi:10.1088/1742-6596/758/1/012011

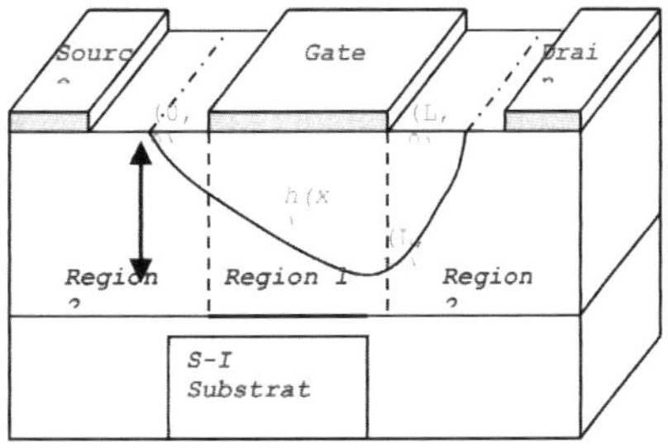

Figure 1: Depletion regions:

(1) Controlled by the gate.

(2) Not controlled by the gate.

The electric potential due to the electrical charge Formed under the gate can given by[2]:

$$V_q(x,y)=\int_0^y \frac{eN_d(x,y)}{\varepsilon} y dy + y\int_y^{h(x)} \frac{eN_d(x,y)}{\varepsilon} dy + Vbi - Vg \quad (1)$$

With *Nd(x,y)* is the density of the donors,*Vbi* is the built in potential of Schottky barrier gate and ε is the permittivity.It should be noted that the approximations in (1) is based on the fact that the depletion layer thickness under the gate h(x) is a slowly varying function in the channel an is giving by:

$$h(x)=\left[\frac{2\varepsilon(V_{bi}-V)}{qNd}\right]^{1/2} \quad (2)$$

The channel potential is obtained by integration limits with y=h(x)

$$V(y)=\frac{q.Nd}{\varepsilon}\left[h.y-\frac{y^2}{2}\right](3)$$

The equation of the potential take a maximum of values in diffusion potential $Vb_i(y=h)$:

$$Vbi = V(y=h) - V(y=0)(4)$$

The dimensional potential of the channel under the gate is given as follows:

$$V(x)=\frac{qN_D h^2(x)}{2\varepsilon}+V_G-V_{bi} \quad (5)$$

To calculate the drain current expression as a function of the drain voltage, we must make some approximations [3]:

One neglects the current flow in the y-direction; this approximation is valid for the components with the length short gate.

- An abrupt Schottky barrier junction.

- A channel with uniform donors doping N_D (x, y) = N_D ,N_D is constant.
- Neglecting edge effects, and the overflow depleted area on the sides of the gate

The density of the current is given by:

$$J_x = \sigma(x,y,z).E_x \qquad (6)$$

$$\sigma(x,y) = \rho(y)\mu_n(E_x)$$

$$J_x = qN_D\mu_n.E_x = -q\mu_n Nd\frac{dV(x)}{dx} \quad (7)$$

μ_n (E_x) is the electron mobility which depends of the electric field.

The drain current Id is considered positive in the drain source is obtained by integrating the current density Jx over the conducting channel section:

$$I_d = -\int_0^Z\int_{h(x)}^a J_x dydz = -Z\int_{h(x)}^a J_x dy \qquad (8)$$

The calculations made above don't take into account the contribution of the space charge region located below the gate contact. In order to consider this contribution we introduce the pinch off voltage V_P at drain voltage equation and the saturation current Ip as following:

$$Ip = \frac{(qNd)^2\mu_n Za^3}{2L\varepsilon} \qquad Vp = \frac{qNda^2}{2\varepsilon} = V_{bi} - Vg$$

.

$$a = \left[\frac{2\varepsilon}{qN_D}\left(V_{bi} - Vg\right)\right]^{1/2} ; \quad \frac{h(x)}{a} = \left[\frac{V_{bi} + V(x) - Vg}{V_{bi} - Vg}\right]$$

Then the expression final of the current Id:

$$I_d(V_d, V_g) = Ip\left\{\frac{Vd}{Vp} - \frac{2}{3}\left[\left(\frac{Vd + V_{bi} - Vg}{V_P}\right)^{3/2} - \left(\frac{V_{bi} - Vg}{V_P}\right)^{3/2}\right]\right\} \qquad (9)$$

3.Effect of temperature .

The current characteristics are strongly related to the temperature. However, most simulations assume that the temperature of the component is constant, usually equal to the ambient temperature (300 ° K). A rigorous thermal model requires solving the equation of heat [4]

$$C_R\rho_R\frac{\partial T_R}{\partial t} = \nabla\left(K_R\nabla T_R\right) + H_S \qquad (10)$$

C_R : specific heat of the network, ρ_R: Network density , K_R : temperature of the network.
T_R : thermal conductivity of the network, H_S: Heat generation network.

The dependence of the carrier mobility with temperature [5]:

$$\mu = \mu_0(300^0 K)\left[\frac{300}{T_R}\right]^{0.6} \tag{11}$$

The saturation velocity varies with temperature as

$$v_S = \frac{2.410^5}{1 + \exp(T / 600)} ms^{-1} \tag{12}$$

according to Conger [3] the dependence of the threshold voltage may be approximately given by:

$$V_{TH} = V_{TH}(300^{\circ} K) - \alpha_{VT} T \tag{13}$$

The value of α_{VT} is in the order of 1.2mV/°C.

4. Results and discussion

To show the effect of temperature on the characteristics (I-V), we perform the numerical simulation with the laws of mobility, velocity saturation and the threshold voltage as a function of temperature:

$$\mu_n = \mu_n(300^0 K)\left[\frac{300}{T_R}\right]^{0.6} \tag{14}$$

$$v_S = \frac{2.410^5}{1 + \exp(T / 600)} \tag{15}$$

As the diffusion voltage of the junction "V_{bi}" varies with temperature as follows:

$$V_{bi}(T) = V_{bi}(300^0 K)\left(\frac{T}{T_0}\right) \tag{16}$$

The structure used for the calculation is shown in Figure (1)

For the mobility we have choice these expressions:

$$\begin{cases} \mu_1(E) = \dfrac{\mu_n}{1+\left(\dfrac{E}{Ec}\right)} \\ \mu_2(E) = \dfrac{\mu_n + v_S\left(\dfrac{E^3}{Ec^4}\right)}{1+\left(\dfrac{E}{Ec}\right)} \end{cases} \quad (17)$$

In table 1.we give the characteristic of our MESFET.

MESFET	*L(μm)*	*a(μm)*	*Z(μm)*	$N_D(Cm^{-3})$	$\mu_n(Cm^2S^{-1}V^{-1})$	V_{bi} *(V)*	R_S *(Ω)*	R_D*(Ω)*
MESFET	*4*	*0.3*	*360*	6.7×10^{16}	*3740*	*0.8*	*16*	*16*

The Figures (2) and (3) present the change of speed of the electrons as a function of electric field for various values of the temperature; the speed is highest when the temperature increases.

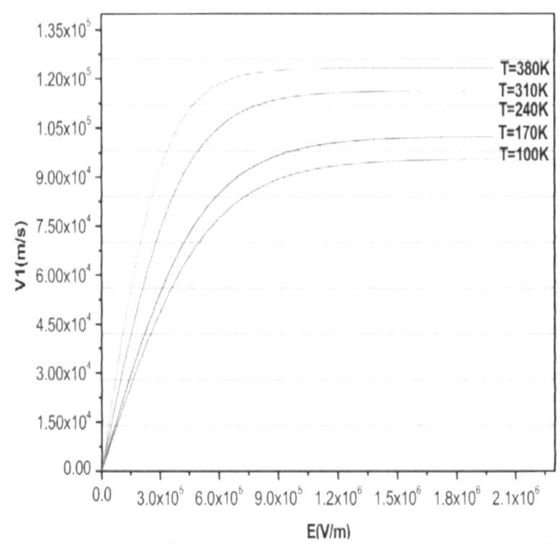

Figure 2. Variation of the velocity V_1 versus electric temperatures field for different values of temperatures.

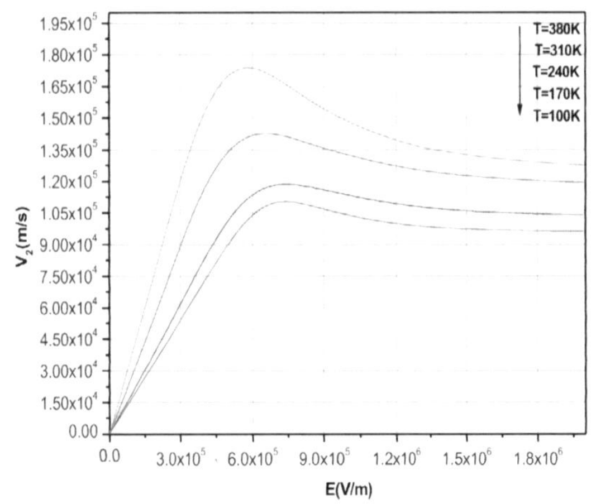

Figure 3. Variation of the V_2 velocity versus electric field for different values of temperatures.

The Figures (4) and (5) shows the variation of the mobility of electrons as a function of electric field for various values of temperature. We find that mobility is higher as the temperature decreases.

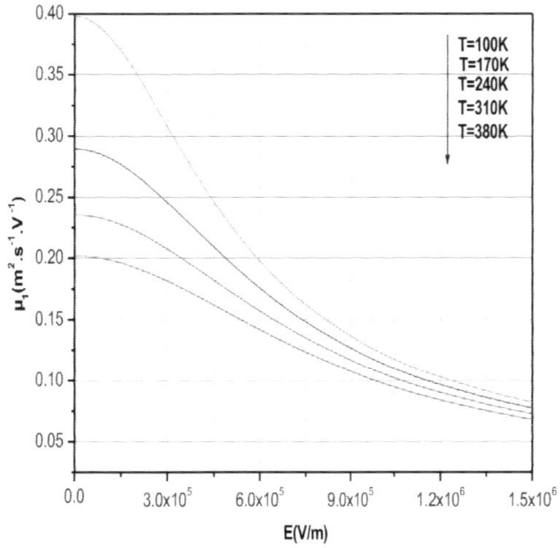

Figure 4.Variation of the μ_1 mobility with electric field Electric field for different values of temperature

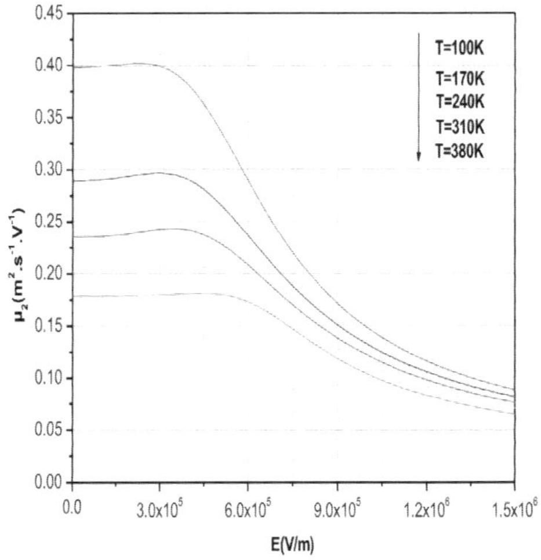

Figure 5: Variation of the μ_2 mobility with for different values of temperature

The figure (6) show the variation of the drain current as a function of the voltage V_{DS} for different values of temperature. The current increases as the temperature T decreases, constant mobility.

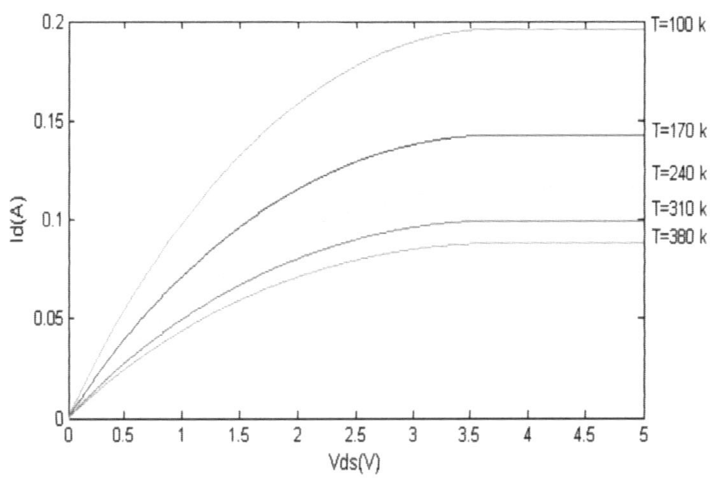

Figure 6 Characteristics (I-V) for different values of temperature T

with constant mobility μ_0 for GAT1

The same manner when the mobility varies, we present in Figures (7) and (8), the influence of temperature on the I-V characteristics of transistors "GAT1». We choose five values of temperature "T = 100 K ," "T = 170K " and "T = 240 K ","T = 310 K ","T = 380 K ",. From these figures, we find that the performance and reliability of the transistors are strongly influenced by temperature. Conduction along the channel is due to the majority carriers (electrons), it will be affected by the temperature variation for certain parameters (electron mobility, the Schottky barrier height, the saturation velocity, the dielectric constant and the same the specific resistance of ohmic contacts). If the temperature increases, thermal motion of the carriers also increases and hence the electron mobility of the channel decreases from expression (5), which causes a decrease in the current "Ids". Similarly the height of the potential barrier increases with increasing temperature from expression (8), therefore the width of the space charge region increases against the conductive channel narrows, and thus the drain current decreases.

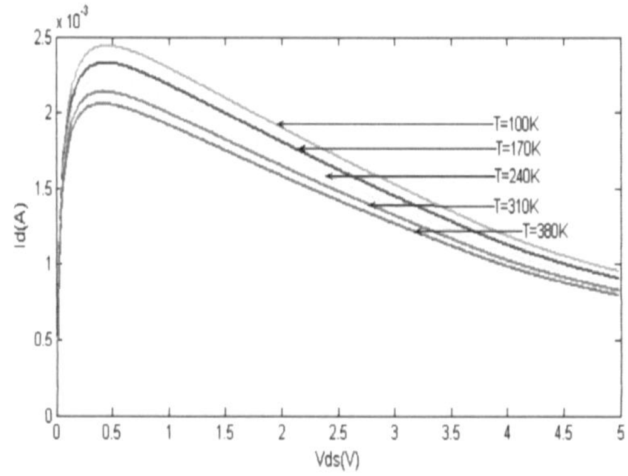

Figure 7. The characteristics (I-V) different values of temperature using the expression of μ_1 mobility for GAT1

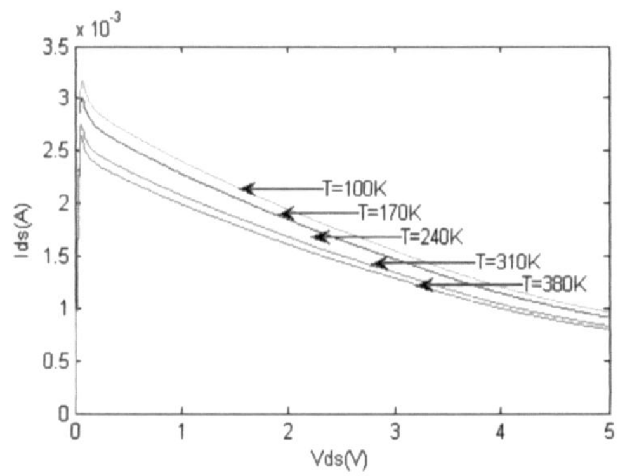

Figure 8. The characteristics (I-V) for different values of temperature using the expression of μ_2 mobility for GAT1

5.Conclusion

In this study we have developed an analytical model to calculate the I-V characteristics of short gate length GaAs MESFET which takes into account the one-dimensional analysis of the charge distribution in the active region and incorporates the effect of temperature on field electron mobility, velocity saturation and effect of this parameter to the temperature expressions. Moreover, comparisons between the analytical models with different values of temperature showed the effect the output characteristics (I-V) of GaAs MESFET,then the MESFET structure allows very significant improvement in performance when it is operated at low temperatures.

6.References

[1] S. Morarka, S. Mishra "has 2-D model for the potential distribution and threshold voltage off fully depleted short-Chanel ion-implanted silicon MESFET' S", newspaper off semiconductor technology and science, Vol 5, NO3, p 173-181, Marsh 2005.

[2] T.A. Fjedly. T. Yterdal, M.S.Shur Introduction to Device Modeling and circuit simulation", Wiley, New " York, 1998.

[3] F.S.Shoucair and Pekka K. Ojala, " High-Temperature Electrical Characteristics of GaAs MEFET's", IEEE uary.19Transactions on Electron Devices. Vol.39.No.7. Febrs

[4] S. P. Chin, G. Y. We, "A new two dimensional model for the potential distribution of short gate lenght MESFET's and its applications," IEEETrans. Elec 5881, Dev, vol **39**, pp. 1928-1937, August 1992.

Journal of Physics: Conference Series **758** (2016) 012012 doi:10.1088/1742-6596/758/1/012012

Structure and Dynamics of a Model Discotic Organic Conducting Material

Mohamed Zbiri[1], Lucas A. Haverkate[2], Gordon J. Kearley[3], Mark R. Johnson[1] and Fokko M. Mulder[2]

[1]Institut Max von Laue-Paul Langevin, CS 20156, Grenoble, France
[2]Faculty of Applied Sciences, Reactor Institute Delft, TU-Delft, Delft, The Netherlands
[3]Australian Nuclear Science and Technology Organisation, Lucas Heights, NSW, Australia

E-mail: zbiri@ill.fr

Abstract. Organic conducting materials exhibit promising functionalities, inducing hence a keen interest due to their potential use as a next generation photoconverters. However, unlike the more expensive inorganic analogues, the underlying properties that give rise to these advantages also cause organic materials to be inherently inefficient as photovoltaics. Understanding their properties at the microscopic level is a major step towards an efficient and targeted design. We probed the morphological and dynamical aspects of a model organic discotic liquid crystal material hexakis(n-hexyloxy)triphenylene (HAT_6) by using neutron-based diffraction and quasielastic scattering techniques to gain deeper insights into structure and dynamics. The neutron measurements are accompanied, in a synergistic way, by molecular dynamics simulations for the sake of the analysis and interpretation of the observations.

1. Introduction

Due to their potential applications as photovoltaic energy-converters and thanks to their low cost, inherent flexibility and relative ease of processing, organic conducting materials offer a better alternative to the traditional, cost-prohibitive, inorganic semiconductors [1, 2, 3]. When disc-shaped poly-aromatic molecules are decorated with aliphatic side chains, they may self-assemble into columnar superstructures with both solid and liquid-like properties [4] arising from the cores and tails, respectively. This leads to properties like self-healing of defects and tunable alignment of the conducting columns [5]. In this context, discotic liquid crystals (DLCs) are a promising class of materials for molecular electronic devices, such as light emitting diodes, field-effect transistors and solar cells. These disk-like molecules with poly-aromatic cores stack on top of one another to form stable columns, providing one-dimensional molecular wires for charge-carrier transport, via a π-π overlap.

The main factor which affects the columnar charge migration in DLCs is the dynamics of the system. That is any conformational change and/or disorder/fluctuations could hinder the π-overlap. Therefore, for a widespread application, the conductivity properties of discotics require optimisation. Indeed, charge transport in these systems requires adequate intermolecular π-orbital overlap of the poly-aromatic cores along the entire columnar wire. The overall conductivity is therefore strongly dependent on the local conformation of the molecules, the presence of structural defects and thermal motions induced by the fluidic side chains [6, 7, 8]. Knowledge of how each of these three factors limits the hopping of charge carriers is valuable for rational design of discotic compounds with optimal device performance.

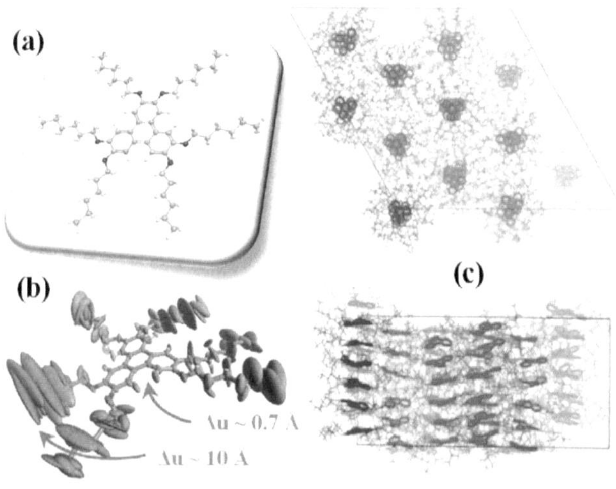

Figure 1. (a) HAT_6 in its starting D_{3h} configuration, (b) anisotropic displacements for a HAT_6 molecule ranging from 0.7 $\AA$ (aromatic core) to 10 $\AA$ (tail end), and (c) snapshot of a model MD simulation (further details on simulation and structure are provided in Section 2). The coulors illustrate the orientational order of the colmuns. The deviation from the initial D_{3h} symmetry is due to the thermally-activated inter and intra molecular motions. This is pictured in terms of fluctuations and deformations occuring between and within the cores, respectively. The geometrical fluctuations reproduced by MD simulations are consistent with the behaviour of the liquid-crystalline phase.

We have studided various aspects (morphological/structural, dynamical, and electronic) of a model organic discotic liquid crystal material consisting of the molecular system hexakis(n-hexyloxy)triphenylene (HAT_6) [8, 9, 10, 11, 12, 13, 14]. In this context, neutron scattering measurements were performed, in the electronic ground state, to gain a deeper and better understanding of the structural and dynamical properties of this material. The neutron measurements were accompanied by Raman and nuclear magnetic resonance probes. At the vibronic level, the low-lying excited states were explored using resonant Raman and UV-vis spectroscopies. Atomistic simulations, either classical via empirical force fields, or first-principles-based via improved density functional theory methods, were used synergistically for the sake of the analysis, interpretation and predictions. However, the presentation here is limited to the application of neutron scattering to probe structure and dynamics of HAT_6. The reader can refer to the indicated literature (and the related references therein) for a complete overview on the other above cited experimental and modelling techniques we have employed to treat this topic.
Neutron scattering is a convenient tool to study the dynamical structure of these systems. Neutron diffraction on deuterated samples generally achieves complementary information to X-ray diffraction. Considering the dynamics, quasi-elastic neutron scattering (QENS) is well suited to probe the picosecond timescale molecular motions of the discoids [15]. We have successfully elucidated the fundamental mechanisms that limit charge transport in discotics, by linking neutron diffraction and QENS experiments directly with classical molecular-dynamics simulation (MD). This enabled us to reveal the dynamic morphology of HAT_6 by determining the effect of dynamical disorder on conductivity [8, 9, 10].

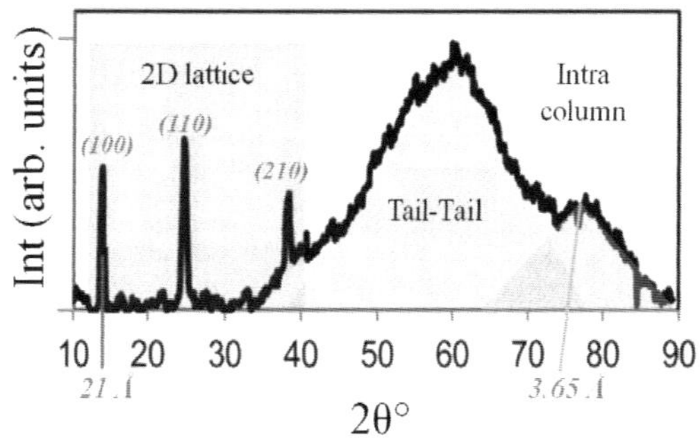

Figure 2. Diffraction pattern of HAT_6 in the columnar phase from neutron measurements. The two-dimensional lattice of columns, and tail-tail and intra-column distances distributions are highlighted.

2. Results and Discussion

HAT_6 (Figure 1(a)) is one of the most basic compounds forming a discotic liquid crystalline phase [7, 16, 17, 18]. The MD simulations were performed on different initial model structures, which were built from a hexagonal superlattice of 12 columns each consisting of six molecules (Figure 1(c)). The initial lattice parameters were determined from the observed reflections in the neutron powder-diffraction pattern (Figure 2), and the measured density of HAT_6. Three regions are distinguishable in the neutron diffraction pattern of a DLC in the columnar phase (Figure 2): (i) a region with three sharp peaks originating from the two-dimensional hexagonal lattice, (ii) the broad distribution of the tail-tail distances form an intermediate region originating, and (iii) a region indicating the intra-columnar distances.

During the MD pre-relaxation of the model structures, the evolution of the simulated diffraction pattern was followed and compared with the experimental data (Figure 3). The crystal structure predicted by the MD simulations reproduces the three regions in the diffraction pattern [9, 10]. The dynamical behaviour of the liquid-crystalline phase is included in the comparison of the MD simulations with the experimental observations. The dynamic behaviour was included by calculating the anisotropic displacements (Figure 1(b)).

We found that different initial models converged towards a comparable overall minimum in configuration space, and consistently resulted in a satisfactory agreement with the experimental diffraction pattern. After the long pre-relaxation run a second MD simulation was performed on the fully relaxed structural models. The diffraction patterns retained their good agreement with experiment during this time evolution, and thermally-averaged structure parameters were obtained (Figure 1(c)).

The experimentally-observed diffraction pattern is intermediate between the simulated patterns of the models. The simulated diffraction pattern of the models TWIST60 and TWIST25 show an excellent agreement with diffraction data. The relaxed structures stem from different starting configurations (the caption of Figure 3 include details of the models). Based on the agreement with the measured diffraction pattern [9, 10], both models converge to comparable minima in configuration space. This happens close to the actual liquid-crystalline structure. Rather than a uniform shift of the core-core distance distribution to higher distances, a high disorder in the core-core distances is actually indcued by the competition between the mutual van der Waals

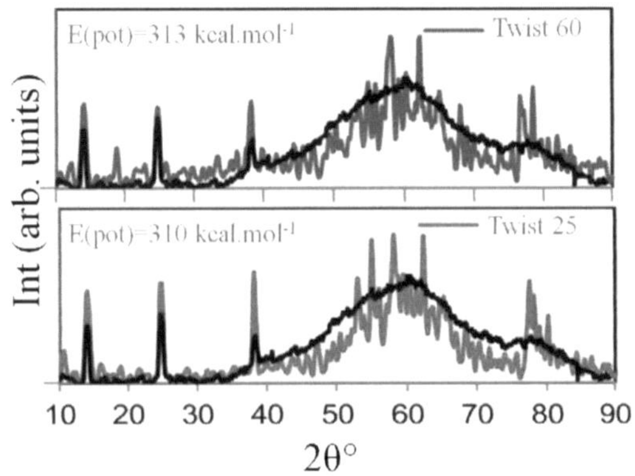

Figure 3. The experimental (black line) and simulated (models TWIST60 and TWIST25) diffraction patterns. The model TWIST60 consists of the rotation of the nearest neighbour molecules in a column by 60° around their principal axis, while a similar orientation is kept for molecules in a two-dimensional layer. On the other hand, the model TWIST25 consists of a unit of three columns built by considering a twist angle of 25° for the first column, then the same mutual rotation angle is applied for the second column, but in the opposite direction. This constrains the twist angle of the third column to 5°, helping hence to avoid superposition of the tails.

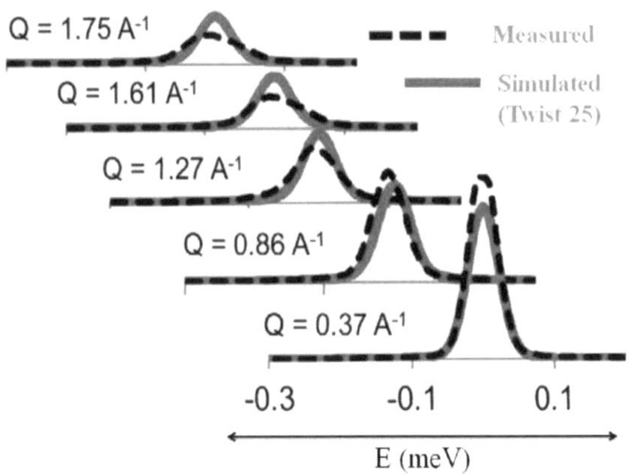

Figure 4. The Q-dependence of the incoherent scattering function from quasielastic neutron scattering measurements (QENS) and molecular dynamics simulations (MD).

interactions of the cores and the steric repulsion of the tails. Indeed, the tails tend to orient toward the open spaces left by these core-core defects. This is supported by the fact that the whole-molecule peak lies at a higher distance in comparison to the core-core peak. The inferred twist-angle distribution values are found to be around 36°. This is close to the DFT-estimated minimum-energy twist angle of 30° [8].

Dynamics occuring on the picosecond timescale can be determined by studying the incoherent-scattering functions which are extracted from MD simulation of the two MD models TWIST25 and TWIST60, and compared directly, for the sake of the validation [9], with QENS experiments

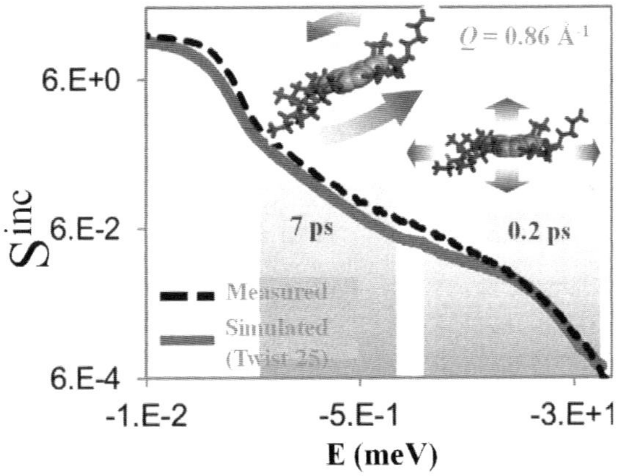

Figure 5. A log scale representation of the simulated and measured incoherent scattering function. The regions of dominant rotational (7ps) and translational (0.2 ps) motions are highlighted.

(Figure 4). The QENS technique offers the opportunity to follow both the temporal and spatial characteristics of the atomic motions, thanks to a well-characterized interaction of the neutrons with the atomic nuclei. Therefore, it is fairly straightforward to estimate the expected spectral profiles by using the atomic trajectories from the MD simulations to assign the time scales of motions to the underlying mechanisms.
The analysis of the experimental data showed that at least two Lorentzian functions (in addition to the elastic peak) are required to perform the fit [9, 15]. Two characteristic timescales of about 0.2 and 7 ps were extracted from the peak widths. Simulations based on these MD models agree well with the measured temporal and spatial dynamics (Figure 5). A closer investigation of the simulation trajectories allowed to characterize the underlying thermal motions, and therefore the typical amplitudes of molecular motions on the 0.2 and 7 ps can be estimated.
The characteristic translational and tilt motions for a single molecule on the 0.2 ps timescale are illustrated in Figure 6. On this timescale the twist-angle deviations are negligible (amplitudes smaller than 0.1°). The tilt degree of freedom of the molecular core seems to be too fast to be followed by the aliphatic tails. The rotational motions are much more pronounced on the 7 ps timescale where the motion of the aliphatic tails is induced by the rotations of the core.

3. Conclusions

We found that molecular translations (in-plane motions) are the dominant contribution to the spectral component of 0.2 ps in the QENS spectrum, whilst the 7 ps component mainly stems from both tilt and twist motions of the whole molecule. In order to determine the effect of local conformation, structural defects and thermal motions on charge transport, the situation for HAT_6 was compared with that of larger discotic molecules exhibiting higher conductivities. It turns out that the large disorder in core-core distances is the major factor limiting the conductivity of HAT_6. The charge hopping rate decreases exponentially as a function of core-core separations . As a result, we consistently found that the structural defects in core-core distances account for a decrease in the mobility by a factor of about 100 in going from larger discotics to HAT_6. The combination of MD simulations with neutron scattering is clearly of a key value for the study of liquid-crystalline phases.

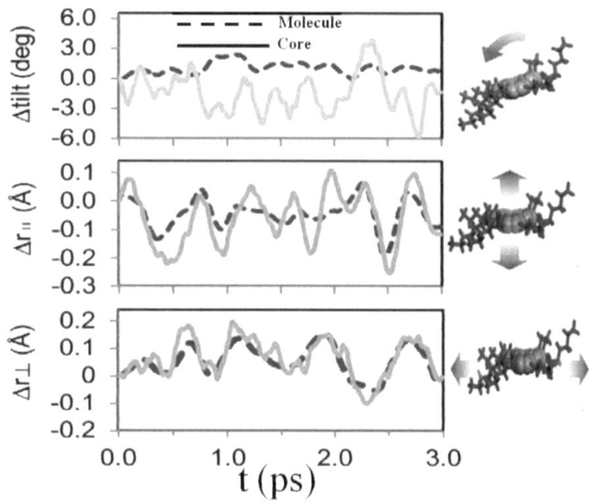

Figure 6. Translational and tilting motions on the sub-picosecond time scale for a typical molecule extracted from the simulated MD trajectory. Both dynamics of the molecular centre of mass (CM) including the tails (dashed lines), and the aromatic core CM alone (solid lines), are shown.

References

[1] Shirakawa H, Louis E J, MacDiarmid A G, Chiang C K and Heeger A J 1977 *J. Chem. Soc. Chem. Commun.* **16** 578
[2] Yu G, Gao J, Hummelen J C, Wudl F and Heeger A J 1995 *Science* **270** 1789
[3] Dreschel J, Mannig B, Kozlowski F, Pfeiffer M, Leo K and Hoppe H 2005 *Appl. Phys. Lett.* **86** 244102
[4] Chandrasekhar S, Sadashiva B K and Suresh K A 1977 *Pramana* **9** 471
[5] Kaafarani B R 2011 *Chemistry of Materials* **23** 378
[6] Feng X L, Marcon V, Pisula W, Hansen M R, Kirkpatrick J, Grozema F, Andrienko D, Kremer K and Mullen K 2009 *Nature Materials* **8** 421
[7] Etchegoin P 1997 *Phys. Rev. E* **56** 538
[8] Zbiri M, Johnson M R, Kearley G J and Mulder F M 2010 *Theoretical Chemistry Accounts* **125** 445
[9] Haverkate L A, Zbiri M, Johnson M R, Deme B, Mulder F M and Kearley G J 2011 *Journal of Physical Chemistry B* **115** 13809
[10] Haverkate L A, Zbiri M, Johnson M R, Deme B, Mulder F M and Kearley G J 2012 *Journal of Physical Chemistry B* **116** 3908
[11] Zbiri M, Johnson M R, Haverkate L A, Mulder F M and Kearley G J 2010 *Aust. J. Chem.* **63** 388
[12] Haverkate L A, Zbiri M, Johnson M R, Deme B, de Groot H J M, Lefeber F, Kotlewski A, Picken S J, Mulder F M and Kearley G J 2012 *Journal of Physical Chemistry B* **116** 13098
[13] Haverkate L A, Zbiri M, Johnson M R, Carter E, Kotlewski A, Picken S, Kearley G J and Mulder F M 2014 *J. Chem. Phys.* **140** 014903
[14] Zbiri M, Haverkate L A, Kearley G J, Johnson M R and Mulder F M 2015 In *Neutron Applications in Materials for Energy* p 109, Ed. Springer
[15] Mulder F M, Stride J, Picken S J, Kouwer P H J, de Haas M P, Siebbeles L D A and Kearley G J2003 *J. Am. Chem. Soc.* **125** 3860
[16] Van de Craats A M, De Haas M P and Warman J M 1997 *J. Synth. Met.* **86** 2125
[17] Kruglova O, Mulder F M, Siebbeles L D A and Kearley G J 2006 *Chem. Phys.* **330** 333
[18] Kearley G J, Mulder F M, Picken S J, Kouwer P H J and Stride J 2003 *Chem. Phys.* **292** 185

Journal of Physics: Conference Series **758** (2016) 012013 doi:10.1088/1742-6596/758/1/012013

Comparative study of energy of particles ejected from coulomb explosion of rare gas and metallic clusters irradiated by intense femtosecond laser field

N Boucerredj [1] and K Beggas[2]

[1,2] Physics department; Faculty of Sciences, Semiconductor Laboratory, Badji Mokhtar Annaba University, B. P. 12, 23000, Annaba, Algeria.

E-mail: boucerredj@yahoo.fr

Abstract. We present our study of high intensity femtosecond laser field interaction with large cluster of Kr and Na (contained 2.10^3 to 2.10^7 atoms). When laser intensity is above a critical value, it blows off all of electrons from the cluster and forms a non neutral ion cloud. The irradiation of these clusters by the intense laser field leads to highly excitation energy which can be the source of energetic electrons, electronic emission, highly charge, energetic ions and fragmentation process. During the Coulomb explosion of the resulting highly ionized, high temperature nanoplasma, ions acquire again their energy. It is shown that ultra fast ions are produced. The goal of our study is to investigate in detail a comparative study of the expansion and explosion then the ion energy of metallic and rare gas clusters irradiated by an intense femtosecond laser field. We have found that ions have a kinetic energy up to 10^5 eV and the Coulomb pressure is little than the hydrodynamic pressure. The Coulomb explosion of a cluster may provide a new high energy ion source.

1.Introduction

High power femtosecond lasers are capable of achieving a light intensity up to 10^{21}-10^{22} W/m^2. In this regime, a novel physics comes in our grasp of modern physics. The production of high temperature plasmas with small scale, short pulse, and high intensity lasers has been actively pursued during the last years. The interaction of clusters with high intensity laser fields has been studied by several numbers of groups [1, 2, 3]. Energetic ions and electrons up to 1 keV are produced as noble gas clusters explode [4, 5]. The ions are stripped to very high charge state [6, 7, 8], and X-ray yield is comparable to that from solid targets [9]. The observation of neutrons from deuterium clusters demonstrated that nuclear fusion can occur in the plasma formed after the explosion of the clusters [10]. Of particular interest, the measurement of the energy absorption efficiency of high density plasma created by intense irradiation of a solid target, have shown that the plasma typically absorbs a large fraction of the laser energy depending upon intensity and laser wavelength [11]. When the laser interacts with clusters it changes the regime of electromagnetic wave propagation and very efficient absorption of radiation has been demonstrated with a formation of very high temperature under dense

[1] Corresponding author: Physics department; Faculty of Sciences, Semiconductor Laboratory, Badji Mokhtar Annaba University, B. P. 12, 23000, Annaba, Algeria. E-mail: boucerredj@yahoo.fr

Journal of Physics: Conference Series **758** (2016) 012013 doi:10.1088/1742-6596/758/1/012013

plasma [12, 13]. It has been noticed that for small diameter clusters, the coulomb explosion plays an important role [14].

Recent studies of individual clusters interaction with intense laser have confirmed that hot electrons (up to 3 keV) are produced during the laser cluster interactions and a greater energy can be deposited in the ions when these hot, highly ionized clusters explode. These studies indicated that the clusters are rapidly heated by the laser, to non-equilibrium, superheated state, due to the passage of free electron density in the cluster through a Mie resonance with the laser field during the cluster expansion [15, 16]. These superheated cluster nanoplasma eject electrons with many keV of energy [17, 18]. After the cluster are heated, charge separation of the hot electrons leads the cluster to explode, and ions acquired energy by the conversion of the energy deposited by the laser in the cluster to ion kinetic energy.

The goal of our paper is to study the comparative behaviour of coulomb explosion and ion energies ejected from the exploding metallic and noble large gas clusters (Na and Kr) respectively irradiated by intense femtosecond laser fields. We further study the temporal comparison between the different parameters of the formed nanoplasma of Na and Kr clusters within the framework of the modified nanoplasma model. After a general introduction on intense laser cluster interaction, we show in section 2 a brief description of the nanoplasma model. In section 3, we discuss the obtained results and finally, we draw our conclusion.

2.Nanoplasma model

In The nanoplasma model developed firstly by T. Ditmire et al [8], and reformulated by F. Megi et al [15] by adding the term of electron collisions with surface in the expression of the electron ion collision frequency, has proved a quite successful in reproducing most of the experimental results concerning expansion, ion and electron emissions from single clusters [16]. This model treats the expanding cluster as a spherical nanoplasma, subjected to the standard processes of a laser heated nanoplasma such as collisional heating, as well as collisional and tunnel ionization. The model suggests that the cluster explosion is driven by an enhancement in the electron heating in the cluster; also treats the formation of high density nanoplasma, ionization and heating in the laser field and its rapid expansion then explosion of the cluster leading to the ejection of energetic electrons and ions. A key aspect was the inclusion of shielding and enhancement of the laser field inside the cluster; which gives rise to a resonance in collisional heating rate. This model relies on three main assumptions: first, the laser field is described within the dipolar approximation; this is valid as long as $\lambda >> R_0$ where λ and R_0 are the laser wavelength and the initial cluster radius respectively. Second, the electronic and ionic densities are supported to be uniformly distributed within the cluster throughout the interaction and the third assumption is that the electrons are assumed to be instantaneously thermalized so their energy distribution is Maxwellian. The nanoplasma model is reduced to a set of differential coupled equations describing the time evolution of the electronic density n_e inside the cluster, which vary due to ionization processes and free streaming, of the populations of the different charge states that evolve through field and collisional ionization, of the net cluster charge, which increases due to free streaming, of the cluster radius, subjected to Coulombian and hydrodynamic pressures, and of the electronic temperature T_e; which results from balancing between energy gains and losses within the cluster volume.

In this model, the external electric field is given by

$$\vec{E}_{ext}(t) = E_0 \exp\left(-2\ln 2\left(\frac{t}{\tau}\right)^2 \right)\vec{e}_z \qquad (1)$$

Where τ is the full width at half maximum (FWHM) and E_0 is the amplitude of the laser field. The internal field is given by

$$E_{int}(t) = \frac{3}{|\varepsilon + 2|} E_{ext}(t). \tag{2}$$

f(t) is a Gaussian shape and ε is a dielectric constant given by

$$\varepsilon = 1 - \frac{\omega_p^2}{\omega(\omega + i\nu)}, \tag{2a}$$

where

$$\omega_p = \sqrt{\frac{n_e e^2}{\varepsilon_0 m}}. \tag{2b}$$

is the electronic plasma frequency, n_e the electron density and ν is the total electron ion collision frequency with the inclusion of the term of electron surface collisions

$$\nu = \nu_{ei} + \nu_s = \nu_{ei} + \frac{v}{R}. \tag{3}$$

Where ν_{ei} is the electron-ion collision frequency given by [19, 20]

$$\nu_{ei} = \frac{1}{2} \frac{n_i \langle Z^2 \rangle e^4 \ln \Lambda_I}{(\pi\varepsilon_0 m_e)^2 v^3}. \tag{4}$$

And the velocity v is given by

$$v = \sqrt{v_{th}^2 + v_{osc}^2}. \tag{4a}$$

V_{th} is the thermal velocity and v_{osc} is the oscillation velocity of the electron in the field. R is the cluster radius, <Z> is the mean ion charge and $\ln\Lambda_I$ the modified coulomb logarithm given by the expression

$$\ln \Lambda_I = \begin{cases} \frac{1}{4} \ln^2\left(\frac{1 + m_e v_{osc}^2}{\hbar w}\right), & \text{for } \hbar w >> k_B T_e \\ \frac{1}{4} \ln^2\left(1 + \frac{v_{osc}^2}{v_{th}^2}\right) + \ln\left(\frac{v_{osc}^2}{v_{th}^2} + \exp(\frac{1}{3}\sqrt{\frac{\pi}{2}})\right) \ln\Lambda & \text{for } \hbar w << k_B T_e \end{cases}. \tag{5}$$

And $\ln\Lambda$ is the classical coulomb logarithm [21].

3.Results and discussion

In the following, we are used a Kr and Na clusters (2.10^7 atoms per cluster) irradiated by a Gaussian laser pulse, pulse duration 200fs (Full width at half maximum), and peak intensity of 10^{17} W/cm^2. We assume no temperature nor density gradients and the electric field inside the cluster is uniform. To explore the aspect of the cluster explosion; we have used the nanoplasma model for numerical calculation of the cluster explosion during the irradiation by the intense femtosecond laser pulse. In the cluster, the high electron temperature created through laser driven heating rapidly strip the ions by collisional ionization to high charge state.

In figure1, we show the temporal evolutions of the cluster radius R for the rare gas Kr and metallic Na clusters. The cluster expands rapidly in the case of Na cluster than the Kr cluster during the pulse once heating of the electrons in the cluster has begun. In the case of the metallic cluster, it expands rapidly due to the rapid ionization of the cluster.

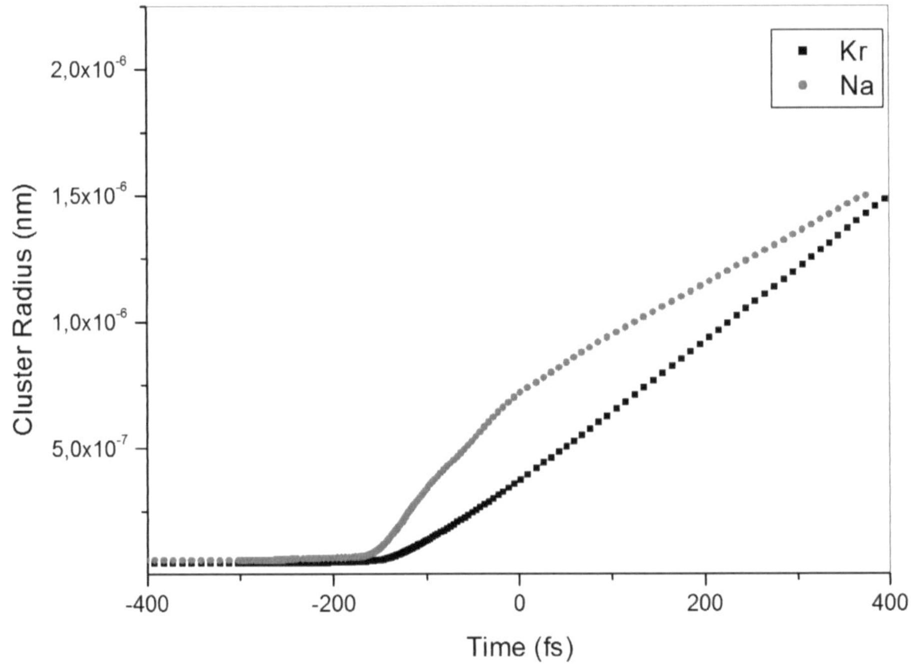

Figure 1. Temporal evolution of the radius of the cluster R for 2.10^7 atoms Kr and Na clusters irradiated by 200 fs pulse with peak intensity of 10^{21} W/m^2 and 390nm wavelength.

The cluster explosion is driven by an enhancement in the electron heating in the cluster; this enhancement occurs when the electron density in the cluster drops to three times the critical density n_{crit}, where the critical electron density is given by:

$$n_{crit} = \frac{\varepsilon_0 m_e \omega^2}{e^2}. \qquad (6)$$

Where ε_0 is the vacuum permittivity and ω is the laser frequency.
During the laser interaction, free electrons are produced by photo ionization early in the laser pulse. The laser collisionally heats these electrons which further ionize the atoms of the cluster through collisional ionization. At around t≈-289,52fs, the first resonance occur in the Na cluster but at -277,92fs it occurs in Kr cluster when electron density rises to reach $3n_{crit}$ (figure2); also the times of the second resonance are different, -94,54fs and -148,39fs for both clusters Na and Kr respectively. The maximum value of the ration $\frac{n_e}{n_{crit}}$ is about 14,54 for Na cluster than 7,32 for Kr cluster; this difference in the time of resonances and the maximum values of the ratio $\frac{n_e}{n_{crit}}$ is caused by the number of electrons in such cluster.

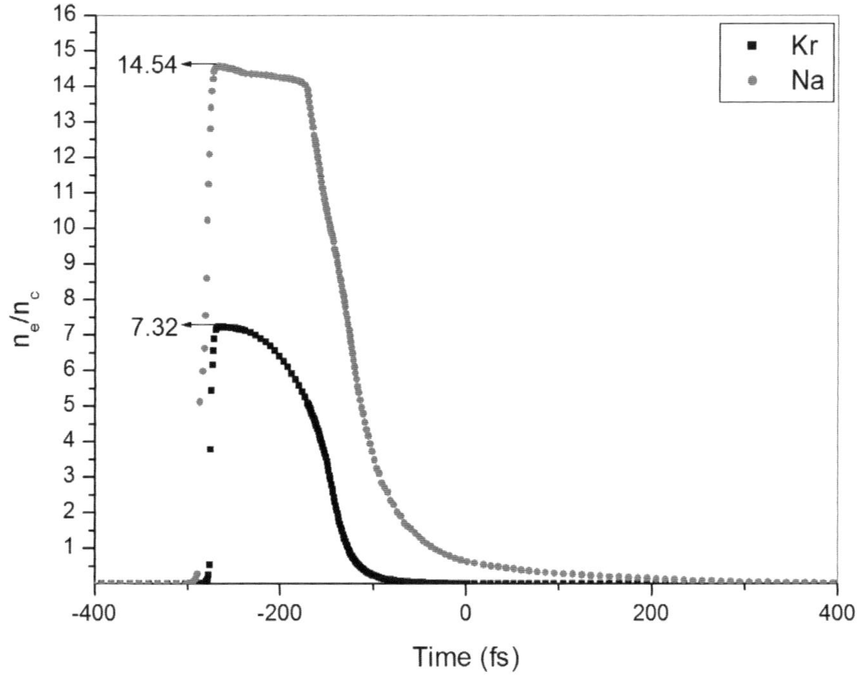

Figure2. Electron density n_e normalized by the critical one n_{crit} as a function of time for 2.10^7 atoms Kr and Na nanoplasma for 200 fs (FWHM) pulse, 300 nm wavelength and peak intensity of 10^{21} W/m^2.

The electron density rises to reach $3n_{crit}$ at time t≈-277,92fs and 289,52fs for Kr and Na clusters respectively. At this point, the field inside the cluster is enhanced (figure3), the cluster expansion velocity increases dramatically; and more electrons are liberated through tunnel, laser driven and thermal ionization. The hot electrons can leave the cluster to the $3n_{crit}$ resonance.

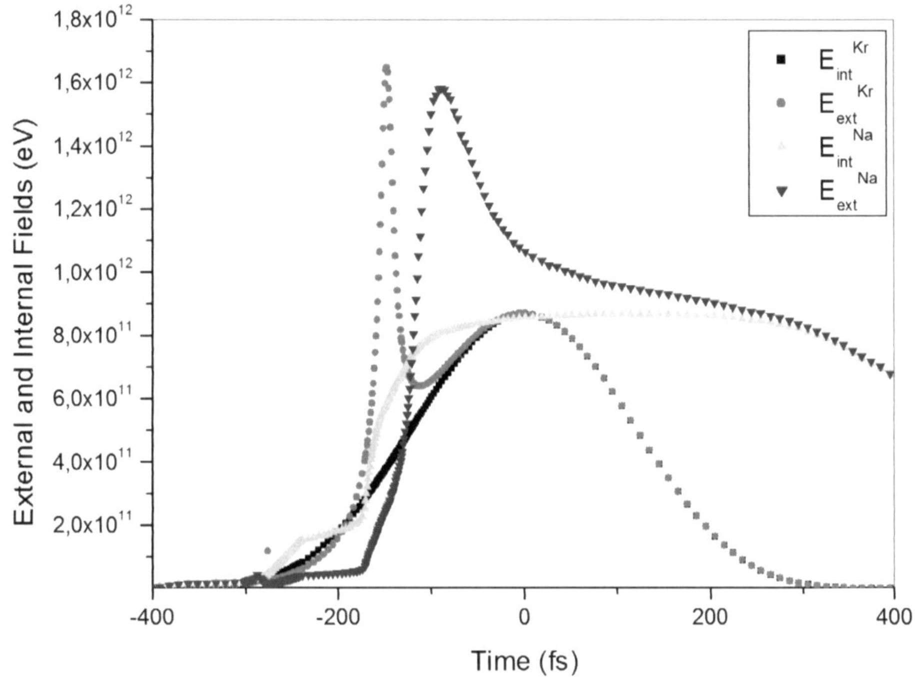

Figure3. Time variation of the amplitudes of the internal and external electric fields for Na and Kr clusters. The laser and cluster parameters are the same as those in Figure 2.

The combined effect of the free streaming of electrons out of the cluster and the hydrodynamic expansion of the cluster is that the electron density starts to fall, after peaking at over 7,32 and 14,54 for both clusters (Kr and Na respectively). The expansion of the cluster lowers the electron density to bring the system into resonance with laser field. The electron density in the cluster drops to $3n_{crit}$ (figure 2); and the field inside the cluster is strongly enhanced and reach the value 1.65 10^{12} V/m and 1,58 10^{12} V/m for Kr and Na clusters respectively (figure 3); then we have a very rapid deposition of the energy into the electrons; at this point, the ions are very rapidly stripped the hot electrons. The inner field $E_{int}(t)$ (which is almost equal to the external field since the electron density n_e is very low.

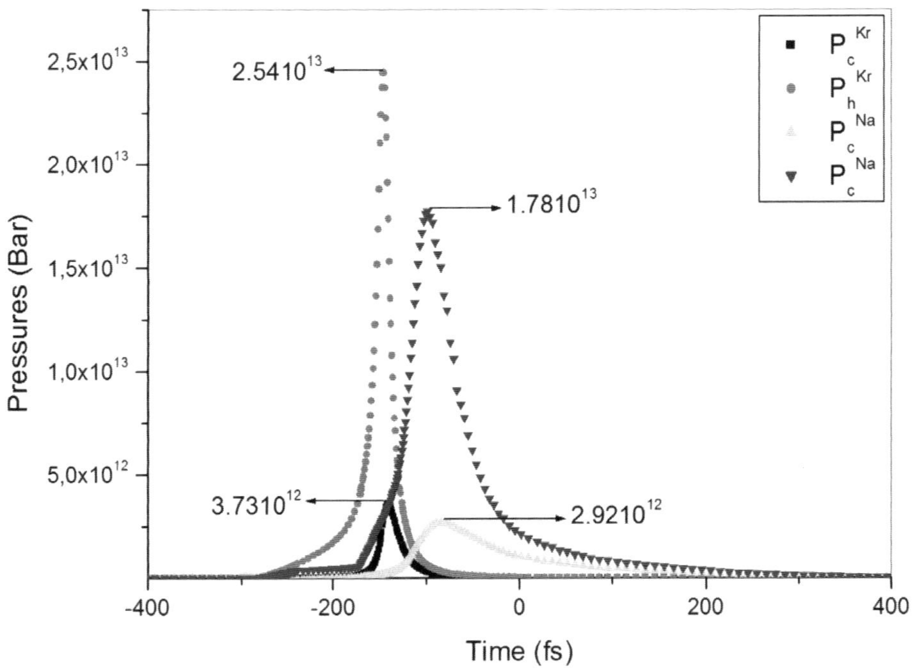

Figure4. Variation of the Coulomb and hydrodynamic pressures as a function of time in the Kr and Na clusters for 2.10^7 atoms irradiated by 200 fs laser pulse, 390 nm wavelength and peak intensity of 10^{21} W/m^2.

The contributions of the hydrodynamic pressure and the coulomb one are shown in figure 4 for the two clusters. The total charge on the cluster increases, then the coulomb pressure increase to $3{,}7310^{12}$ Bar and $2{,}9210^{12}$ Bar for Kr and Na clusters respectively; However this is small compared to the hydrodynamic pressure due to the hot electrons of $2{,}5410^{13}$Barand $1{,}7810^{13}$ Bar for the rare gas (Kr) and metallic (Na) clusters respectively. This pressure causes a sharp increase in the cluster expansion velocity. During the majority of time, the dominant pressure is the hydrodynamic pressure with a little contribution of the coulomb explosion force; then the hydrodynamic pressure is dominant in driving the explosion. It has been noticed that for small diameter clusters, the coulomb explosion force plays a key role [4]. Our investigation of the coulomb explosion of large Kr and Na clusters shows that the hydrodynamic pressure plays an important role than the coulomb one which is in good agreement with previous results [22].

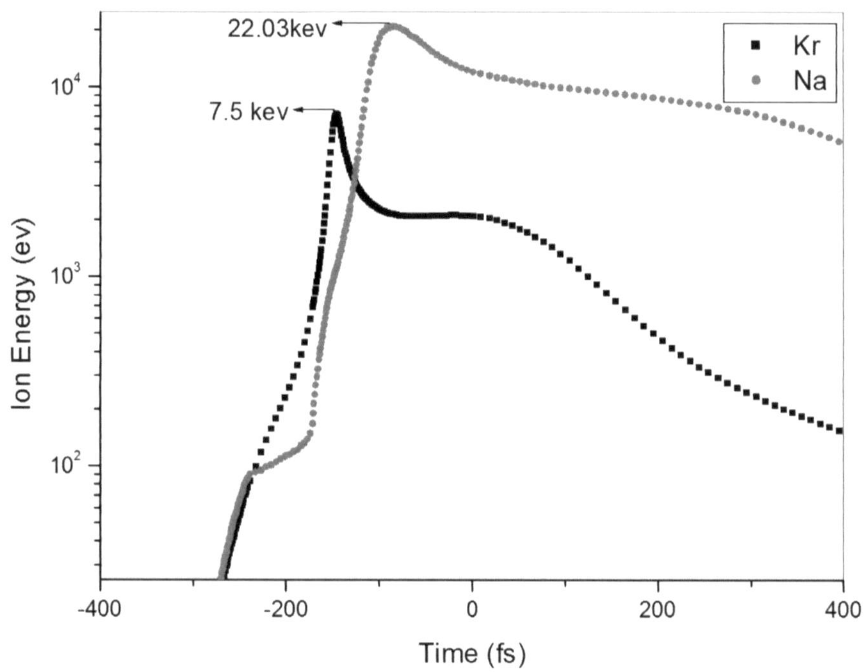

Figure5. Ion energy as a function of time for Kr and Na nanoplasma with 200 fs laser pulse, 2.10^7 atoms, 390 nm wavelength and peak intensity of 10^{21} W/m^2.

When the electrons mainly gain energy through inverse bremsstrahlung, the ion energy (figure 5) roughly follows the evolution of the internal field. The increase of the electronic temperature leads to higher ionized states, and then electrons with high energy can leave the cluster. The coulomb pressure induced by this loss of the plasma neutrality combined with the hydrodynamic pressure induces the expansion of the nanoplasma. The electron density n_e also decreases to reach the $3n_{crit}$ resonance condition at different time (second resonance) mentioned above. The enhancement of the internal field (figure3) then leads to a very efficient absorption of the laser energy, resulting to the production of high charge states and high energetic ions 7,5 keV for Kr cluster and 22,03keV for Na cluster three times than ion energy of Kr cluster . Then the combined effect of free streaming and the coulombian and hydrodynamic pressures leads to the final explosion of the cluster.

4. Conclusion

We have studied the comparison behaviour of the different physics parameters between rare gas and metallic clusters (Kr and Na) using the modified nanoplasma model to examine the cluster explosion dynamics and scaling of ion energies; similar behaviour was found for the two cases metallic and rare gas clusters. We have found that ions have energy up to 10^3eV, ions of Na cluster have energy up to three times than the ion energy of Kr cluster, rapid ionization and expansion of Na metallic cluster than Kr rare gas cluster. The coulomb pressure is little than the hydrodynamic one for both cases (metallic and rare gas clusters); then we conclude that the hydrodynamic pressure is responsible for the dynamic of ionization, expansion and explosion of the clusters.

5. References

[1] Reinhart P G and Suraud E 2004, Introduction to cluster dynamics, Wiley-VCH Verlag. Weinheim.

[2] J Nishihara K, Hamitani H, Murakami M et al 2001, High energy ions generated by laser driven coulomb explosion of cluster, *Nuclear Instruments and Methods in Physics Research (Elsevier)* A **464** 98-102.

[3] Perry M D and Mourou G 1994 *Science* **32** 917

[4] Ditmire T , Tish J W G, Springate E et al 1997 *Nature (London)* **386** 54

[5] Springate E, Hay N, Tish J W G et al 2000, Explosion of atomic clusters irradiated by high intensity laser pulses: Scaling of ion energies with cluster and laser parameters, *Phys. Rev.* A**61** 063201

[6] McPherson A, Thompson B D, Krainov A B et al 1994 *Nature (London)* **370** 631

[7] Lezius M, Dobosz S, Normand D and Schmidt M 1998 *Phys. Rev. Lett.* **80** 261

[8] Ditmire T, Donnelly T, Rubenchikh A M et al 1996 *Phys. Rev.* A **53** 3379

[9] Ditmire T, Zweiack J, Yonovsky V P et al 1999 *Nature (London)* **398** 489

[10] Ditmire T, Smith R A, Tish J W G and Hutchinson M H R 1997, High intensity laser absorption by gases of atomic clusters, *Phys. Rev. Lett.* **78**, N° 6 3121-3124

[11] Heidenreich A, Last I and Jortner J 2008, Nanoplasma dynamics in Xe clusters driven by ultraintense laser fields, *Eur. Phys. J. D* **46** 195-202

[12] Micheau S, Jouin H and Pons B 2008, Modified nanoplasma model for laser cluster interaction, *Phys. Rev*. A **77** 053201

[13] Last I, Schek I and Jortner J 1997 *J. Chem. Phys.* **107** 6685

[14] Ditmire T 1998, Simulation of exploding clusters ionized by high intensity femtosecond laser pulses, *Phys. Rev.* A **57** N° 6

[15] Megi F, Belkacem M, Bouchen M, Suraud E and Zweicknagel 2003, On the impotance of damping phenomena in clusters irradiated by intense laser fields, *J. Phys. B: At. Mol. Opt. Phys.* **36** 273-282

[16] Ditmire T, Springate E, Tisch J W G et al 1998, Explosion of atomic clusters heated by high-intensity femtosecond laser pulse, *Phys. Rev.* A. **57** N° 1 369

[17] Boucerredj N, Beggas K and Djabri I 2009, Coulomb explosion of Kr clusters irradiated by intense laser field, *2nd International Meeting on Materials for Electronic-Applications (IMMEA)* May 8-10, Hammamet, Tunisia

[18] Boucerredj N, Beggas K and Djabri I 2011, Study on the interaction of intense laser pulse with large metallic clusters of Pb using modified nanoplasma model, *Global Journal of Physical Chemistry* Vol. **2** 83-87

[19] Djaoui A and Offenberger A A 1994, Heating of underdense plasmas by intense-short pulse lasers, *Phys. Rev.* E **50** 4961-4968

[20] Polishchuk A Ya and Vehn J M T1994, Electron-ion relaxation in a plasma interacting with an intense laser field, *Phys. Rev.* E **49** 663-666

[21] Tohnston T W and Dawson J M 1973, Correct values for high-frequency power absorption by inverse bremstrahlung in plasmas, *Phys. Fluids* **16** 722

[22] Tish J W G, Hay N, Mendham K J et al 2003, Interaction of intense laser pulses with atomic clusters: Measurements of ion emission, simulations and applications, *Nuclear Instruments and Methods in Physics Research (Elsevier)* B **205** 310-323

Investigation of ionic mobility in NASICON-type solid electrolytes

A Vyalikh[1,4], V Vizgalov[2], D M Itkis[2,3], D C Meyer[1]
[1] Institut für Experimentelle Physik, Technische Universität Bergakademie Freiberg, Leipziger Str. 23, D-09599 Freiberg, Germany
[2] Department of Materials Science, Moscow State University, Leninskie gory, 119991 Moscow, Russian Federation
[3] Department of Chemistry, Moscow State University, Leninskie gory, 119991 Moscow, Russian Federation

Abstract. Impedance spectroscopy and ^{7}Li NMR have been applied to characterize the lithium conducting glass-ceramics membranes of the $Li_{1.5}Al_{0.5}Ge_{1.5}(PO_4)_3$ composition with the NASICON-type structure. The ^{7}Li NMR spectra and T_1 relaxation times have been compared for the precursor glass and two glass-ceramics annealed for 2 and 6 hours, and analysed with respect to the ionic conductivity in these materials. The ^{7}Li static NMR spectra reveal two components in the glass-ceramics samples: A quadrupole pattern with C_Q of 38.7 kHz and 32.5 kHz, and a narrow signal of the Lorentzian or Gaussian lineshape for the samples annealed for 2 and 6 hours, respectively. Variation of the lineshape and the deconvolution parameters point out to the modification of the NASICON framework in the former, which affects the conductivity channels towards improved movement of lithium ions. The NMR data correlate with the conductivity measurements demonstrating enhanced ionic mobility in the glass-ceramics annealed for 2 hours. The ^{7}Li NMR relaxation data seem to be very sensitive to the species with different mobility and reveal the presence of an additional minor component, which can be responsible for decrease of conductivity at longer thermal treatment.

Introduction

Highly demanded high-voltage and safe lithium-based energy storage is associated today with all-solid state lithium-ion batteries. All-solid-state batteries utilize solid electrolytes, which demonstrate a sufficiently high ionic conductivity (> 10^{-4} S/cm), good chemical stability and an electrochemical stability window, which is significantly broader than in liquid electrolytes. Due to an unmatchable combination of their chemical and electrical properties, materials with the NASICON-type structure have been proposed as optimum candidate structures for ion migration in Na- or Li-based all-solid-state batteries. [1] By a partial replacement of the structural elements and a total charge balance, one can vary carrier concentration and therefore tailor the electrical properties. In the present work we focus on the $Li_{1+x}Al_xGe_{2-x}(PO_4)_3$ composition (LAGP) with a stoichiometric value of x=0.5, as this NASICON-type material is reported to demonstrate particular stability against lithium metal. [2, 3]

Since conducting properties are related to the crystalline phase and microstructures of the materials, development of improved ceramics-based electrode materials is focused today on optimising the morphology of their microstructure, which is influenced by the composition, the choice of nucleating

[4] To whom any correspondence should be addressed, email: Anastasia.Vyalikh@physik.tu-freiberg.de.

agents and the heat treatment processes used. As sintering behaviour has a strong effect on phase formation and thus on electrical characteristics, a detailed characterisation of the amounts of the phases and their ionic dynamics on an atomic level is of great importance for gaining understanding of structure-property relationships.

Among the existing analytical tools able to characterize the complex morphology of ceramic materials, NMR has an important advantage as it can probe each phase directly through differences in chemical shift and relaxation behaviour. [4,5] Moreover, NMR as being reflecting short-to-medium-range-order is more sensitive to the very local structure, and is particularly useful in the cases, where X-Ray and other diffraction techniques are limited due to a lack of long-range order.

The NMR relaxation times of nuclear spins are sensitive to a number of parameters including degree of crystallinity, presence of impurity phases, morphology of the ceramics system and molecular mobility. Depending on the relaxation kind (spin-spin or spin-lattice relaxation in the laboratory or rotating frame) under study, different motional time scales can be investigated. [6] The spin-lattice or longitudinal relaxation characterized by the time constant T_1 describes the recovery of the magnetisation along the axis of the magnetic field and has been frequently applied to correlate the structure [7, 8] and morphology [9, 10] of solid electrolyte materials with their conducting properties.
In the present work, we apply solid-state nuclear magnetic resonance (NMR) in order to characterize LAGP glass and glass-ceramics as a function of sintering conditions and to reveal a relationship between the microstructure and ionic conductivity towards further development of materials for sustainable energy.

Experimental part

1.1. Sample preparation

For preparation of $Li_{1.5}Al_{0.5}Ge_{1.5}(PO_4)_3$ glass ceramics 1.2420 g of Li_2CO_3 (99.9%, Sigma-Aldrich), 0.5713 g of Al_2O_3 (99.9%, Sigma-Aldrich), 3.5162 g of GeO_2 (99.9%, SMI Ltd.), 8.8801 g of $(NH_4)_2HPO_4$ (99,9%, Sigma-Aldrich) and 0.3189 g (5 vol. % or 12 mol. %) of Y_2O_3 (99,99%, Sigma-Aldrich) were mixed in required molar ratios in an agate mortar without any additional purification. Stoichiometric mixtures of precursors were preannealed at 500°C for 5 hours. Resulting powder was placed into a platinum crucible and melted at 1300°C for 2 hours. The melt was quenched between two steel plates preheated at 450°C, forming 500 or 1000 μm thick discs, and then cooled down to room temperature for 3 hours. The glass discs were then tempered at 500°C and 400°C subsequently. For crystallisation, glass was rapidly heated to 650°C and annealed for 2 hours and 6 hours with subsequent cooling in air yielding the samples LAGP-2h and LAGP-6h, respectively. Finally, glass-ceramic membranes were obtained by rapid heating and crystallization at 750°C. For impedance spectroscopy, as-obtained glass-ceramic membranes were polished down to 300 – 350 μm using Struers Tegrapol-11 polishing machine and DiaDuo abrasives. For NMR measurements the samples were ground to powder and filled into zirconia rotors.

1.2. Impedance spectroscopy

Ionic conductivity of the resulting samples was characterized by electrochemical impedance spectroscopy (EIS). Nickel blocking electrodes were sputtered on both sides of a glass-ceramic membrane using Quorum Technologies Q150T Turbo-Pumped Sputter Coater / Carbon Coater. Electric contacts were fixed on sputtered electrodes using EPO-TEK H20S conductive epoxy. The ac impedance measurements were conducted in the frequency range of 7 MHz to 100 Hz with an amplitude of 50 mV using Bio-Logic SP-300 with a frequency response analyzer. The obtained Niquist plots generally consist of one or two semicircles in a higher frequency region typical for bulk and grain boundary resistance of electrolyte and a spike at the lower frequency corresponding to diffusion due to the blocking electrodes.

1.3. NMR measurements
Static ^{7}Li NMR measurements were performed on a BRUKER AVANCE III 400 MHz spectrometer operating at a resonance frequency of 155.5 MHz employing a 2.5 mm MAS probe head. Static ^{7}Li NMR spectra were recorded at a single pulse of 1 μs pulse duration ($\pi/12$), a recycle delay of 1 s and 4096 repetitions. The spectra were referenced to 1M solution of LiCl at 0 ppm. The spectra were fitted using Dmfit. [11] ^{7}Li NMR spin-lattice relaxation times were recorded with the standard saturation recovery pulse sequence employing a saturation pulse train of 50 pulses separated by 0.0002 s. The duration of the 90° pulse of 5 µs, 128 repetitions and the repetition delay of 1 s were applied. 16 increments in the range from 0.05 s to 150 s and 22 increments in the range from 0.002 s to 50 s were used for the studies of glass and glass-ceramics samples, respectively. The magnetization decay profiles were analysed with a double and triple exponential functions with Bruker TopSpin 3.2 software.

Results and Discussion
Figure 1(a) shows the static ^{7}Li NMR spectra of LAGP glass and glass-ceramics annealed at 2 hours and 6 hours. The spectrum of glass is represented by a single line inhomogeneously broadened due to a long-range disorder typical for the glass structure. The spectra of both glass-ceramics demonstrate the characteristic I=3/2 quadrupolar lineshape formed by the central line, those linewidth is significantly narrower as compared to glass, and the satellite transitions. [6] Fit shows two overlapping spectral components for both LAGP-2h and LAGP-6h spectra, their fit parameters are given in Table 1. The major spectral component in both cases shows the quadrupolar pattern and can be attributed to Li ions surrounded by six oxygens (component I) [8], while the minor component does not show a quadrupolar splitting and can be described by a Lorentzian or Gaussian lineshape (component II). Such lineshape for quadrupolar nuclei arises, either in the symmetrical local environment, or when quadrupolar interactions are averaged out on the timescale of the NMR experiment due to fast atomic motion. We assume that the component II is attributed to more mobile Li ions accommodated within the interstitial channels formed by the NASICON framework.

Analysis of the deconvolution parameters helps to correlate the local structure with the conducting properties. A drastic change is observed for the lineshape of the component II, which transforms from the pure Lorentzian in LAGP-2h to the pure Gaussian one in LAGP-6h, which is known to be responsible for slower motions with correlation times. A similar lineshape transformation has been observed for $LiHf_2(PO_4)_3$ with the NASICON structure and associated with a transition from the triclinic phase characterized by very low ionic conductivity to the rhombohedral symmetry with significantly higher conductivity. [7] It is worth mentioning that such a transformation in $LiHf_2(PO_4)_3$ has been accompanied by an enormous increase of the T_1 relaxation time. Considerable increase of the linewidth values in LAGP-2h as compared to LAGP-6h is observed for both spectral components. Such behaviour has been reported previously for LAGP glass-ceramics of the same composition, where Li_2O was used as a nucleating agent to considerable promote crystallisation during heat treatment that resulted in a dense microstructure with a uniform grain size and therefore in an enhanced ionic conductivity. [12] The main cause of the linewidth increase is likely to be an increase in the strength of the dipole-dipole interactions (most likely Li-P) associated to the incorporation of a part of excess lithium into the crystal lattice of the NASICON-type structure. Variations of the other deconvolution ^{7}Li spectral parameters such as chemical shift, quadrupole coupling constant C_Q and asymmetry parameter at different thermal treatment points out to changes in the local environment around ^{7}Li, which are not directly related to the ionic conductivity, but suggests a modification of the conducting channels that favour the movement of lithium ions.

Characterization of the NMR relaxation times may give independent estimation on the number of the components with different mobility. Figure 1(b) shows the experimental relaxation curves for three samples under study. It is obviously that the relaxation time for LAGP-glass deviates strongly from

(a) (b)

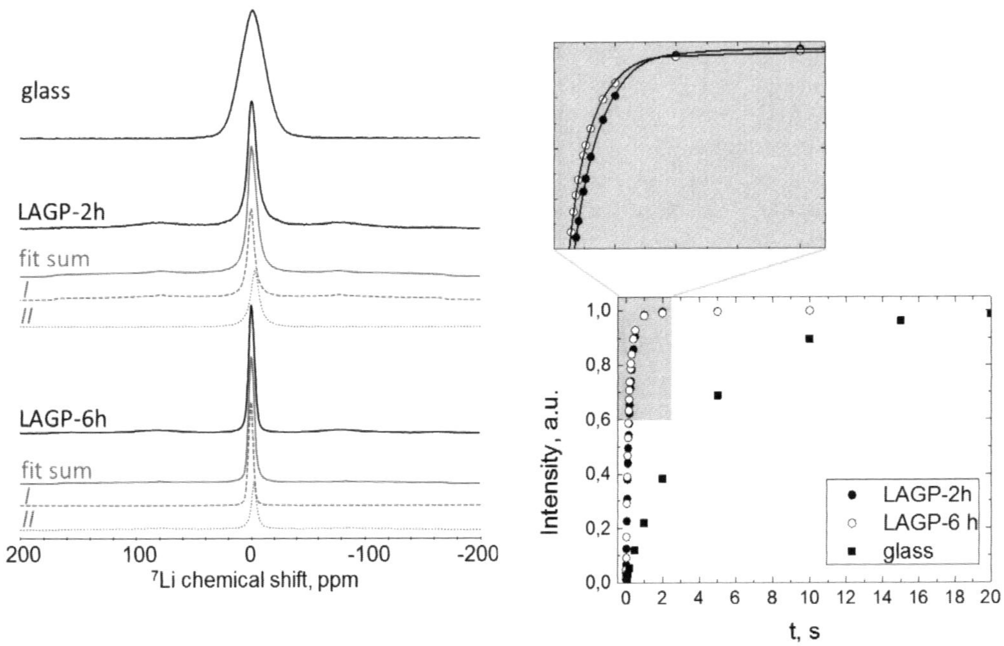

Figure 1. (a) Static ^{7}Li NMR spectra of LAGP-glass (top), LAGP glass-ceramic annealed at 2 hours (middle) and LAGP glass-ceramic annealed at 6 hours (bottom) with their fit sum and fit components; (b) ^{7}Li relaxation curves for three samples. The insert shows the best fits to LAGP-2h and LAGP-6h in the zoomed region.

those for LAGP glass-ceramics. The fit parameters of the experimental data are indicated in Table 2. While the relaxation in glass dominates by a contribution (97 %) with a long T_1 of 4.38 s characteristic for the glassy structure, in glass-ceramics it shows a multicomponent character reflecting a heterogeneous nature of ionic dynamics in $Li_{1+x}Al_xGe_{2-x}(PO_4)_3$. In LAGP-2h two relaxation components of 0.23 s (81%) and 0.06 s (19%) are found, their fractions are in good agreement with the quantitative relationship of both spectral components reported in Table 1. In LAGP-6h, while two relaxation components deviate only slightly in magnitude (0.19 s and 0.05 s) from those in LAGP-2h, but their fractions change (68 % and 30 %, respectively) and an additional slow component (7.45 s, 2 %) appears. Although, the fractions of former two contributions agree with the deconvolution data in Table 1, the fraction of the latter is too small to be resolved in the ^{7}Li NMR spectrum as well as to be detected by other characterization techniques. The formation of the phase with the slow T_1 component at longer crystallization time can be associated with the grain growth and appearance of cracks, in which a new phase is formed, which interrupts the conduction pathways and therefore suppresses ionic conductivity in these materials.

Although the chemical composition and morphology of the phase, which suppresses conductivity, cannot be determined from the data available, the systematic study in the LAGP glass-ceramics prepared at different annealing conditions using the complementary methods will help in understanding the factors that influence the ionic conductivity in the materials with the NASICON-type structure and therefore contribute to the design of new solid electrolyte materials.

Table 1. ^{7}Li NMR fit parameters with their experimental errors.

Sample	component I					component II			
	$\delta(^{7}Li)$ (±0.1, ppm)	C_Q (±0.5, kHz)	η, (±0.05)	width (±0.2, ppm)	%, (±2)	$\delta(^{7}Li)$ (±0.1, ppm)	G/L ratio, (±0.1)	width (±0.3, ppm)	%, (±2)
LAGP-2h	0.1	38.7	0.35	6.8	78	-3.2	0	9.0	22
LAGP-6h	0.4	32.5	0.20	4.5	64	-0.9	1	6.4	36

Table 2. ^{7}Li spin-lattice relaxation T_1 parameters. Fraction of each component is indicated in parentheses relative to the sum. The experimental uncertainty is ±0.01 s for T_1 and ±1 % for the fraction.

Sample	**conductivity (mScm^{-1})**	**I component (s)**	**II component (s)**	**III component (s)**
Glass	$<10^{-9}$	4.38 (97%)	0.49 (3%)	-
LAGP-2h	0.5	-	0.23 (81%)	0.06 (19%)
LAGP-6h	0.3	7.45 (2%)	0.19 (68%)	0.05 (30%)

Conclusions

The structure and mobility in NASICON-like glass-ceramics electrolyte membranes of a $Li_{1+x}Al_{0.5}Ge_{1.5}(PO_4)_3$ composition have been studied and analyzed in respect to ionic conductivity. We have shown that NMR spin-lattice relaxation time measurement is a simple non-destructive analytical tool, which is sensitive to morphological changes that occur in glass-ceramics due to thermal treatment. Quantitative analysis of the relaxation data has shown a multicomponent character of the relaxation behavior associated with different mobility of lithium ions accommodated within interstitial channels formed by the NASICON framework. In the sample with the longer crystallization time an additional slow relaxing component appears, which is indistinguishable in the ^{7}Li NMR spectrum, and which is believed to account for the decrease of the ionic conductivity.

References

[1] Masquelier C, Croguennec L. Polyanionic (Phosphates, Silicates, Sulfates) Frameworks as Electrode Materials for Rechargeable Li (or Na) Batteries. 2013 *Chem. Rev.* **113** 6552–91

[2] Adachi G, Imanaka N, Aono H. Fast Li-ion Conducting Ceramic Electrolytes. 1996 *Adv. Mater.* **8** 127–35

[3] Leo CJ, Chowdari B, Subba Rao GV, Souquet JL. Lithium conducting glass ceramic with Nasicon structure. 2002 *Mater. Res. Bull.* **37** 1419–30

[4] Granger P, Harris R. 1990 *Multinuclear Magnetic Resonance in Liquids and Solids — Chemical Applications* (Netherlands Springer)

[5] McKenzie K, Smith ME. 2002 *Multinuclear Solid-State NMR of Inorganic Materials* (Pergamon)

[6] Abragam A. 1961 *The principles of nuclear magnetism.* (Oxford: Oxford Univ. Press)

[7] París MA, Martínez-Juárez A, Iglesias JE, Rojo JM, Sanz J. Phase Transition and Ionic Mobility in $LiHf_2(PO_4)_3$ with NASICON Structure. 1997 *Chem. Mater.* **9** 1430–6

[8] Arbi K, París MA, Sanz J. Li mobility in Nasicon-type materials $LiM_2(PO_4)_3$, M = Ge, Ti, Sn, Zr and Hf, followed by ^{7}Li NMR spectroscopy. 2011 *Dalton Trans.* **40** 10195–202

[9] Kuhn A, Wilkening M, Heitjans P. Mechanically induced decrease of the Li conductivity in an alumosilicate glass. 2009 *Solid State Ionics* **180** 302–7

[10] Dunst A, Epp V, Hanzu I, Freunberger SA, Wilkening M. Short-range Li diffusion vs. long-range ionic conduction in nanocrystalline lithium peroxide Li_2O_2—the discharge product in lithium-air batteries. 2014 *Energy Environ. Sci.* **7** 2739–52

[11] Massiot D, Fayon F, Capron M, King I, LeCalve S, Alonso B et al. Modelling one- and two-dimensional solid-state NMR spectra. 2002 *Magn. Reson. Chem.* **40** 70–6

[12] Xu X, Wen Z, Wu X, Yang X, Gu X. Lithium Ion-Conducting Glass–Ceramics of $Li_{1.5}Al_{0.5}Ge_{1.5}(PO_4)_{3-x}Li_2O$ (x=0.0–0.20) with Good Electrical and Electrochemical Properties. 2007 *J. Am. Ceram. Soc.* **90** 2802–6

Acknowledgments

We thank the Ministry of Education and Science of the Russian Federation (agreement no. 14.616.21.0007, identificator RFMEFI61614X0007) and the Bundesministerium für Bildung und Forschung (project no. 05K2014) for financial support in the framework of the joint Russian–German research project "SYnchrotron and NEutron STudies for Energy Storage (SyNeSTeSia)". V.V. gratefully acknowledges the support by The German-Russian Interdisciplinary Science Center (G-RISC) in the framework of the Grant no. P-2015a-13 and by the President of the Russian Federation in the framework of the Grant no. MK-7167.2015.3. We are grateful to Erica Brendler for numerous and fruitful discussions.

Journal of Physics: Conference Series **758** (2016) 012015 doi:10.1088/1742-6596/758/1/012015

Vibrational properties of noncovalently oligothiophene-functionalized graphene nanomaterials

M Boutahir, A H Rahmani, H Chadli and A Rahmani
Laboratoire d'Etude des Matériaux Avancés et Applications (LEM2A), Université Moulay Ismail,FSM-ESTM, BP 11201, Zitoune, 50000 Meknes, Morocco
E-mail: a.rahmani@fs-umi.ac.ma

Abstract. In this paper, the non resonant Raman spectra of hybrid nanostructures obtained by confinement of oligothiophene derivatives in two sheets of graphene are calculated in the framework of spectral moments method, together with a bond-polarizability model. Minimum energy calculations are performed, using a convenient Lennard-Jones expression of the van der Waals intermolecular potential, to derive the optimum configurations of oligothiophene with graphene. Parallel configuration of thiophene molecules with respect to the graphene plane was considered.

1. Introduction

The worldwide demand for energy has grown enormously over the recent years and it is likely to grow even more in the near future [1, 2, 3]. Because of the limited supply of our main energy sources (fossil fuels)[4], the development of renewable energy sources such as solar energy is a great priority.

Organic solar cells are a promising low-cost alternative to silicon solar ones [5], but the main handicap is the low power conversion efficiency of these devices. One of the alternatives to solve this handicap is the introduction of carbon nanotubes (NTs) or graphene to form an interpenetrating blend with the polymer. Since the discovery of photoinduced charge transfer between organic conjugated polymers (as donor) and nanotubes or graphene (as acceptor), these latter have been used to fabricate photovoltaic devices in combination with different polymers to the aim of increasing the power conversion efficiency [6, 7, 8].

The process consists of combining species with carbon nanotubes and graphene to create a new hybrid system. Three methods are currently used for that purpose. The first one is a functionalization of polymers at the surface of the tubes [9]. The second method is the noncovalent adsorption or wrapping of various functional molecules [10]. The third one consists of encapsulating oligomers inside nanotubes or between two sheets of graphene in order to keep their original physical properties [11, 12, 13]. The advantage of the last method is that the nanotubes and graphene protect the internal organic molecules from photo degradation and attacks by external

Journal of Physics: Conference Series **758** (2016) 012015 doi:10.1088/1742-6596/758/1/012015

reactive species, there by preventing undesirable alteration of the organic molecules [14].

Thiophene and its polymerized derivatives are considered as the reference compounds of conjugated polymers widely used in organic electronic devices and recently in novel hybrid graphene materials for solar cell. In fact, new 1D hybrid nano-systems are elaborated with oligomers either encapsulated into the hollow core of single-wall carbon nanotubes or non-covalently functionalized at the surface of graphene sheet. The non-covalent functionalization of graphene with conjugated polymers molecules provides an elegant approach to elaborate hybrid systems with original physical properties. Several experimental works dealt with the encapsulation of oligothiophene derivatives (oTh) into NTs(oTh@NT) [15, 13, 16]. Indeed, these systems are particularly interesting as the π-conjugation of the thiophene molecules provides nonlinear optical properties and intrinsic electrical conductivity [17, 18, 19, 20, 21].

In this work, we investigate the non resonant Raman spectra of hybrid nanostructures obtained by confinement of oligothiophene derivatives in multi-layered graphene. We study two hybrid systems, bithiophene in two sheets of graphene (2T@G) and quater-thiophene between two sheets of graphene (4T@G). The calculations have been performed in the framework of the bond-polarization theory using the spectral moment's method (SMM)[22].

2. Models and computational method

The dynamical matrix of our hybrid system is built block by block from the dynamical matrices of each subsytems: The graphene and oligothiophene forms the tridiagonal block of the dynamical matrix of the hybrid systems. While the interactions of graphene-oligothiophene and graphene-graphene forms the off-diagonal block.

The dynamical matrix of the oligomers molecules is calculated using DFT as implemented inside the Spanish initiative for Electronic Simulations with Thousands of Atoms (SIESTA) package [23].

The C-C interactions in graphene sheets are represented by the force constant model introduced by Saito et al [24]. This model was previously used by our group to study the nonresonant Raman spectra of SWNTs as a function of their lengths[22]. Whereas, the interactions between oligomer-graphene are described according to the van der Waals potential. We choose a Lennard-jones type

$$U_{LJ}(r) = 4\varepsilon \left[\left(\frac{\sigma}{r}\right)^{12} - \left(\frac{\sigma}{r}\right)^{6} \right] \tag{1}$$

with parameters: $\epsilon = 2.964$ meV and $\sigma = 0.3407$ nm for carbon atoms. $\epsilon = 10.7$ meV and $\sigma = 0.3560$ nm for sulfur atoms. $\epsilon = 0.67$ meV and $\sigma = 0.1069$ nm for hydrogen atoms.

The Sulfur-Carbon and Carbon-Hydrogen parameters were calculated using the rule of mixtures of Lorentz -Berthelot: $\varepsilon_{AB} = \sqrt{\varepsilon_A \varepsilon_B}$ et $\sigma_{AB} = (\sigma_A + \sigma_B)/2$.

3. Results

The Raman spectra of graphene are dominated by the tangential mode (TM) located at $1588cm^{-1}$. Many experimental and theoretical Raman studies have shown that the TM frequency follows a dependence as a function of the number of sheet of graphene. This dependence is well reproduced by Saito empirical force field[24].

The Raman spectrum of a free bithiophene(2T) molecule have been intensively studied in the literature [25]. Our calculated frequencies are close to those obtained in Ref[25]. From the low frequency region of the spectra of 2T(resp. 4T) molecule (Figure2(resp. Figure3)), we found a double peaks around 8 and $20cm^{-1}$(resp. $8cm^{-1}$) corresponding to a libration modes. The spectra is dominated by the region between 600 and $1600cm^{-1}$ with intense bands associated to the C-S-C deformation ($677cm^{-1}$(2T) and $697cm^{-1}$(4T)), the C-H bending ($1050cm^{-1}$(2T) and $1055cm^{-1}$(4T)), the C-C inter-ring stretching($1271cm^{-1}$(2T) and $1284cm^{-1}$(4T)), and C=C stretching($1504cm^{-1}$(2T) and $1503cm^{-1}$(4T)).

These results support that the dynamical matrices of graphene and 2T or 4T are expected to be reliable and can be used to build the tridiagonal block of the dynamical matrix of the new hybrid systems (2T@G and 4T@G).

In this preliminary work, an isolated molecule (nT) is inserted between two sheets of graphene with a supercell contains more than 192 atoms each. A supercell size close to $19.7\AA$, along OX axis, and $25.5\AA$, along OY one, is considered in order to avoid the interactions between molecules when we made the periodic conditions (figure1). Minimum energy calculations are performed, using a convenient Lennard-Jones expression of the van der Waals intermolecular potential, to derive the optimal distances between molecule (nT) and graphene sheets. In this parallel configuration, we found that the graphene-nT distance is close to$3.49\AA \pm 0.05$ and graphene-graphene distance is close $6.88\AA \pm 0.05$ (table1). The graphene-graphene distance in graphite is close to $3.4\AA$ which is smaller than the graphene-nT one in multi-layered system graphene-nT-graphene.

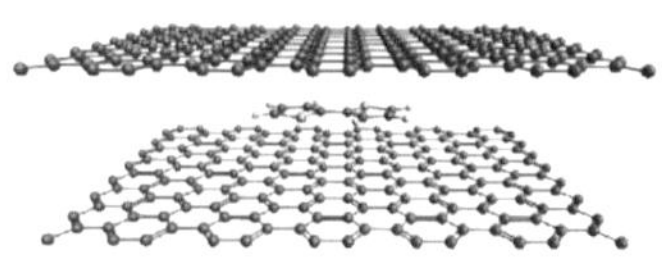

Figure 1. Structure of 2T@G.

Table 1. optimized structural of 2T and 4T between two sheets of graphene.

nT	nT-graphene distance (nm)	graphene-graphene distance (nm)
2T	0.349±0.005	0.688±0.005
4T	0.347±0.005	0.687±0.005

Now we investigate the Raman spectra of the two hybrid systems 2T@G and 4T@G obtained after minimizations. The figure2 displays the ZZ-polarized Raman spectra of 2T molecule (bottom) and 2T@G (top). Raman lines can be divided into three frequency ranges; (i)below $600cm^{-1}$ (panel a), (ii) an intermediate range between $600cm^{-1}$ and $1500cm^{-1}$ (panel b), (iii) above $1500cm^{-1}$ where the tangential modes are located (panel c).

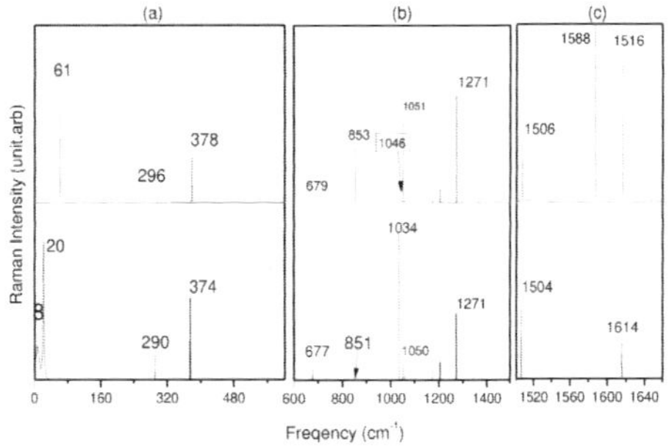

Figure 2. The ZZ-polarized Raman spectra of 2T molecule (bottom) and 2T@G hybrid (top).

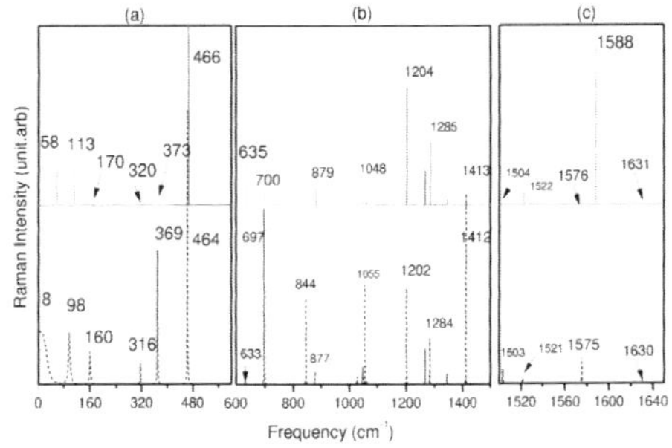

Figure 3. The ZZ-polarized Raman spectra of 4T molecule (bottom) and 4T@G hybrid (top).

The modes located in the lower frequency range are very sensitive to the confinement of thiophene into graphene, because major changes of frequencies and intensities are observed. First, we observe that the libration modes ($8cm^{-1}$ and $21cm^{-1}$) of 2T disappear in the Raman spectrum of 2T@G. We observe also a frequency shift of the other modes of 2T@G with respect to the corresponding modes of 2T.

In the intermediate range, Raman spectra of 2T@G show a several lines which originate from the splitting of the 2T degenerate modes due to van der Waals interactions from 677 to $679cm^{-1}$ and from 1050 to $1051cm^{-1}$ as well as from 1504 to $1506cm^{-1}$. Moreover, a new mode appears at $1046cm^{-1}$ which is associated to C-H bending as shown in the Figure4, where we presente the eigendisplacement vector of this coupled mode. These behaviours could be associated to the competition between intramolecule interactions and molecule-graphene ones.

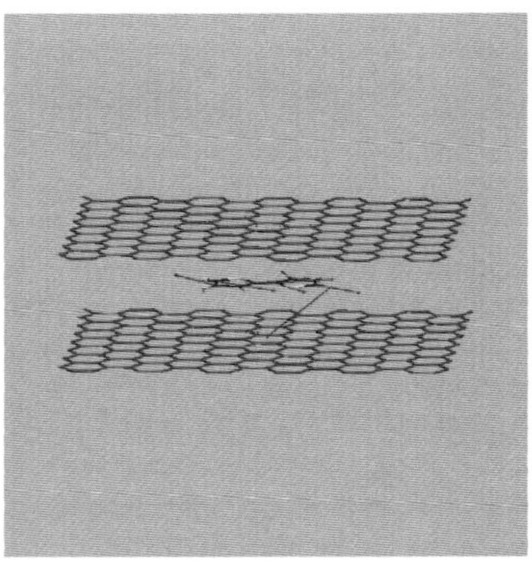

Figure 4. Calculated atomic motions of $1046cm^{-1}$ mode.

In the TM range, we observe an upshift of the modes in 2T@G with respect to the corresponding modes of 2T. The G-band of graphene is the same in 2T@G around $1588cm^{-1}$.

For the other system (4T@G), the figure3 displays the ZZ-polarized Raman spectra of 4T molecule (bottom) and 4T@G (top). As in (2T@G) the mode of libration ($8cm^{-1}$) of 4T molecule disappears in the Raman spectrum of 4T@G, and we find an upshift of the other modes of 4T@G with respect to those of 4T molecule.

In the intermediate range, like in 2T@G Raman spectra of 4T@G show several lines due to van der Waals interactions. An upshift of the all modes by ($2cm^{-1}$ or $3cm^{-1}$) in 4T@G with respect to those in 4T molecule.

We note that the G-band of graphene, around $1588cm^{-1}$, is not affected by the encapsulation in 4T@G hybrid.

4. Conclusion

In this study, the optimal configuration of the isolated nT-oligomers inside two sheets of graphene are derived using a convenient Lennard-Jones potential. In these multilayer graphene-nT-graphene hybrids, we found that the nT-graphene distance is greater than the graphene-graphene one in graphite. For the obtained configurations of 2T@G and 4T@G, the nonresonant Raman spectra have been calculated using the bond-polarizability model combined with the spectral moment's method. We showed that the major changes is obtained in the low frequency range where the libration modes in the 2T and 4T are disappeared in the Raman spectra of the hybrids. The Study with nT-nT interactions inside graphene multilayers is in progress.

5. ACKNOWLEDGMENT

The work was supported by Moulay Ismail University Research Support (13-16).

References

[1] Lave L 2008 The economics of energy options *MRS Bull* **33** 291-6.
[2] Holditch S and Chianelli R 2008 Factors that will influence oil and gas supply and demand in the 21st century *MRS Bull* 33 317-23.
[3] Agency IE World Energy Outlook 2007 Edition 2007.
[4] IEA World Energy demand to grow briskly to 2030 oil gas J 2002 **100**(42) 36-8.
[5] Sariciftci N S 2004 Plastic photovoltaic devices Mater Today **7**(9) 36-40.
[6] Nogueira A F, Lomba B S, Soto-Oviedo M A, Correia G R D, Corio P and Furtado G A 2007 Polymer solar cells using single-wall carbon nanotubes modified with thiophene pedant groups *J Phys Chem C* **111**(49) 18431-8.
[7] Berson S, de Bettignies R, Bailly S, Guillerez S and jousselme B 2007 Elaboration of P3HT/PCBM composites for organic photovoltaic cells *Adv Funct Mater* **17**(16) 3363-70.
[8] Landi B J, Castro S L, Ruf H J, Evans C M, Bailey S G and Raffaelle R P 2005 CdSe quantum dot-single wall carbon nanotube complexes for polymeric solar cells *Sol Energy Mater Sol cells* **87**(1-4) 733-46.
[9] Babaa M R, Bantignies J L, Alvarez L, Parent P, Le Normand F, Gulas M, Mane J M, Poncharal P and Doyle B P J 2007 *Nanosci Nanotechnol* **7** 3463-3467.
[10] Roquelet C, Lauret J S, Alain-Rizzo V, Voisin C, Fleurier R, Delarue M, Garrot D, Loiseau A, Roussignol P, Delaire J A and Deleporte E 2010 *Chem Phys Chem* **11** 1667-1672.
[11] Loi M A, Gao J, Codella F, Blondeau, P, Menna E, Bartova B, Hebert C, Lazar S, Botton G A, Milko M and Ambrosch-Draxl C 2010 *Adv Mater* **22** 1635-1639.
[12] Malard L M, Nishide D, Dias L G, Capaz R B, Gomes A P, Jorio A, Achete C A, Saito R, Achiba Y, Shinohara H and Pimenta M A 2007 *Phys Rev B* **76** 233412.
[13] Kalbac M, Kavan L, Gorantla S, Gemming T and Dunsch L 2010 *Chem-Eur J* 16 11753-11759.
[14] Yanagi K, Miyata Y and Kataura H 2006 Highly Stabilized Beta-Carotene in Carbon Nanotubes *Adv Mater* **8** 437-441.
[15] Loi M A, Gao J, Cordella F, Blondeau P, Menna E, Bartova B, Hebert C, Lazar S, Botton G A, Milko M and Ambrosch-Draxl C 2010 Encapsulation of Conjugated Oligomers in Single-Walled Carbon Nanotubes: Towards Nanohybrids for Photonic Devices *Adv Mater* **22** 1635-1639.

Journal of Physics: Conference Series **758** (2016) 012015 doi:10.1088/1742-6596/758/1/012015

[16] Alvarez L, Almadori Y, Arenal R, Babaa R, Michel T, Le Parc R, Bantignies J L, Jousselme B, Palacin S, Hermet P and Sauvajol J L 2011 Charge Transfer Evidence between Carbon Nanotubes and Encapsulated Conjugated Oligomers *J Phys Chem C* **115** 11898-11905.
[17] Brdas J L and Street G B 1985 Polarons, Bipolarons and Solitons in Conducting Polymers *Acc Chem Res* **18** 309-315.
[18] Swager T M 1998 The Molecular Wire Approach to Sensory Signal Amplification *Acc Chem Res* **31** 201-207.
[19] McQuade D T, Pullen A E and Swager T M 2000 Conjugated Polymer-Based Chemical Sensors *Chem Rev* **100** 2537-2574.
[20] Kertesz M, Choi C H and Yang S J 2005 Conjugated Polymers and Aromaticity *Chem Rev* **105** 3448-3481.
[21] Chen Z F, Wannere C S, Corminboeuf C, Puchta R and Schleyer P V 2005 Nucleus-Independent Chemical Shifts as an Aromaticity Criterion *Chem Rev* **105** 3842-3888.
[22] Rahmani A, Sauvajol J L, Rols S and Benoit C 2002 *Phys Rev B* **66** 125404.
[23] Soler J M, Artacho E, Gale J D, Garca A, Junquera J, Ordejn P and Snchez-Portal D 2002 The SIESTA method for ab initio order-N materials simulation *Journal of Physics Condensed Matter* **14**(11) 2745.
[24] Saito R, Dresselhaus G and Dresselhaus M S 1998 *Imperial Colege Press* London.
[25] Hermet P, Izard N, Rahmani A and Ghosez Ph 2006 *J Phys Chem B* **110** 24869-24875.

Buffer layer optimization for high efficiency CIGS solar cells

N Severino[1], N Bednar, N Adamovic
Technische Universität Wien, Institute of Sensor and Actuator Systems,
Gusshausstrasse 27-29, 1040 Vienna, Austria

E-mail: noemi.severino@tuwien.ac.at Phone: +43-1-58801-76674

Abstract. This work presents a study concerning the numerical optimization of a buffer layer for high efficiency CIGS solar cells. The dependence of the solar cell properties on the buffer layer material, the layer thickness, the type and density of defects within the same layer were numerically investigated and analysed. Promising results were obtained with alternative Cd-free buffer layers (ZnSnO, InS and ZnS) in place of the standard CdS.

1. Introduction

Among thin-film solar cells based on the I-III-VI semiconductor compound: the polycrystalline Cu(In,Ga)Se_2 (CIGS) absorber layer, is one of the most promising [1,2]. CIGS with its chalcopyrite structure is a direct band gap semiconductor characterized by a high absorption coefficient that allows decreasing the absorber layer thickness in relation to crystalline and multi-crystalline Si solar cells resulting in a reduction of production costs; in fact an active layer with a thickness between 1-2 μm is already enough to guarantee high stability and high efficiencies of the device [3]. Moreover, the use of flexible substrates, in the production flow of these thin film solar cells (e.g. polymer films or stainless steel), have opened new market possibilities. CIGS solar cells grown on polymer film have already demonstrated, at laboratory scale, promising results [4].

In order to enhance the performances of a CIGS solar cell, not only a deep control of the absorber properties is required, but also an optimization of the buffer layer properties is fundamental. Thin film buffer layers grown with different deposition techniques have been extensively studied in the last years. The use of a buffer layer in CIGS solar cell has demonstrated superior performances compared to solar cells without it [5-7]. Several advantages were observed like an evident decrease of the recombination at the interfaces and a reduction of the damages caused by the sputtering of the ZnO layer directly on the absorber [8]. Moreover, since the buffer has usually a refractive index comparable to the one of the ZnO and CIGS layer the reflection losses at the front of the device are kept insignificant.

In order to support the technology development and allow the exploration of new cell/module concepts [9, 10], while meeting specific requirements, we investigated the characteristics of different buffer layers in order to produce highly efficient flexible CIGS solar cells. The chemical bath deposited CdS film is usually employed as buffer layer in a high efficiency CIGS solar cell [11, 12]. The reason of this choice is due to the fact that CdS can cover entirely the rough surface of the CIGS film avoiding the formation of shunt paths [13]. Moreover an efficient heterojunction in ZnO/CdS/CIGS is created; in fact a favorable band alignment can be observed when a CdS buffer layer is used. This work

[1] N Severino E-mail: noemi.severino@tuwien.ac.at

investigates the dependence of the solar cells performances on the change of the CdS thickness. Moreover, the influence of the increase of the defects density in the CdS layer was analyzed in order to study its effects on the open circuit voltage (V_{OC}) and the short circuit current (J_{SC}) of the final device. Furthermore we investigated, through numerical modelling, the possibility to use alternative buffer layers in place of the standard CdS. The replacement of the CdS by other materials is related to environmental concerns, due to the toxicity of Cd. Additionally, the aim is to increase the short circuit current density of the device by enhancing the photon collection in the region of low wavelength (between 350-550 nm). Since the band gap of CdS is 2.42 eV [14], which corresponds to the wavelength of around 512 nm, only the incident light with a wavelength higher than 512 nm can be transmitted to the absorber layer, resulting in a drop of the quantum efficiency in the high energy region. For this reason the application of a buffer layer with higher energy gap and lower electron affinity, compared to the one of CdS, could result in an enhancement of the solar cell performances reducing the losses in the buffer layer.

2. Simulations

The simulations took account following working conditions: on room temperature, using the solar spectrum AM1.5G as light source incidents on the device. The simulations are based on 1-D electrical model and were carried out using SCAPS, a one-dimensional software package freely available to the PV community [15].

The structure of the simulated CIGS solar cell consists of 200 nm thick n-type ZnO:Al, 80 nm thick i:ZnO, n-type buffer layer (different thicknesses and buffer layer materials were taken account) and a layer of p-type CIGS 1800 nm thick (figure 1).

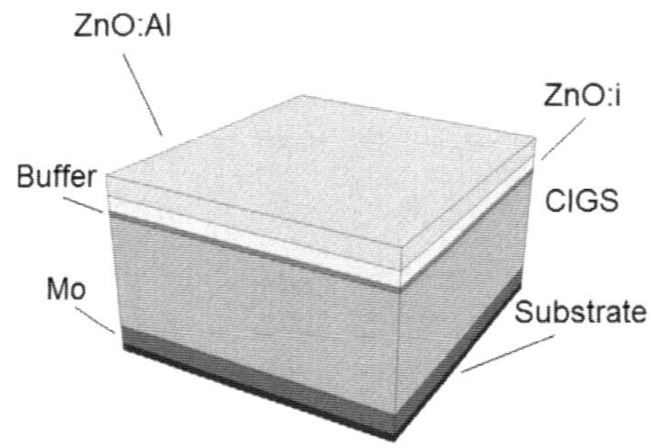

Figure 1. Schematic structure of the simulated CIGS solar cell.

The baseline set of the parameters used for the simulations was obtained from in-house measurements performed on each separate layer of the samples and from a study of the data found in literature [16-18]. The numerical model was calibrated with a reference measured CIGS solar cell grown on flexible substrate (reported in [4]). A graded composition of Ga throughout the absorber layer was considered in the simulation. Increased solar cell efficiency was obtained with a double grading profile with a minimum amount of Ga in the middle region of the layer and an increased amount toward the back and the front contacts. In the simulations all the properties, except the ones of the buffer layer, were kept unchanged in order to be able to analyse only the influences of the changes in the buffer layer and their effects on the solar cell parameters. The buffer layer optimization, performed through the simulations, was done in the following steps:

1) variation of the CdS thickness to find out the optimal value taking into consideration the technological constraints, such as the difficulty to grow a uniform thin CdS layer using chemical bath deposition process, properly covering the rough surface of the absorber layer,

2) variation of the defects properties in the CdS layer in order to identify how much the performances are influenced by the change of the defects density and the defects energy level,

3) replacement of the CdS layer with a Cd-free buffer layer in order to investigate the possibility to use alternative buffer enhancing the device performances.

2.1 CdS layer thickness

The CdS layer thickness was varied from 10 nm to 80 nm. The results of the simulated current density vs. voltage (I-V) and external quantum efficiency (EQE) curves for different CdS thicknesses are reported in figure 2.

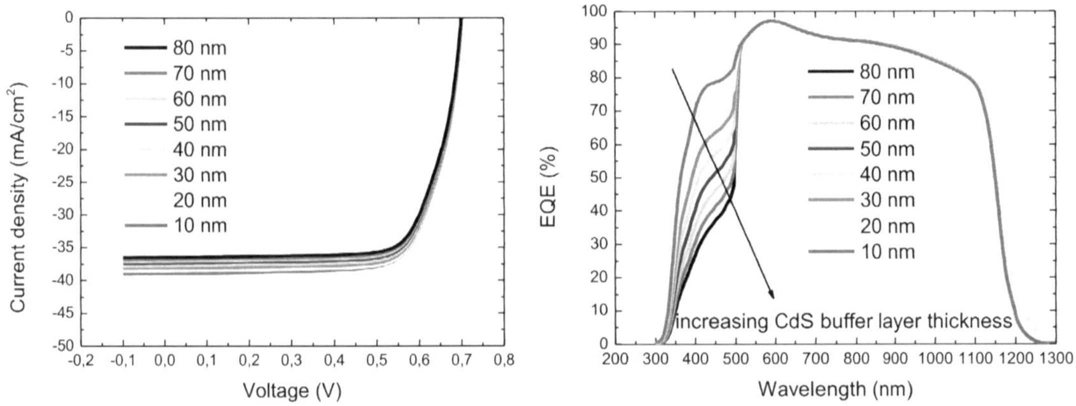

Figure 2. Simulated I-V (left) and EQE (right) curves for different CdS layer thickness. The buffer layer thickness was varied from 10 nm to 80 nm.

As shown in the figure 2, the influence of the change of the CdS thickness is more evident from the EQE curves. It can be observed that a rise of the buffer layer thickness causes a drop of the blue response of the EQE due to the increased absorption in the CdS layer. Moreover, it should be considered that a thickness of around 10 nm represents a limit for the buffer layer thickness [19]. Under this thickness the buffer will not compensate the degradation effects of the rough surface of the polycrystalline CIGS layer and these effects will not be any longer negligible. In figure 3 the solar cell parameters at different CdS thicknesses are shown. It is evident from the results that the open circuit voltage (V_{OC}) and the fill factor (FF) are just slightly influenced by the change on the CdS layer. On the contrary its influence on the short circuit current (J_{SC}) and consequently on the efficiency (η) is more accentuated. An exponential decrease of the J_{SC} was observed and shown by an exponential fitting line in figure 3. A too thick CdS layer will increase absorption in the buffer layer itself causing a decrease of the incident photons reaching the active layer and an increment of the recombination of the minority carriers that causes the reduction observed in EQE curves between 350 and 500 nm. The simulations reported on this paper showed that the preferred thickness of CdS layer is around 20 nm, the future challenge for the material developers will be to grow such a thin, uniform, adherent, and reproducible large area CdS layer with a good quality using chemical bath deposition or other techniques.

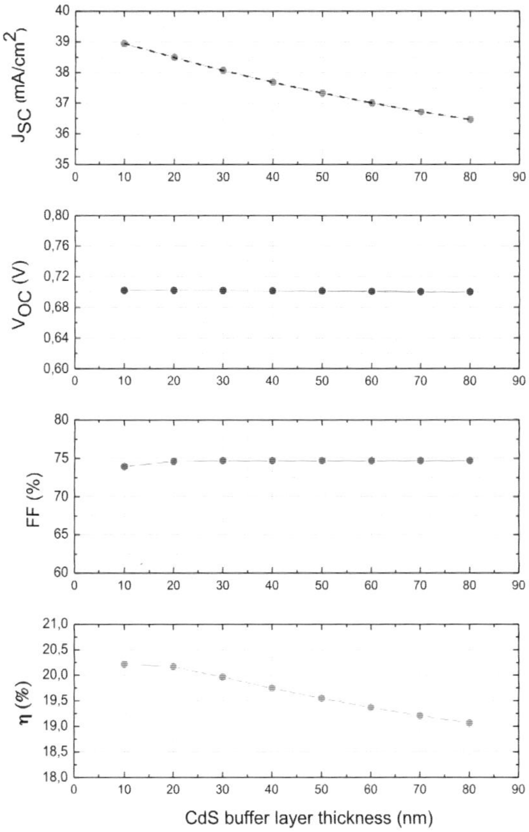

Figure 3. Effects of the CdS thickness change on the solar cell performances.

2.2 CdS defects density and defects energy level

The impacts of the traps properties in the CdS layer on the solar cell defects density and defects energy level in the bulk or at the interface with the CIGS layer was evaluated. As the first approach the defects density in the bulk region of the buffer layer 50 nm was varied in order to investigate its effects on the solar cell parameters. The results can be observed in the I-V and EQE curves simulated and presented in figure 4.

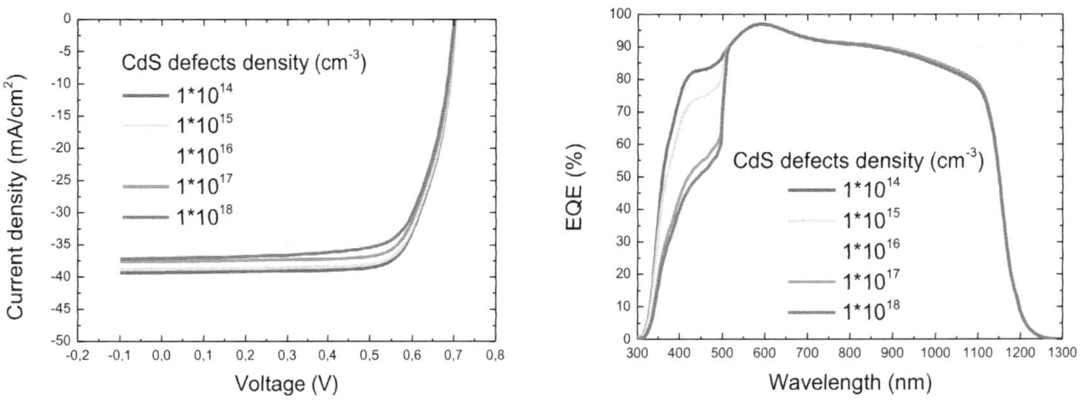

Figure 4. Simulated I-V (left) and EQE (right) curves for different CdS defects density in the bulk region. The total density of traps was varied from 10^{14} to 10^{18} cm^{-3}.

From figure 4 it is evident that the rise of the defects density brings to a deterioration of the solar cell performances as shown in the simulated I-V curves and in the spectral responses. An increase of the total trap density of the acceptor-like mid-gap defects in the bulk region of the CdS layer will cause an enhancement of the recombination processes. In order to evaluate also the impacts of the interface defects on the solar cells, the same simulations were carried out considering variable defect densities at the interface with the absorber layer instead of the middle gap defects. For the simulation of the interface defects a surface defect layer (SDL) with optical properties similar to the CIGS layer was considered. The SDL band-gap, the electron affinity and the defects concentration were varied. The obtained results are presented in figure 5. The performed simulations show that the changes in the defects density have more influence on the solar cell performances than the position of the traps in the buffer layer. In fact the interface and bulk defects present the same behaviour and effects on the solar cell performances in the majority of the simulated range of the traps densities. Only for the defect densities over 10^{17} cm^{-3} the defects located in the bulk of buffer layer have more impact on the deterioration of the solar cell parameters than the defects on the buffer/absorber interface.
The rise of the interface or bulk defects density in the buffer layer deteriorates the J_{SC} without significantly affecting the V_{OC}.

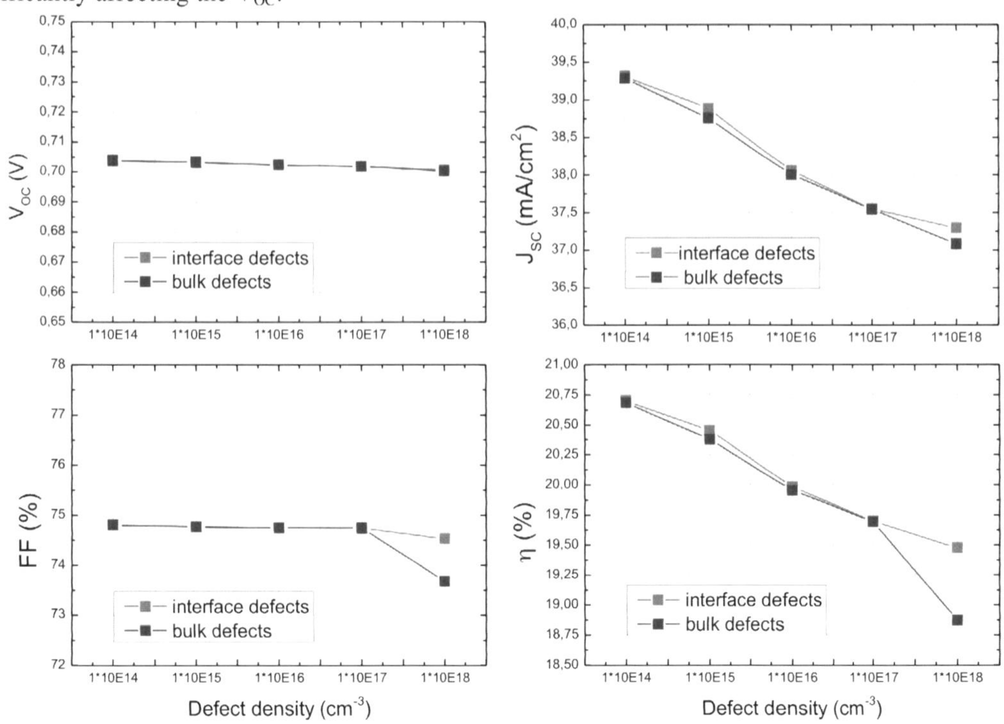

Figure 5. Effects of different trap densities at the interface, between buffer and absorber layer, and in the bulk region on the performances of the solar cell.

2.3 New buffer layer

Three different alternative buffer layers (ZnSnO (ZTO), InS and ZnS) were modelled in order to check the possibility to replace the CdS with a non-toxic material characterized by higher energy gap in order to overcome the photocurrent losses that characterize the use of the CdS layer. Different technique of growth can be used to produce and test new buffer layer material and in this work, the transmittance data of the three different sputtered buffer layers were considered. The transmittance curves of the alternative buffer layers, are presented in figure 6 and were used as input data for the calculation of the absorptions at different wavelengths used in the model.

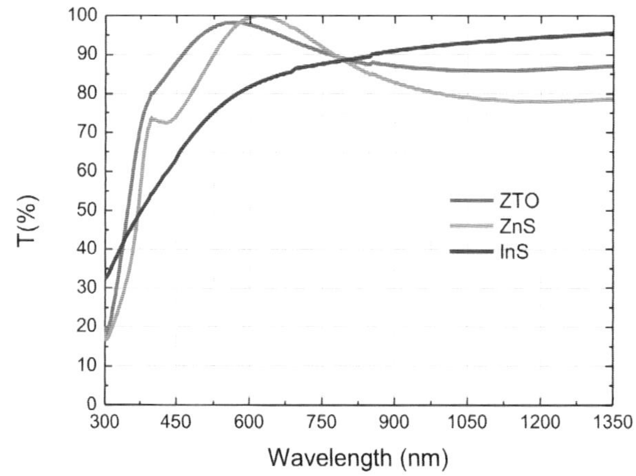

Figure 6. Transmittance measurements of three alternatives Cd-free layers used as input in the simulations.

In the model three new materials used as buffer layer were simulated considered a higher electron affinity and energy gap compared to the CdS. In particular a 3.0 eV energy gap was used and 100 nm thick buffer layers. The resulted I-V and EQE curves obtained from simulations are reported in figure 7.

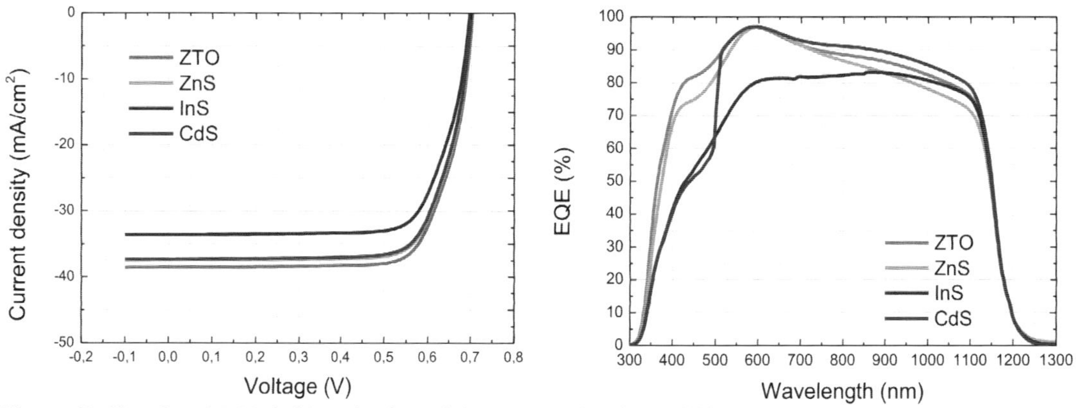

Figure 7. Simulated I-V (left) and EQE (right) curves for three different sputtered new buffer layers: ZTO, ZnS and InS.

As shown in the simulations results the most promising Cd-free buffer is the ZTO, having the best I-V curve and the highest spectral response. A comparison among the simulated solar cells performances using different buffer layers and the reference cell with CdS is presented in table 1.

Table 1. Solar cell parameters obtained from simulations using three different sputtered buffer layer and a CdS layer for comparison.

Buffer layer used in the simulation	V_{OC} (mV)	J_{SC} (mA/cm^2)	FF (%)	η (%)
CdS	702	37.3	74.7	19.5
ZTO	704	38.6	75.0	20.3
ZnS	702	37.5	75.0	19.7
InS	698	33.6	74.9	17.6

As reported in table 1 using ZTO and ZnS as buffer layer enhances the short circuit current and consequently the efficiency. On the contrary the open circuit voltage and the fill factor are just slightly influenced by the change of the buffer layer.

3. Conclusions

In this work we simulated the effect of the different buffer layer properties (thickness, defects, materials) on the performance of the CIGS solar cell in order to find the optimal configuration. The simulations show that the best performances in a CIGS solar cell, using a CdS layer, are obtained for a 20 nm thick buffer. The simulation results suggested that the interface and bulk defects in the CdS layer have similar influence on the solar cell performace. In both cases a deterioration of the short circuit current is observed while just a slightly change in the open circuit voltage is showed. Moreover simulations have also demonstrated the possibility to use alternative buffer layer instead of CdS, obtaining promising results especially with the ZTO layer. Further experimental work will be carried out in order to confirm the presented simulation work.

4. Acknowledgments

This work has been supported by the Austrian Research Promotion Agency FFG through the project SynerCIS, project No. 840706, and also by the European Commission FP7-NMP Programme project SolarDesign, under the grant agreement No. 310220.

5. References

[1] Green M A, Emery K, Hishikawa Y and Warta W 2011 *Prog. Photovolt.* **19** 84-92
[2] Repins I, Conteras M A, Egaas B, DeHart C, Scharf J, Perkins C L, Noufi R 2008 *Prog. Photovolt.* **16** 235-239
[3] Green M A, Emery K, Hishikawa Y, Warta W and Dunlop E D 2015 *Prog. Photovolt: Res. Appl.* **23** 1-9
[4] Chirilă A et al. 2011 *Nat. Mater.* **10** 857-861
[5] Schock H W and Noufi R 2000 *Prog. Photovolt. Res. Appl.* **8** 151-160
[6] Naghavi N et al. 2010 *Prog. Photovolt. Res. Appl.* **18** 411-433
[7] Klenk R 2001 *Thin Solid Films* **387** 135-140
[8] Witte W, Spiering S and Hariskos D 2014 *Vakuum in Forschung und Praxis* **26** 23-27
[9] Severino N, Bednar N, Acciarri M and Adamovic N 2014 *Proc. in E-MRS Spring Meeting Conf.* (Lille) pp.18
[10] Bednar N, Severino N and Adamovic N 2015 *J. of Renewable and Sustainable Energy* **7** 011201
[11] Chirilă A et al. 2013 *Nat. Mater.* **12** 1107-1111
[12] Jackson P, Hariskos D, Lotter E, Paetel S, Wuerz R, Menner R, Wischmann W and Powalla M 2011 *Prog. Photovolt: Res. Appl.* **19** 894-897
[13] Hashimoto Y, Kohara N, Negami T, Nishitani N and Wada T 1998 *Sol. Energ. Mat. Sol. C.* **50** 71-77
[14] Kittel C 2003 *Introduction to solid state physics 7th ed.* (Singapore) Wiley, pp.201
[15] Burgelman M, Nollet P and Degrave S 2000 *Thin Solid Films* **361-362** 527-532
[16] Krc J, Cernivec G, Campa A, Malmström J, Edoff M, Smole F and Topic M 2006 *Opt. Quant. Electron.* **38** 1115-1123
[17] Ray J R, Panchal C J, Desai M S and Trivedi U B 2011 *J. Nano- Electron. Phys.* **3** 747-754
[18] Werner J H, Mattheis J and Rau U 2005 *Thin Solid Films* **480-481** 399-409
[19] Contreras M A, Romero M J, To B, Hasoon F, Noufi R, Ward S and Ramanathan K 2002 *Thin Solid Films* **403-404** 204-211

Journal of Physics: Conference Series **758** (2016) 012017 doi:10.1088/1742-6596/758/1/012017

Effect of particle size on the photocatalytic activity of $BiNbO_4$ under visible light irradiation

S P Adhikari[1, 2] **and A Lachgar**[1, 2]

[1]Department of Chemistry, Wake Forest University, Winston Salem, NC 27109, USA
[2]Center for Energy, Environment and Sustainability, Wake Forest University, Winston Salem, NC 27109, USA

Email: lachgar@wfu.edu

Abstract
Numerous attempts have focused on developing more efficient visible-light active photocatalysts. Bismuth niobate ($BiNbO_4$) was reported as a potential candidate for photocatalytic water splitting and for the degradation of organic molecule under visible-light irradiation because of its small band gap and appropriate band positions. Here, three $BiNbO_4$ samples with different, uniform particle sizes were synthesized. The effect of $BiNbO_4$ particle size on its photocatalytic activity was studied by using the reduction of 4-nitroaniline as model reaction. Results show that the photocatalytic degradation of 4-nitroaniline varies systematically with the change in particle size and surface areas. To better understand the correlation between structure, composition, morphology, and optical properties, the catalysts were characterized by transmission electron microscopy (TEM), scanning electron microscope (SEM), Powder X-ray diffraction (PXRD), Brunauer-Emmett-Teller (BET) surface area, UV-vis diffuse reflectance spectroscopy (DRS), and photoluminescence spectroscopy (PL).

Key Words: Photocatalysis, Particle size, Surface area, Bismuth niobate

1. Introduction

Semiconductor based photocatalytic processes are widely recognized as viable solutions to current environmental problems and clean energy generation [1]. After the discovery of titanium dioxide (TiO_2) as an ultraviolet light active photocatalyst for water splitting and degradation of organic compounds in the early 1970's, a large number of semiconductors have been reported to be active for energy generation and environmental applications [2,3]. Most semiconductors explored to date have large band gaps, thus require UV light irradiation to induce catalytic activity. However, solar energy consists mainly of visible (~45 %) and infrared light (~50 %) with a small contribution of UV light (~5 %). To more efficiently utilize solar energy irradiation, photocatalysts must show good activity under visible light irradiation (400 nm $< \lambda >$ 800 nm) with a band gap less than 3 eV [4–7]. In addition, the quantum efficiency is negatively affected by thermodynamically favored recombination of photogenerated electron-hole pairs. Also, many semiconductor photocatalysts, chalcogenides in particular, undergo degradation under solar energy due to the photo-corrosion [8]. Therefore, there is a significant need to develop visible-light active photocatalysts with high activity and good chemical and physical stability.

To overcome the limitation that a majority of oxide-based photocatalysts are only UV responsive, many methods have been developed to extend their light absorption range into the visible region, such as doping with metal or non-metal ions or forming semiconductors heterojunctions by combining two or three semiconductors [9–17]. Other methods for enhancing the activity of given catalyst by allowing better separation of the photogenerated carriers have been explored. Some of these approaches include the addition of sacrificial reagents, cocatalyst loading, and preparing materials with higher active sites densities [1,4,18–20]. Another interesting approach is the reduction of grain size to decrease the time necessary for the charge carriers to migrate to the surface of the particle and be involved in electrochemical reactions. It is well known that for randomly generated charge carriers, the average diffusion time from bulk to the surface is given by $\tau = r^2/\pi^2 D$, where r is the grain radius and D is the diffusion coefficient of the carrier [21]. Therefore, when the grain radius decreases, the diffusion time is reduced and the migration of the charge careers to the surface of the particles is facilitated. The preparation of particles with small and uniform size is one of the key issues to address the photocatalytic activity of different materials [4,18,22–24]. Conventional high temperature solid-state synthetic methodologies generally lead to rather large particle size due to sintering, and small surface area. Synthesis methods using mild temperatures and solvents, on the other hand, has significant advantages in terms of homogeneity, low temperature and control over particle size, leading to higher surface area that can enhance the overall photocatalytic activity [4,18,25,26].

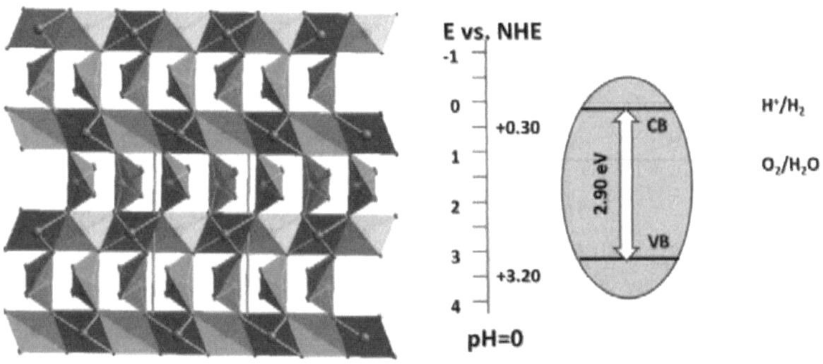

a B

Figure 1: (a) Crystal structure representation of $BiNbO_4$. (b) Band gap and calculated band positions of $BiNbO_4$. The conduction band (CB) bottom (E_{CB}) was calculated empirically according to the formula, $E_{CB} = X - 0.5E_g + E_0$, where E_g is the band gap energy, E_0 is a scale factor relating the reference electrode's redox level to the absolute vacuum scale (E_0 =4.5 eV for normal hydrogen electrode), and X is the electronegativity of the semiconductor, which can be expressed as the geometric mean of the absolute electronegativity of the constituent atoms. The value of X for $BiNbO_4$ was taken as 6.26 eV for the calculation)

Bismuth niobate ($BiNbO_4$) was reported to be photocatalytically active for water splitting and organic molecule degradations using sunlight [27–29]. Figure 1 show the crystal structure of orthorhombic phase of $BiNbO_4$, where both the Bi and Nb are coordinately bonded to the six neighboring oxygen atoms. Several band-gap engineering approaches such as anion and cation doping and dye sensitization were investigated to make $BiNbO_4$ photocatalytically active in the visible light region [29–33]. Zou et al. reported a novel series of Solid Photocatalysts, $BiTa_{1-x}Nb_xO_4$ ($0 \leqq x \leqq 1$) for water splitting [34]. The same group demonstrated the potential of Bi_2MNbO_7 (M= Al, Ga, In) as catalyst to split water into hydrogen and oxygen under both ultraviolet and visible light [28]. Most photocatalysts based on these studies are synthesized by the solid-state synthetic methodology at high temperature, leading to large particle size and small surface areas. Still, some recent papers have reported a wet chemical route to synthesize bismuth and niobium metals based oxides [35–39]. Here, we report the preparation of $BiNbO_4$ with different particle sizes and surface areas using different wet chemistry methods, and the study of their photocatalytic activity towards the reduction of 4-nitroaniline using visible light irradiation ($\lambda \geq 400$ nm). The results are compared with that of the compound synthesized by solid-state reaction method to correlate the activities with particle sizes and surface areas.

2. Experimental Section

2.1 Synthesis

All solvents and chemicals used were reagent-grade quality and were used without further purification. Bismuth Oxide (99.8%) and Niobium Chloride (99.9%) were purchased from Alfa Aesar (USA). Niobium Oxide (99.9%) and Triblock Copolymer Surfactant Pluronic P123 ($HO(CH_2CH_2O)_{20}(CH_2CH(CH_3)O)_{70}(CH_2CH_2O)_{20}H$) were obtained from Sigma Aldrich (USA). Three different preparation methods were performed including high temperature solid state reaction method. In a typical solid-state reaction method (**S1**), 2.5 mmol Bismuth oxide (Bi_2O_3) and 2.5 mmol Niobium oxide (Nb_2O_5) were mixed and ground manually in an agate mortar to obtain fine powder[40]. The mixture was heated to 900 °C at a rate of 5 °C/min and the temperature was held at 900 °C for 24 hours then cooled to room temperature at the rate of 10 °C/min to produce light yellow microcrystalline powder of pure $BiNbO_4$ as determined by powder X-ray diffraction (PXRD). $BiNbO_4$ was also synthesized by solution method using P123 as surfactant (**S2**)[41]. In a typical synthesis, 2.5 mmol Bismuth nitrate pentahydrate and 2.5 mmol Niobium chloride were separately dissolved in 10 ml of ethanol and stirred for 0.5 hr. A solution of 1.0 gm of P123 dissolved in 10 ml of ethanol was also prepared. The three solutions were mixed and stirred to obtain a colorless transluscide solution, which was aged for 48 hours at room temperature then dried at 80 °C to remove excess ethanol. The dry mass was calcined at 750 °C at the rate of 5 °C/min and held at this temperature for 12 hours. The temperature of calcination was determined from thermogravimetric analysis (TGA).

$BiNbO_4$ was also synthesized by two-step hydrothermal process (**S3**). The first step was carried out to prepare niobic acid (Nb_2O_5 .nH_2O) by hydrothermal route [38,39]. In a typical procedure, 2.5 mmol of Nb_2O_5 was added to 18 ml of 3 M KOH solution, which was loaded into 23 mL teflon-lined stainless steel cylindrical autoclave and then heated at 200 °C for 5 hours, a clear solution was obtained. The pH of the solution was made acidic (around 5.5) by dropwise addition of concentrated HCl to precipitate niobic acid, which was filtered and washed several times with distilled water to remove any potassium ions. In the second step freshly prepared niobic acid was added to an aqueous solution of oxalic acid (1M) at 80 °C with constant stirring to obtain milky-white suspension. Then, 2.425 g of bismuth nitrate (5 mmol) was then added to that suspension and the pH of the solution was adjusted to 12 to precipitate as hydroxide by adding 3M KOH solution. The slurry was transferred to 100 mL Teflon-lined stainless steel autoclave up to 80 % of total volume (slurry volume 80 mL). The sealed autoclave was heated at 140 °C for 12 hours and the yellow product was collected by filtration followed by washing with distilled water after natural cooling. Finally the sample was dried in an oven at 80 °C to obtain crystalline $BiNbO_4$.

2.2 Structural Characterization

A Bruker-AXS D2 phaser powder diffractometer equipped with a Ni-filtered Cu Kα sealed X-ray tube (λ=1.54184 Å) and a Lynxeye position-sensitive detector was used for X-ray analysis. The PXRD pattern was collected at room temperature and analyzed using the Bruker-AXS EVA software package. The samples were scanned in the 2θ range of 10-70° with a step-size of 0.01° and collection time 0.3 sec at each step. The indexing and refinement were done using TOPAS software. A Perkin Elmer Pyris 1 thermogravimetric analyzer was used to determine the thermal stability and thermal behavior of the compounds prepared. The samples prepared by solution method were heated up to 1000 °C with a temperature ramp of 5 °C/ min under a flow of air. The percent weight loss was calculated relative to the initial weight of the sample to determine the calcination temperature and the thermal stability of the material. Surface areas and pore-size distribution of the samples were determined by nitrogen adsorption–desorption isotherms at liquid nitrogen temperature (77K) using a Quantachrome autosorb–IQ instrument. Prior to the measurements, the samples were outgassed at 120 °C for 2 h under dynamic vacuum. The pore sizes were determined using ASiQwin software. A JEOL 6330 cold field emission Scanning Electron Microscope (SEM) and JEOL 1200 EX Transmission Electron Microscope (TEM) were used to determine the particle size of each of the $BiNbO_4$ materials. UV-vis diffuse reflectance spectra (DRS) were collected on an Ocean Optics FLAME-S-XR1-ES spectrophotometer equipped with an integrating sphere. Photoluminescence (PL) spectra were measured at room temperature on fluorescence spectrometer (LS55, PerkinElmer) with an excitation wavelength of 300 nm.

2.3 Photocatalytic test

The reduction of 4-nitroaniline (4-NA) was used as a model reaction (scheme 1) to evaluate the photocatalytic activity of $BiNbO_4$ catalysts prepared by different methods. The reduction of 4-NA has been chosen as model reaction because of its long-term impact to environment. It is found in wastewater discharges from industries where it is either manufactured or used as an intermediate such as in the synthesis of dyes, antioxidants, pharmaceuticals, in gum inhibitors, poultry medicines, and as a corrosion inhibitor.[42]

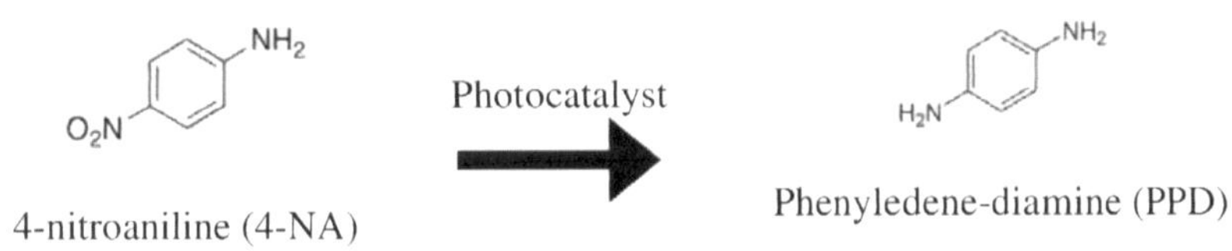

Scheme 1: Photocatalytic reduction of 4-nitroanilne (4-NA) to phenyledene-diamine (PPD)

The photocatalytic reaction was carried out in 80 ml photocatalytic quartz glass cell with 100 mg of $BiNbO_4$ dispersed in 50 mL of an aqueous solution containing 10 ppm of 4-NA. After adding 10 mg of ammonium oxalate, the suspension was kept under stirring for 1 h in the dark to establish the adsorption–desorption equilibrium. The suspension was then irradiated with visible light ($\lambda \geq 400$ nm). The experiment was carried out under nitrogen atmosphere. Aliquots of 3 mL were taken out every 30 min and centrifuged to remove the solid catalyst particles. Then, the clear solution was analyzed by UV–vis spectroscopy. The 4-NA and PPD concentrations as function of time were determined by measuring the absorbance at 380 nm and 238 nm. The absorbance was converted to concentration using Beer–Lambert law. The photolysis of 4-NA in the presence of hole scavenger ammonium oxalate was also conducted without photocatalyst under visible light irradiation to determine self-induced photodegradation.

3. Results and Discussion

3.1 Characterization:

The phase composition and crystallinity of as-prepared $BiNbO_4$ were analyzed using PXRD.

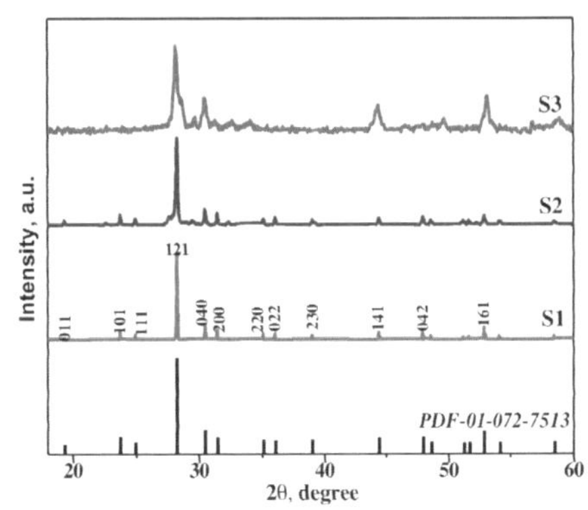

Figure 2: Powder X-ray diffraction analysis of $BiNbO_4$.

Journal of Physics: Conference Series **758** (2016) 012017 doi:10.1088/1742-6596/758/1/012017

Figure 2 shows the PXRD patterns of $BiNbO_4$ catalysts synthesized by different methods. The major peaks have been indexed in the orthorhombic phase of $BiNbO_4$ in the *Pnma* space group (PDF-01-072-7513). The main diffraction peaks in all samples occur at almost the same 2θ positions. It was reported that $BiNbO_4$ synthesized at temperatures between 750-1030 °C crystallizes in orthorhombic phase. When heated at temperatures higher than 1030 °C, the orthorhombic phase converts to the triclinic phase [40,43]. In the present study, $BiNbO_4$ was synthesized at temperatures less than 900 °C, which suggests that all samples must correspond to orthorhombic phase. The PXRD of the compound prepared by the hydrothermal method (S3) shows mostly broad diffraction peaks, suggesting less crystallinity and smaller particle size. A summary of refined lattice parameters along with synthetic methods is provided in Table 1.

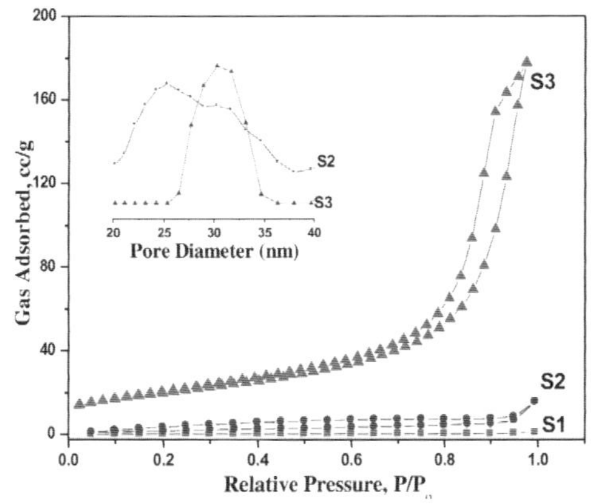

Figure 3: Nitrogen adsorption-desorption isotherms for S1, S2 and S3. The inset picture shows pore size distribution curves for S2 and S3.

Table 1: Textural properties of S1, S2 and S3.

Sample	Synthesis Method	BET Surface Area (m^2g^{-1})	Average Pore Size (nm)	Average Particle Size (nm)	Refined Lattice Parameters (Å)
S1	Solid-State Reaction	0.63	-	>1000	a = 5.6821(2) b = 11.7173(4) c = 4.9558(8)
S2	Solution Method with P123	5.8	>100	114	a = 5.649(5) b = 11.5182(2) c = 4.912(4)
S3	Hydrothermal Method	69.85	37	23	a = 5.655(4) b = 11.6769(3) c = 4.97113(7)

The nitrogen adsorption–desorption isotherms were studied to evaluate the surface area and pore size of all $BiNbO_4$ samples. Typical isotherms for each sample are provided in Figure 3. The surface area and pore size of different samples are provided in Table 1. Typical adsorption-desorption isotherm for sample S3 is of Type IV indicating the mesoporous nature of the sample. In addition, when compared to S1 and S2, the isotherms for S3 show higher absorption at relatively higher pressure ranges ($P/P_0 \sim 1$) suggesting the formation of large meso-pores. On the other hand, the isotherms indicate that the amount adsorbed by the different samples at low P/P_0 (P/P_0<0.8) decreases from S3 to S1 at, indicating that pore size increases and specific surface area decreases. The pore size distribution curves for S2 and S3 are shown in the inset in Figure 3. It indicates that sample S3 mainly consists of meso-pores with a peak pore diameter of ca. 37 nm.

To further examine the nature of the particles in each sample, the morphology and microstructure were investigated by electron microscopy. SEM and TEM images of the sample synthesized by solid-state reaction (S1) clearly showed large irregularly shaped particles with sizes in the micrometer range (Figure 4a and d). Similarly, particles size for sample 2 (S2, Figure 4b and d) varies from 90 nm to 138 nm (with average size ~114 nm). The larger particles in samples S1 and S2 can be explained on the basis of agglomeration of particles due to the high temperature used in the synthesis of these samples. In contrast the sample synthesized by the hydrothermal method (S3, Figure 4c and f) consists of smooth, spherically shaped particles with average size of ~23 nm.

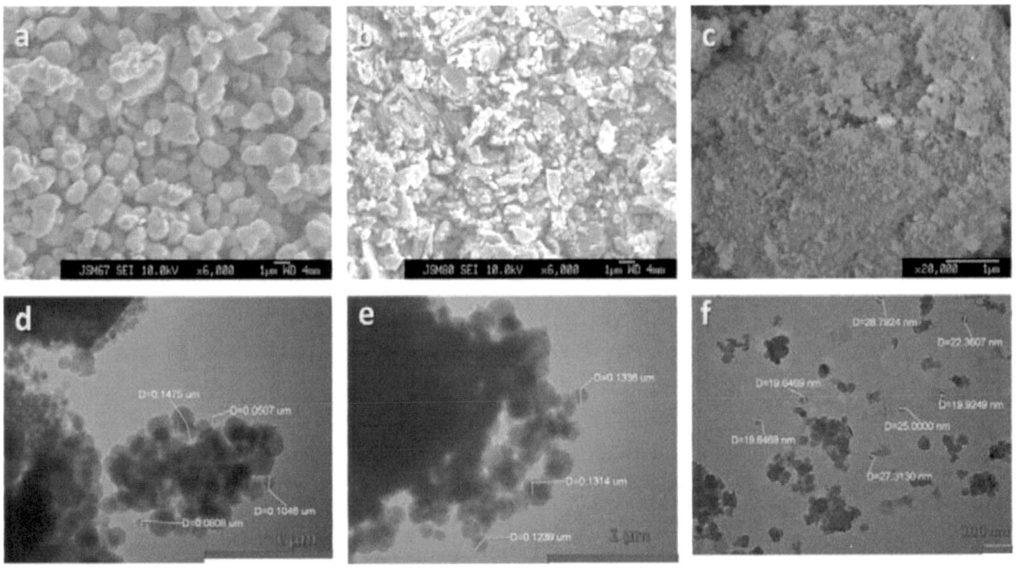

Figure 4: SEM (a, b and c) and TEM (d, e and f) images for $BiNbO_4$ powders from different synthetic methods.

Figure 5 shows the TGA data of different samples of $BiNbO_4$ synthesized by wet chemistry routes. The major weight loss (~ 75%) observed for sample 2 (S2) around 300 °C may correspond to the loss of carbon from P123 surfactant. The combination of TGA and PXRD study of S2 clearly show that the crystalline phase of $BiNbO_4$ is formed only after 700 °C. In contrast, TGA of S3 shows negligible weight loss suggesting the presence of pure crystalline phase after the hydrothermal treatment at 140 °C.

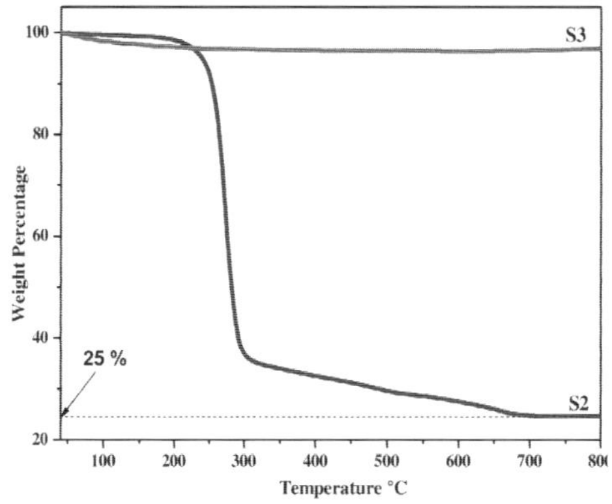

Figure 5: Thermogravimetric analysis of different samples of $BiNbO_4$ synthesized by wet chemistry route.

Diffuse reflectance UV-vis spectra (Figure 6) show photo-absorption at wavelength shorter than 425 nm, 420 nm, and 414 nm for S1, S2 and S3, respectively. The UV–Vis absorption spectra were transformed from the corresponding diffuse spectra according to the Kubelka–Munk theory to obtain the corresponding band gaps as 2.91 eV, 2.95 eV, and 2.99 eV for S1, S2 and S3, respectively [17]. These results demonstrate that all catalysts in this study may be active in the visible light region ($\lambda \geq 400$ nm).

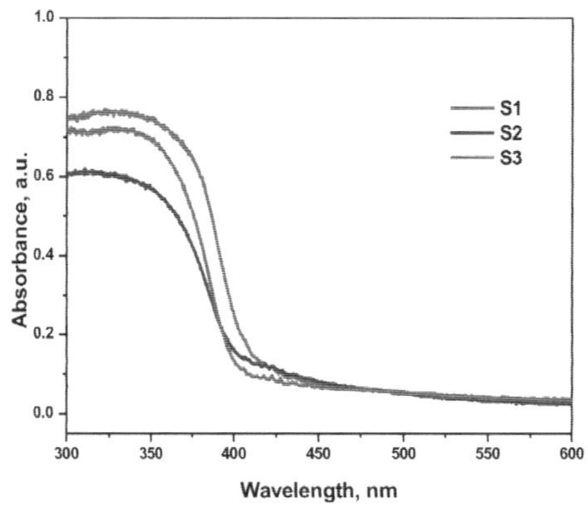

Figure 6: UV-vis diffuse reflectance spectra of different $BiNbO_4$ samples

3.2. Photocatalytic test

Photocatalytic activities of all samples were evaluated under visible light illumination ($\lambda \geq 400$ nm) using the photocatalytic reduction of 4-NA. For all samples, a decrease of 4-NA absorption intensity at wavelength around 380 nm is observed. As the absorption due to 4-NA decreases, absorptions at 238 and 305 nm corresponding to p-Phenyledene-diamine (PPD) increases. Figure 7 shows the photocatalytic degradation of 4-NA using different catalysts under visible light irradiation. The photocatalytic decomposition of 4-NA with time depends on the catalyst used. Control experiment (4-NA and $(NH_4)_2C_2O_4$) without catalysts exhibits negligible photocatalytic reduction of 4-NA, which indicates that photolysis of 4-NA can be ignored. The catalysts with higher surface area and small particle size were more catalytically active. Thus S3 reduced all of the 4-NA within 4 hours of visible light irradiation, while only 50 % and 30 % of 4-NA are decomposed by S2 and S1 respectively after the same irradiation time. To check the recyclability of the best active catalyst (S3), we exchanged the solution (after four hours of photocatalytic reaction) with fresh 4-NA solution and performed the similar photocatalytic experiment for the next 4 hours. It has been found that the degradation efficiency of 4-NA is similar for the second run too. This confirms that those catalysts are very recyclable.

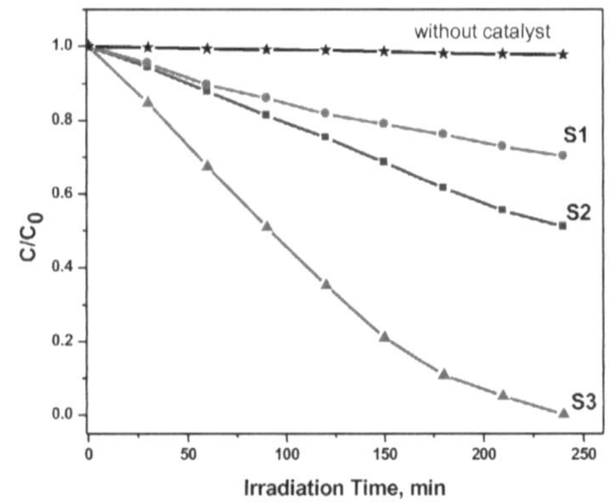

Figure 7: Photoreduction efficiencies of 4-NA as a function of irradiation time by different samples. The absorbance intensity at around 380 nm was used to monitor the concentration of 4-NA. C_o is the initial concentration of 4-NA, C is the concentration of 4-NA at time t.

The photocatalytic activity results are further supported by photoluminescence (PL) experiments by considering the recombination of photogenerated electron-hole pairs. High photoluminescence intensity generally indicates a high recombination rate of photogenerated charge carriers leading to low efficiency of a given catalyst [16,17,44,45]. Figure 8 shows the PL spectra for different $BiNbO_4$ samples with an excitation wavelength of 300 nm. The main emission peak is centered at about 325-440 nm for all samples. The PL emission intensity of S3 is significantly smaller than that of S2 and S1, suggesting lower recombination rate of photogenerated charge carriers in S3, which may explain enhanced photocatalytic performance of S3 compared to S1 and S2.

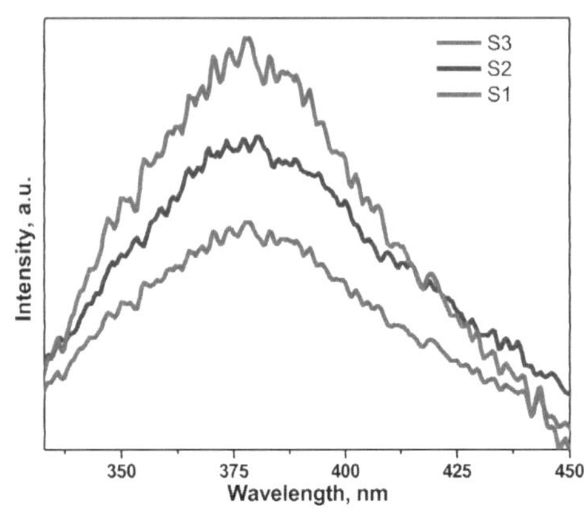

Figure 8: Photoluminescence (PL) spectra for different $BiNbO_4$ samples.

In the photocatalytic processes, a series of photo-induced reactive species, including h^+, $^{\bullet}OH$ or $^{\bullet}O_2^-$, take part in the redox process after the electron–hole pairs are generated. In the degradation of 4-NA, the generation of the superoxide radical ($^{\bullet}O_2^-$) is unlikely because all photocatalytic activity processes were conducted under nitrogen flow. Hence, the photogenerated holes (h^+) or $^{\bullet}OH$ are solely responsible for 4-

NA degradation. The holes (h^+) can produce active carbon-dioxide anion radial ($^{\bullet}CO_2^-$) in the presence of oxalate ($C_2O_4^{2-}$) ion. These $^{\bullet}CO_2^-$ radical anions have a strong reducing ability [46–48].

The photocatalytic degradation of 4-NA was found to follow pseudo-first order kinetics, and can be described by Equation 1.

$$\ln (C_o/C) = -kt \qquad (1)$$

where C_o is the initial concentration of 4-NA, C is the concentration of 4-NA at time t, and k is the rate constant. The corresponding rate constants (k) were calculated and are reported in Table 2. Under similar experimental conditions, the rate constant of 4-NA degradation with S3 was determined to be 3 times larger than that of S2 and 5 times larger than that of S1.

Table 2: Pseudo first order rate constants for photocatalytic degradation process of 4-NA under visible light irradiation (λ ≥ 400 nm).

Sample	**4-NA degradation Rate constants (min^{-1})**
S1	0.020
S2	0.032
S3	0.098

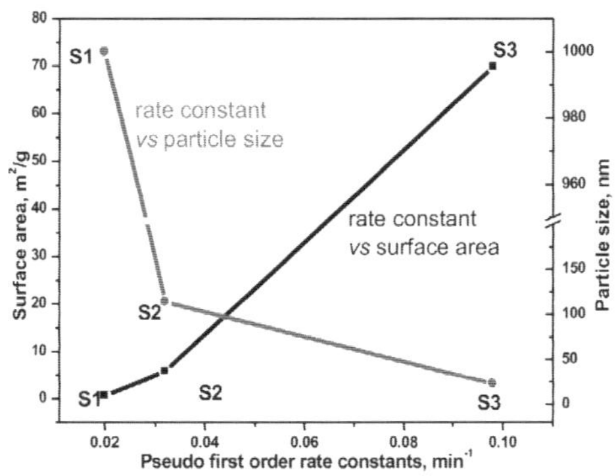

Figure 9: Activity as a function of surface area and particle size

The photocatalytic activity of $BiNbO_4$ depends on particle size and surface areas as shown in Figure 9. The rate constants of the photocatalytic reduction of 4-NA increases with increasing surface area and decreased with increasing with particle size. S3, which was prepared by hydrothermal method, has the largest surface area (69.85 m^2/g) and smallest particle size (23 nm) shows the highest catalytic activity. In contrast S1, which was obtained by high temperature solid state reaction, has the smallest surface area and largest particle size showed the lowest activity. It can be concluded that the size of the particles of $BiNbO_4$ can affect its photocatalytic efficiency. These results are consistent with previous experimental studies with other type of catalysts [26,49–51]. This may be explained by higher density of active sites as the surface area increases. Moreover, the photogenerated carriers in smaller particles are also likely to transfer more quickly to the sample surface without recombination [24,51].

4. Conclusion

In this work, different bismuth niobate samples with particle sizes ranging from 23-1000 nm were successfully prepared by using different synthetic methods. The photocatalytic activity of each sample was evaluated using the reduction of 4-nitroaniline as model system. The relationship between particle size/ surface area and photocatalytic activity were determined. Samples with smaller particles obtained through hydrothermal methodology have been found to exhibit the highest photocatalytic activity for the degradation of 4-NA under visible-light irradiation. The current study enables an in-depth understanding of the structure–performance relationship of bismuth niobate photocatalyst, which can be applied to other photocatalyst systems.

Acknowledgements

Financial support for this work was provided by the WFU Science Research Fund. The authors would like to thank Dr. Scot Geyer, Wake Forest University, Department of Chemistry, for his help with UV-vis DRS data. The help from Dr. Cynthia Day, Dr. Michael Gross, Zachary Hood, and Baxter Mcguirt is also acknowledged.

References

[1] Maeda K 2011 Photocatalytic water splitting using semiconductor particles: History and recent developments *J. Photochem. Photobiol. C Photochem. Rev.* **12** 237–68

[2] Fujishima A and Honda K 1972 Electrochemical Photolysis of Water at a Semiconductor Electrode *Nature* **238** 37–8

[3] Osterloh F E 2013 Inorganic nanostructures for photoelectrochemical and photocatalytic water splitting *Chem Soc Rev* **42** 2294–320

[4] Ni M, Leung M K H, Leung D Y C and Sumathy K 2007 A review and recent developments in photocatalytic water-splitting using for hydrogen production *Renew. Sustain. Energy Rev.* **11** 401–25

[5] Asahi R, Morikawa T, Ohwaki T, Aoki K and Taga Y 2001 Visible-Light Photocatalysis in Nitrogen-Doped Titanium Oxides *Science* **293** 269–71

[6] Kanhere P and Chen Z 2014 A Review on Visible Light Active Perovskite-Based Photocatalysts *Molecules* **19** 19995–20022

[7] Liao C-H, Huang C-W and Wu J C S 2012 Hydrogen Production from Semiconductor-based Photocatalysis via Water Splitting *Catalysts* **2** 490–516

[8] Zhang H, Zong R and Zhu Y 2009 Photocorrosion Inhibition and Photoactivity Enhancement for Zinc Oxide via Hybridization with Monolayer Polyaniline *J. Phys. Chem. C* **113** 4605–11

[9] Pelaez M, Nolan N T, Pillai S C, Seery M K, Falaras P, Kontos A G, Dunlop P S M, Hamilton J W J, Byrne J A, O'Shea K, Entezari M H and Dionysiou D D 2012 A review on the visible light active titanium dioxide photocatalysts for environmental applications *Appl. Catal. B Environ.* **125** 331–49

[10] Takata T, Hitoki G, Kondo J N, Hara M, Kobayashi H and Domen K 2007 Visible-light-driven photocatalytic behavior of tantalum-oxynitride and nitride *Res. Chem. Intermed.* **33** 13–25

[11] Kudo A and Miseki Y 2009 Heterogeneous photocatalyst materials for water splitting *Chem Soc Rev* **38** 253–78

[12] Ibhadon A O and Fitzpatrick P 2013 Heterogeneous Photocatalysis: Recent Advances and Applications *Catalysts* **3** 189–218

[13] Marschall R 2014 Semiconductor Composites: Strategies for Enhancing Charge Carrier Separation to Improve Photocatalytic Activity *Adv. Funct. Mater.* **24** 2421–40

[14] Wang H, Zhang L, Chen Z, Hu J, Li S, Wang Z, Liu J and Wang X 2014 Semiconductor heterojunction photocatalysts: design, construction, and photocatalytic performances *Chem Soc Rev* **43** 5234–44

[15] Tahir M and Amin N S 2013 Advances in visible light responsive titanium oxide-based photocatalysts for CO2 conversion to hydrocarbon fuels *Energy Convers. Manag.* **76** 194–214

[16] Adhikari S P, Dean H, Hood Z D, Peng R, More K L, Ivanov I, Wu Z and Lachgar A 2015 Visible-light-driven Bi2O3/WO3 composites with enhanced photocatalytic activity *RSC Adv* **5** 91094–102

[17] Adhikari S P, Hood Z D, More K L, Ivanov I, Zhang L, Gross M and Lachgar A 2015 Visible light assisted photocatalytic hydrogen generation by Ta2O5/Bi2O3, TaON/Bi2O3, and Ta3N5/Bi2O3 composites *RSC Adv* **5** 54998–5005

[18] Abe R 2010 Recent progress on photocatalytic and photoelectrochemical water splitting under visible light irradiation *J. Photochem. Photobiol. C Photochem. Rev.* **11** 179–209

[19] Rajeshwar K 2011 Solar Energy Conversion and Environmental Remediation Using Inorganic Semiconductor–Liquid Interfaces: The Road Traveled and the Way Forward *J. Phys. Chem. Lett.* **2** 1301–9

[20] Zaleska A 2008 Doped-TiO2: A Review *Recent Pat. Eng.* **2** 157–64

[21] Hagfeldt A and Graetzel M 1995 Light-Induced Redox Reactions in Nanocrystalline Systems *Chem. Rev.* **95** 49–68

[22] Amano F, Ishinaga E and Yamakata A 2013 Effect of Particle Size on the Photocatalytic Activity of WO3 Particles for Water Oxidation *J. Phys. Chem. C* **117** 22584–90

[23] Xu N, Shi Z, Fan Y, Dong J, Shi J and Hu M Z-C 1999 Effects of Particle Size of TiO2 on Photocatalytic Degradation of Methylene Blue in Aqueous Suspensions *Ind. Eng. Chem. Res.* **38** 373–9

[24] Hagfeldt A and Graetzel M 1995 Light-Induced Redox Reactions in Nanocrystalline Systems *Chem. Rev.* **95** 49–68

[25] Colmenares J C, Luque R, Campelo J M, Colmenares F, Karpiński Z and Romero A A 2009 Nanostructured Photocatalysts and Their Applications in the Photocatalytic Transformation of Lignocellulosic Biomass: An Overview *Materials* **2** 2228

[26] Zhang Z, Wang C-C, Zakaria R and Ying J Y 1998 Role of Particle Size in Nanocrystalline TiO2-Based Photocatalysts *J. Phys. Chem. B* **102** 10871–8

[27] Zou Z, Ye J and Arakawa H 2001 Optical and structural properties of the BiTa1−xNbxO4 ($0 \leqq x \leqq 1$) compounds *Solid State Commun.* **119** 471–5

[28] Zou Z, Ye J and Arakawa H 2003 Photocatalytic water splitting into H2 and/or O2 under UV and visible light irradiation with a semiconductor photocatalyst *CANCUN-01* **28** 663–9

[29] Zhai H-F, Li A-D, Kong J-Z, Li X-F, Zhao J, Guo B-L, Yin J, Li Z-S and Wu D 2013 Preparation and visible-light photocatalytic properties of BiNbO4 and BiTaO4 by a citrate method *J. Solid State Chem.* **202** 6–14

[30] Nisar J, Wang B C, Pathak B, Kang T W and Ahuja R 2011 Mo- and N-doped BiNbO4 for photocatalysis applications *Appl. Phys. Lett.* **99**

[31] Wang B C, Nisar J, Pathak B, Kang T W and Ahuja R 2012 Band gap engineering in BiNbO4 for visible-light photocatalysis *Appl. Phys. Lett.* **100**

[32] Muktha B, Darriet J, Madras G and Guru Row T N 2006 Crystal structures and photocatalysis of the triclinic polymorphs of BiNbO4 and BiTaO4 *J. Solid State Chem.* **179** 3919–25

[33] Ding K, Chen B, Li Y, Zhang Y and Chen Z 2014 Comparative density functional theory study on the electronic and optical properties of BiMO4 (M = V, Nb, Ta) *J Mater Chem A* **2** 8294–303

[34] Zou Z, Ye J, Sayama K and Arakawa H 2001 Photocatalytic and photophysical properties of a novel series of solid photocatalysts, BiTa1−xNbxO4 ($0 \leqslant x \leqslant 1$) *Chem. Phys. Lett.* **343** 303–8

[35] Wang L, Wang W, Shang M, Sun S, Yin W, Ren J and Zhou J 2010 Visible light responsive bismuth niobate photocatalyst: enhanced contaminant degradation and hydrogen generation *J Mater Chem* **20** 8405–10

[36] Ullah R, Ang H M, Tadé M O and Wang S 2012 Synthesis of doped BiNbO4 photocatalysts for removal of gaseous volatile organic compounds with artificial sunlight *Chem. Eng. J.* **185–186** 328–36

[37] Zhang G, Yang J, Zhang S, Xiong Q, Huang B, Wang J and Gong W 2009 Preparation of nanosized Bi3NbO7 and its visible-light photocatalytic property *J. Hazard. Mater.* **172** 986–92

[38] Min Y, Zhang F-J, Zhao W, Zheng F, Chen Y and Zhang Y 2012 Hydrothermal synthesis of nanosized bismuth niobate and enhanced photocatalytic activity by coupling of graphene sheets *Chem. Eng. J.* **209** 215–22

[39] Hou J, Cao R, Wang Z, Jiao S and Zhu H 2012 Hierarchical nitrogen doped bismuth niobate architectures: Controllable synthesis and excellent photocatalytic activity *J. Hazard. Mater.* **217–218** 177–86

[40] Dunkle S S and Suslick K S 2009 Photodegradation of BiNbO4 Powder during Photocatalytic Reactions *J. Phys. Chem. C* **113** 10341–5

[41] Yang P, Zhao D, Margolese D I, Chmelka B F and Stucky G D 1998 Generalized syntheses of large-pore mesoporous metal oxides with semicrystalline frameworks *Nature* **396** 152–5

[42] Gautam S, Kamble S P, Sawant S B and Pangarkar V G 2005 Photocatalytic degradation of 4-nitroaniline using solar and artificial UV radiation *Chem. Eng. J.* **110** 129–37

[43] Zhou D, Wang H, Yao X, Wei X, Xiang F and Pang L 2007 Phase transformation in BiNbO4 ceramics *Appl. Phys. Lett.* **90**

[44] Du Y, Zhao L and Zhang Y 2014 Roles of TaON and Ta3N5 in the visible-Fenton-like degradation of atrazine *J. Hazard. Mater.* **267** 55–61

[45] Mayer M T, Du C and Wang D 2012 Hematite/Si Nanowire Dual-Absorber System for Photoelectrochemical Water Splitting at Low Applied Potentials *J. Am. Chem. Soc.* **134** 12406–9

[46] Wu W, Lin R, Shen L, Liang R, Yuan R and Wu L 2013 Visible-light-induced photocatalytic hydrogenation of 4-nitroaniline over In2S3 photocatalyst in water *Catal. Commun.* **40** 1–4

[47] Imamura K, Iwasaki S, Maeda T, Hashimoto K, Ohtani B and Kominami H 2011 Photocatalytic reduction of nitrobenzenes to aminobenzenes in aqueous suspensions of titanium(iv) oxide in the presence of hole scavengers under deaerated and aerated conditions *Phys Chem Chem Phys* **13** 5114–9

[48] Wu W, Liu G, Liang S, Chen Y, Shen L, Zheng H, Yuan R, Hou Y and Wu L 2012 Efficient visible-light-induced photocatalytic reduction of 4-nitroaniline to p-phenylenediamine over nanocrystalline PbBi2Nb2O9 *J. Catal.* **290** 13–7

[49] Dodd A C, McKinley A J, Saunders M and Tsuzuki T 2006 Effect of Particle Size on the Photocatalytic Activity of Nanoparticulate Zinc Oxide *J. Nanoparticle Res.* **8** 43–51

[50] Amano F, Nogami K, Tanaka M and Ohtani B 2010 Correlation between Surface Area and Photocatalytic Activity for Acetaldehyde Decomposition over Bismuth Tungstate Particles with a Hierarchical Structure *Langmuir* **26** 7174–80

[51] Li X, Li Q and Wang L 2013 The effects of NaNbO3 particle size on the photocatalytic activity for 2-propanol photodegradation *Phys Chem Chem Phys* **15** 14282–9

Synthesis and Characterization of Luminescent Amorphous Porous Silicon (ap-Si) Nanoparticles via unconventional Stain Etching

M R.Tchalala*, J K El-Demellawi*, A J Mughal and S Chaieb[1]
Division of Physical Sciences and Engineering, King Abdullah University of Science and Technology, 23955 Thuwal, KSA

E-mail: mohammed.tchalala@kaust.edu.sa, jehad.eldemellawi@kaust.edu.sa, sahraoui.chaieb@kaust.edu.sa

Abstract. Starting from crystalline silicon we synthesised bright suspensions of amorphous porous silicon nanoparticles through unconventional stain etching. Upon excitation with UV light, this novel nanostructured material gives rise to an intense red photoluminescence (PL) which resembles that of some silicon nanostructures. We studied the properties of the prepared nanoparticles using a number of cutting-edge characterization techniques such as TEM, SEM and EDX. The complete crystalline-to-amorphous phase transition, confirmed by the morphological studies, seems fortuitous.

1. Introduction
Taking into account its abundance (second most abundant material on earth) and prominence (most widely used semiconductor in electronic and optoelectronic applications) [1-9], silicon in many respects is one of the most important elements in both nature and chemistry. It has however poor optical qualities. The indirect band gap of bulk crystalline silicon used in the semiconductor industry is too low (1.1 eV) to be used in optical applications [10]. Nonetheless, the desire for integration of optoelectronic devices with silicon microelectronics has stimulated an intense effort to search Si-based materials that emit light efficiently [11]. As they approach the nanometer scale, silicon structures, like those of other elements, exhibit unique optical, electronic and physical properties [12-14]. Their properties are interesting from both a fundamental perspective and a practical viewpoint [13]. In the early 1990s, nanostructured silicon, also known as porous silicon (p-Si), was reported to exhibit a surprising visible emission when excited with UV light. Ever since this remarkable discovery by Canham [15], p-Si spurred a great deal of interest among the science community, particularly for its staggering optical properties [15].

Even though the mechanism for its light emission remains rather controversial, fascination towards porous silicon continues to grow. Given its compatibility with the existing electronic platform and their unique optical properties, p-Si has been extensively studied as a promising optoelectronic material and a

1 mohammed.tchalala@kaust.edu.sa

more efficient photovoltaic material. Furthermore, the ability to select the emission wavelength i.e. red, green and blue emission, of p-Si by adjusting the erosion process of porous silicon [16-19], is appealing for many applications.

In this work, we will review the fabrication as well as the obtained properties of silicon nanostructures derived from p-Si films via an optimized chemical etching process.

2. Experimental

Because of the air- and moisture-sensitive chemicals, we performed all experiments at room temperature under controlled atmosphere. All sample handling was performed in an argon-filled glovebox (MBraun) to avoid sample contamination and to prevent further oxidation.

All chemical reagents were used as received without further purification. For amorphous porous silicon formation, hydrofluoric acid (HF, 49%, Sigma-Aldrich), hydrochloric acid (HCl, 35%, Sigma-Aldrich) and iron (III) chloride ($FeCl_3$, 98%, Alfa Aeser) were used.

Colloidal silicon nanoparticles (SiNPs) were fabricated through electroless stain etching process. Single crystal silicon wafers (boron-doped, (100), 1-10 Ω.cm, Addison electronics) were cleaved into 2.5 x 4.5 cm strips and immersed in a 50:1 solution of deionized water and hydrofluoric acid (49%, Fluka) for 3-4 min for native oxide removal (Scheme 1*). Cleaned strips were then submerged into an etching solution composed of HF, HCl and $FeCl_3$, for 13 hours (Scheme 1**). This solution creates an electroless chemical reaction in which the metal ion oxidant (Fe^{3+}) injects as a hole into the valence band of the silicon atoms, which initiates the reaction and stimulates p-Si formation [20]. The addition of HCl produced much stable and reproducible samples than etchants composed of just Fe^{3+} and HF. Following etching, wafer strips were washed out through brief dipping in deionized water followed by rinsing in absolute toluene then dried under N_2 stream (Scheme 1***). In last, the NPs were dispersed in an organic solvent (toluene, Chromasolv grade, Sigma-Alrich) after breaking-up the formed porous film by sonicating the strips for 45 min.

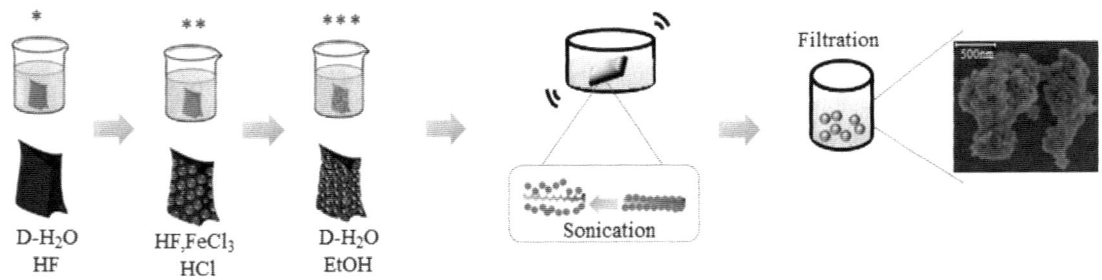

Scheme 1. Schematic diagram for the synthesis route of amorphous porous silicon nanoparticles. The SEM micrographs is for one nanoparticle after ultra-sonication.

The surface morphology of ap-Si nanoparticles was investigated by scanning electron microscope (SEM) and transmission electron microscopy (TEM). SEM was performed in an FEI Nova Nano-SEM 50, at voltage 2 keV. TEM were recorded using FEI Krios CT-TEM at an operating voltage of 120 kV. Fluorescence spectra were obtained using a Horiba Fluoromax- 4 spectro-fluorometer. Absorbance measurements were carried out using a Genysis 10S UV-Vis spectrometer (Thermos-Scientific). We conducted Brunauer-Emmet-Teller (BET) porosimetry measurements to measure the surface area and pore size distributions of the samples, using an ASAP 2420 area and porosimetry system (Micrometrics). Composite homogeneity of the optimized stain etching procedure as well as the chemical analysis of the produced nanoparticles were investigated by energy dispersive X-ray spectroscopy (EDX) in conjunction with SEM. All characterization measurements were performed under standard conditions while maintaining low air exposure.

3. Results and Discussion

A wide spectrum of synthesis methods have been developed for the fabrication of nanostructured silicon. These include chemical stain etching, chemical vapor etching, laser-induced etching, and metal-assisted etching as well as spark processing and reactive ion etching. Each method could produce a nanostructure of specific features [21-25]. Normally anodic etching and stain etching are the most used techniques [26-27]. The setup and the process of anodic etching are however complex and expensive. Meanwhile, stain etching is rapid, simple, and cost-effective with large product yield [28-31].

In this work, starting from p-type crystalline silicon wafers, we used electroless colloidal stain etching (Scheme 2) to synthesize bright red-emitting suspensions of colloidal amorphous porous silicon nanoparticles. This etching process yielded a highly luminescent porous silicon film with a unique dual layer structure.

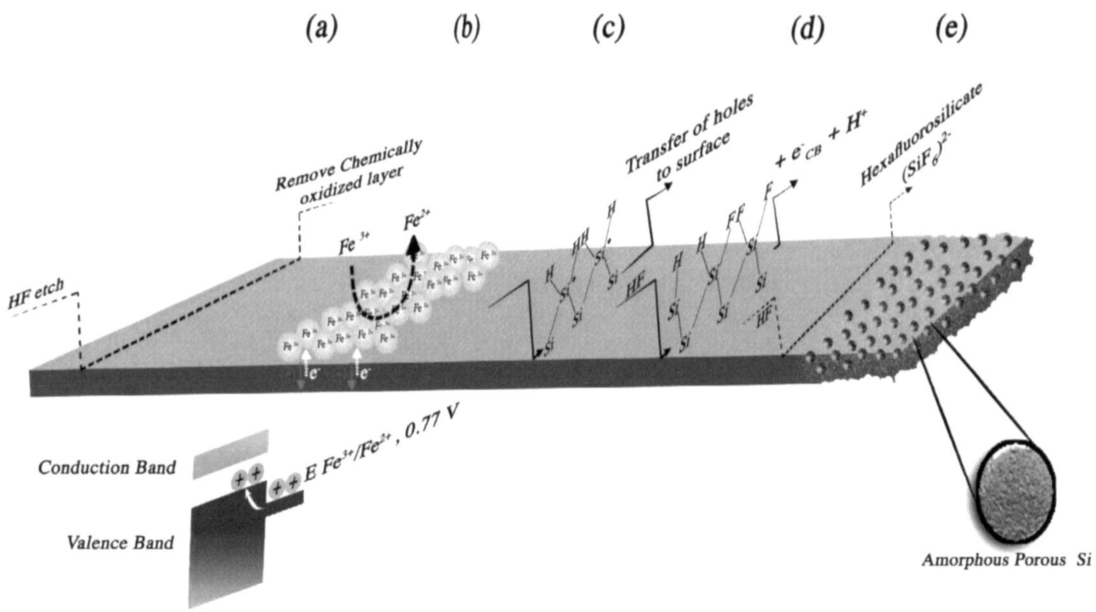

Scheme 2. An illustration of stain etching process using $FeCl_3$: (a) HF etching removes chemically oxidized layer and leave the surface H-terminated, (b) generation of the holes generated during the reduction reaction, (c) electron removal from nearby Si-H bonds, (d) activation of Si-H passivated bonds, (d) etching resulted in the formation of SiF_6^{2-}, and (e) evolution of porous film of amorphous nanostructured silicon.

The morphology of the etched silicon surface was examined by SEM. The interaction between the sample and the impinging electron beam provides insight into the morphology, topography, and composition. Fig. 1 shows that the surface of the etched film is covered with nodules of 10 μm in diameter and 6 μm in height with a gap between them and the bottom surface (shown in Fig. 2). The upper and lower interfaces of the 6 μm thick film were flat and well-defined in shape. The top portion of the film is mainly composed of nodules while the bottom layer consists of a film that seemed to be more adherent to the substrate; moreover, the thermos-capillary forces cracked and partially exfoliated the top layer of the film (see Fig. 2).

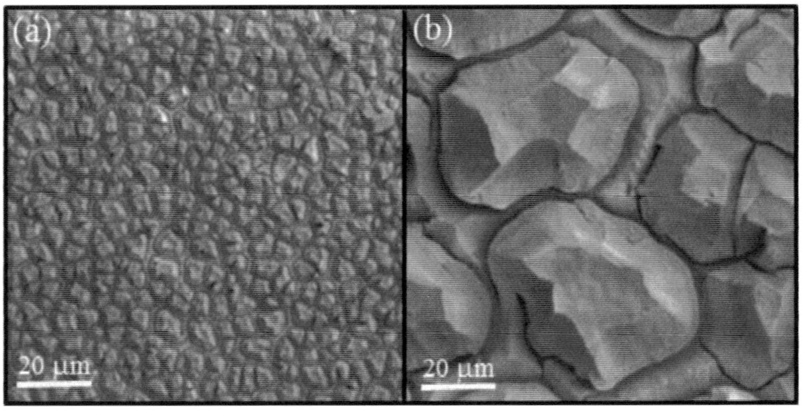

Figure 1. SEM micrographs of the (a) top view of the surface, and the (b) blow-up of the top surface of an etched silicon substrate.

Figure 2. Cross-sectional SEM image for the produced film of p-Si. The exfoliated region exceeded 2μm in thickness whereas the lower layer remained intact and exceeded 6 μm in thickness.

TEM studies carried out on the surface of etched silicon showed that the films are typically characterized by a graded transition from bulk to a shallow high porosity region. It is clear that the surface of the etched silicon becomes porous (see Fig. 3). TEM micrographs, depicted in Fig. 3, confirm that the laterally etched film (stain film) consists of two layers: a fragmented coarse-grained upper layer exhibiting a very high porosity that looks like a damaged surface layer. Whereas the bottom layer shows flat fine uniform texture (Fig. 3c). The film mostly composed of a homogenous distribution of pores. We suggest that the upper layer of the film is mechanically less stable because it has higher porosity than the bottom layer. Diffraction patterns (inset of Fig. 3c) obtained from both films showed them to be amorphous. (The diffraction spots are blurred because the size of the coherent region is smaller compared to single crystal silicon.)

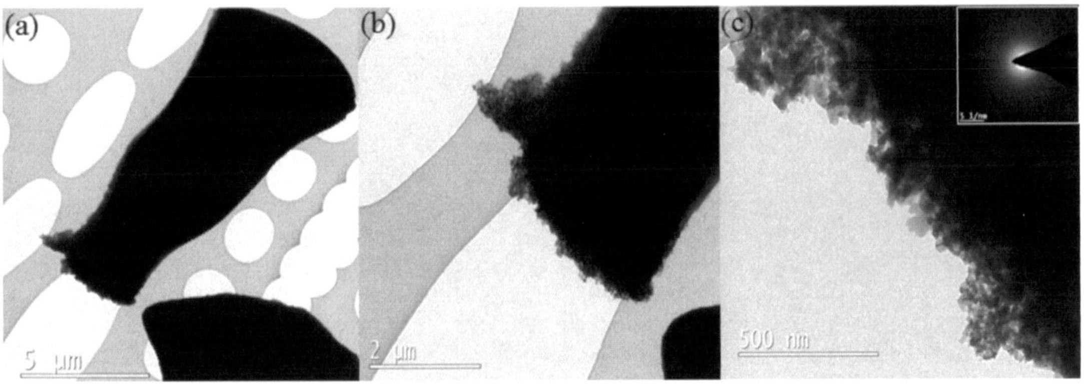

Figure 3. TEM micrographs of porous silicon film etched for 13 hours. (a) The upper and lower interfaces were both sharp. (b) The top layer exfoliated and separated from the bottom part along an irregular crack. (c) TEM image of amorphous porous silicon NPs. FFT pattern is shown in the inset.

The uniform flat films of etched substrates (strips) demonstrated a shiny appearance (Fig. 4b) under UV illumination, before being dried under N_2 stream. During drying the films took on a rough golden brown appearance as seen in Fig. 4c. Such a color change indicates the portion in which the stain etched film is formed, where surface discoloration occurred. The upper portion of the film appeared brighter, indicating a higher degree of charging because of its higher porosity.

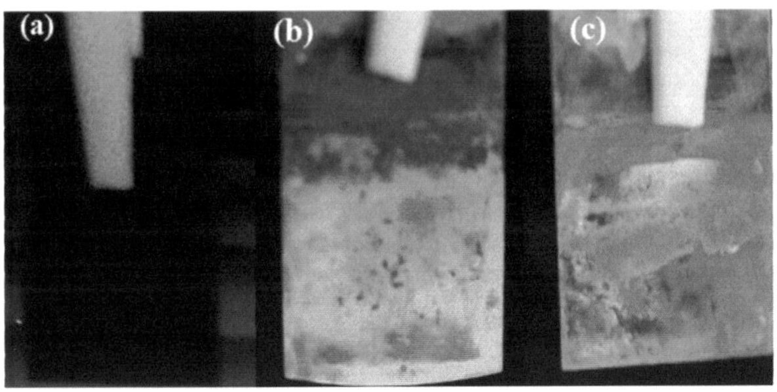

Figure 4. Surface of cleaved silicon strips under UV illumination (a) before etching, (b) after texturing through stain etching and before sonication, and (c) after drying and sonicating the textured strips.

Sonicating the stain etched strips removed the nodules from the surface layer as a consequence of the mechanical interaction of sonication-generated cavitation bubbles with the porous surface layer. The suspended particles were in the order of 10 μm in size as shown in Fig. 5. Fig. 5b shows that the amorphous porous silicon nanoparticles exhibited a cauliflower structure of hundred nanometers in size. These particles were further milled trough additional sonication step. Ultracentrifugation was used to remove the clusters that are larger in size, and that would not remain suspended.

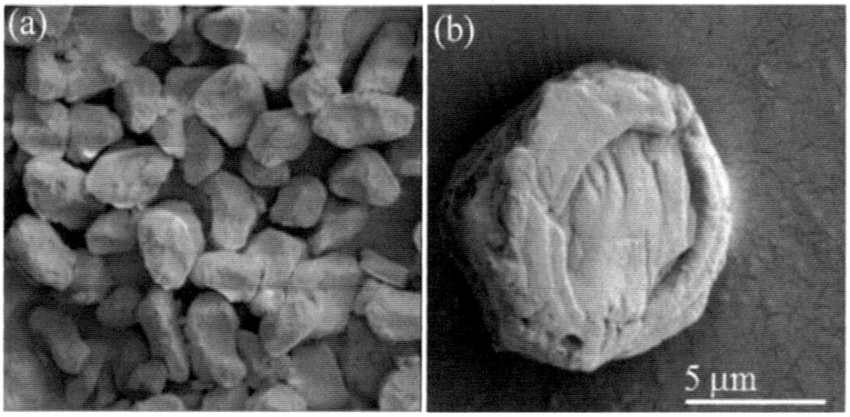

Figure 5. (a) SEM micrograph of $FeCl_3$-assisted etched ap-Si nanoparticles after sonication, and (b) a high magnification SEM image of one nodule.

To quantitatively determine the surface area of our ap-Si NPs, a BET porosimetry measurement were conducted. BET indicated that the ap-Si nanoparticles have a surface area of ~ 300 $m^2.g^{-1}$, with an average pore size of 3.5 nm. This large amount of surface area may be the driving force behind the crystalline-to-amorphous phase transition during the etching process [32]. Gravimetric analysis of the synthesized ap-Si NPs revealed that the average yield was approximately 0.4 mg of ap-Si per cm^2 of the silicon wafer or 10 mg per 10 ml of the solvent (toluene). Dispersing nanoparticles in a non-polar organic medium, such as toluene, is a major challenge as the particles tend to aggregate easily. Toluene, however, does not contain any oxygen that could react with the sample and was therefore used as a dispersing medium to avoid further oxidation.

TEM investigation showed the presence of roughly spherical particles (Fig. 6), which tend to aggregate during preparation of samples for TEM. High-resolution TEM together with the corresponding fast Fourier diffractogram (FFT) of the processed nanoparticles revealed no indication of crystallinity in these nanospheres. They are completely amorphous, and the absence of lattice plane fringes in the former, and the broad diffuse ring in the latter are characteristic signatures of amorphous structures.

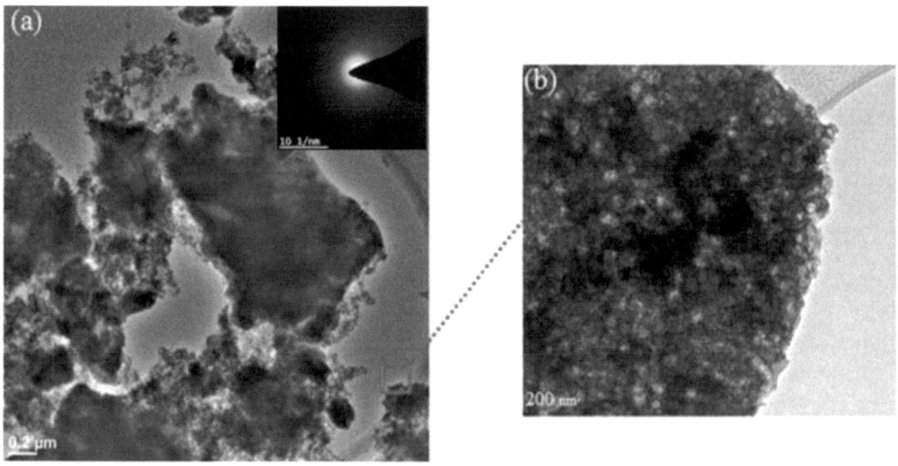

Figure 6. (a) A high-resolution TEM image of the processed ap-Si NPs, and (b) a higher magnification image. The inset is the corresponding diffraction pattern obtained by Fourier transformation showing amorphous structure.

We have performed EDX measurements in conjunction with SEM to analyze the chemical composition of this novel nanostructured material. The result of this study is presented in Fig. 7. From the spectra, it can be seen that silicon is clearly the dominant element in the powder. The small trace of oxygen is attributed to air exposure during sample transport as silicon is readily oxidized. The Cu peak is arising from cupper-made sample holder.

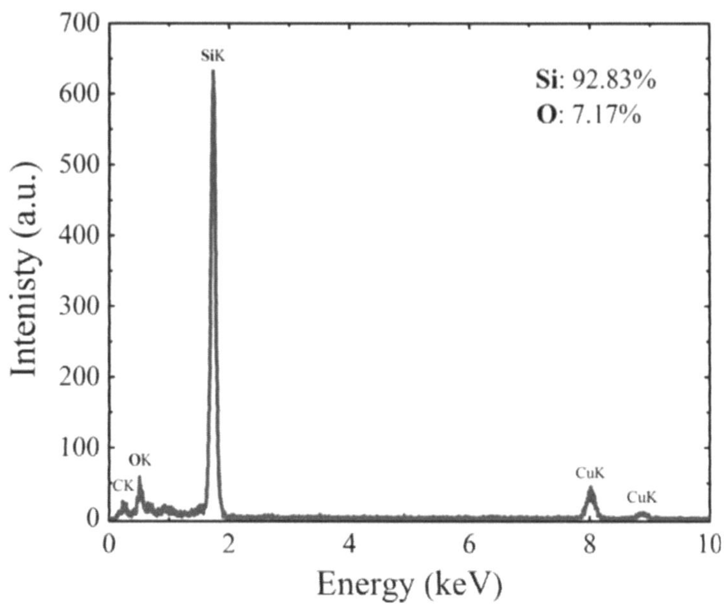

Figure 7. EDX spectrum of amorphous porous silicon nanoparticles.

Journal of Physics: Conference Series **758** (2016) 012018 doi:10.1088/1742-6596/758/1/012018

Fabricated ap-Si nanoparticles exhibited a bright orange-yellow color under white light. This is also considered as an indication of lack of crystallinity, as crystalline silicon has a metallic luster with a grayish color. Qualitatively amorphous porous silicon samples, prepared under sonication, displayed a more uniform luminescence and less pitted porous silicon surface to the naked eye than non-sonicated samples. Fig. 8 shows the UV-Vis absorbance, PL excitation, and PL emission spectra of ap-Si NPs. Our samples exhibit strong UV-Vis absorbance at wavelengths below 400 nm.

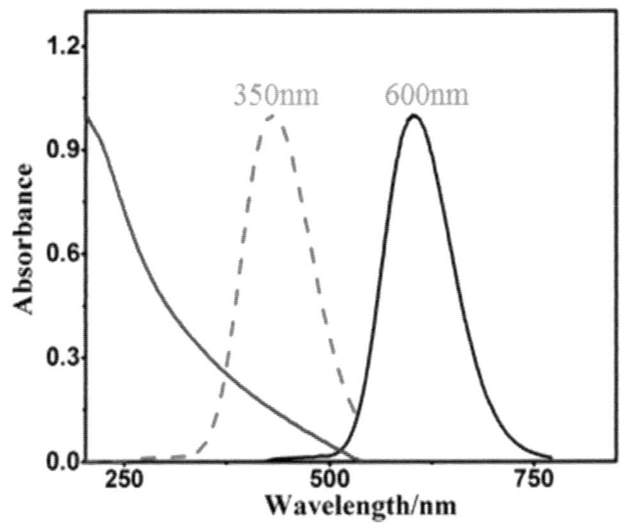

Figure 8. UV-Vis Absorbance (blue), PL excitation (red), and PL emission (black) spectra of ap-Si nanoparticles dispersed in toluene.

The recorded photoluminescence spectra exhibited an excitation peak at 350 nm while the emission peak was centered at 600 nm. It is worth mentioning that while the upper layer is fragile, its PL is not, even after being removed from the substrate. Moreover, samples made up of toluene and kept under N_2 atmosphere were able to maintain their emission for more than one month.

Nevertheless, the etched film and the sonicated one not only exhibited different structures, but they also had different PL behavior [33, 34], as shown in Fig. 9. The shift in PL peak position is mainly due to the difference in the porosity of the top and bottom layers. It could also be a result of interactions between the porous silicon layers and the bulk silicon underneath. It is remarkable that the colloidal nanostructures in the suspension, regardless of their size, produce homogeneous emission.

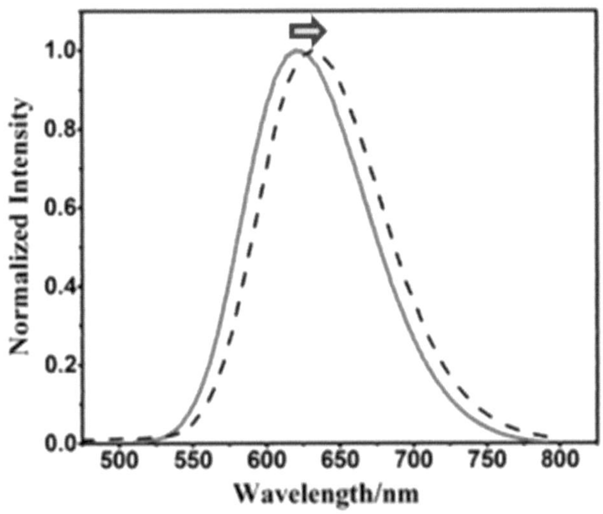

Figure 9. PL emission spectra of ap-Si etched film (black curve peaking at 638 nm) and the underlying substrate from which they were extracted (red curve peaking at 620 nm). The ultra-bright PL emission, seen in and measured for the suspension of amorphous porous nanostructured silicon, renders it an interesting material to study.

4. Conclusion

In conclusion, we succeeded in obtaining high visible photoluminescence from originally non-luminescent crystalline silicon via unconventional stain etching method. This was attained by developing new formulations of stain etchants inducing rapid, efficient and scalable process to fabricate amorphous porous silicon nanoparticles with widely reproducible outcomes. The high visible photoluminescence of the treated nanoparticles is attributed to porosity, and the particular property might be useful for the realization of light-emitting silicon-based optoelectronic devices. The peculiar hollow structure of the ap-Si nanoparticles could also be useful for trapping or storing molecules. It is important to note the passivity of the produced ap-Si nanoparticles as demonstrated by the EDX spectrum, which indicates that the pristine film contained nothing but hydrogen-terminated silicon elements. These colloidal nanostructures can be used in many applications, such as liquid LEDs, liquid filters [35] and solar energy harvesting [1] as well as oxygen detection in solution and on surfaces.

5. References

[1] Canham L 2000 *Nature* **408** 411
[2] Rossi A M and Bohn H G 2005 *Physica Status Solidi (a)* **202** 1644
[3] Ramiro-Manzano F, Fenollosa R, Xifré-Pérez E, Garín M and Meseguer F 2011 *Advanced Materials* **23** 3022
[4] de-Leon S B-T, Oren R, Spira M E, Korbakov N, Yitzchaik S and Saar A 2005 *Physica Status Solidi (a)* **202** 1456
[5] Salonen J, Paski J, Vähä-Heikkilä K, Heikkilä T, Björkqvist M and Lehto V-P 2005 *Physica Status Solidi (a)* **202** 1629
[6] Coffer J L, Whitehead M A, Nagesha D, Mukherjee P, Akkaraju G, Totolici M, Saffie R and Canham L 2005 *Physica Status Solidi (a)* **202** 1451

[7] Litvinenko S, Alekseev S, Lysenko V, Venturello A, Geobaldo F, Gulina L, Kuznetsov G, Tolstoy V, Skryshevsky V, Garrone E and Barbier D 2010 *International Journal of Hydrogen Energy* **35** 6773
[8] Dimova-Malinovska D 2006 *Functional Properties of Nanostructured Materials* 323
[9] Shin H-C, Corno J A, Gole J L and Liu M 2005 *Journal of Power Sources* **139** 314
[10] Al-Khayat W M S and Wilde G 2012 *American Journal of Materials Science* **6** 210
[11] Vij D R 2004 Series on optics and optoelectronics *Handbook of Electroluminescent Materials. Inst. of Physics Publishing Ltd., Bristol and Philadelphia Press* p 394
[12] Alivisatos A P 1996 *Science* **271** 933
[13] Zou J, Sanelle P, Pettigrew K A and Kauzlarich S M 2006 *J. Clust. Sci.* **17** 565
[14] Chopra N, Gavalas V G and Hinds B J 2007 *Anal. Lett.* **40** 2067
[15] Canham L T 1990 *Appl Phys Lett* **57** 1046
[16] Chin R P, Shen Y R and Petrova-Koch V 1995 *Science* **270** 776
[17] Lin V S-Y, Motesharei K, Dancil K-P S, Sailor M J and Reza Ghadiri M 1997 *Science* **278** 840
[18] Li Y Y, Cunin F, Link J R, Gao T, Betts R E, Reiver S-H, Chin V, Bhatia S N and Sailor M J 2003 *Science* **299** 2045
[19] Yae S, Kawamoto Y, Tanaka H, Fukumuro N and Matsuda H 2003 *Electrochem. Commun.* **5** 632
[20] Dudley M E and Kolasinski K W 2009 *Phys. Status Solidi A* **206** 1240
[21] Cullis A G, Canham L T and Calcott P D J 1997 *J Appl Phys* **82** 909
[22] Kalem Ş, Göbelek D, Kurtar R, Misirli Z, Aydinli A and Ellialtioğlu R 1995 *Nanostructured Materials* **6** 847
[23] Vázsonyi É, Szilágyi E, Petrik P, Horváth ZE, Lohner T, Fried M and Jalsovszky G 2001 *Thin Solid Films* **388** 295
[24] Saadoun M, Bessaïs B, Mliki N, Ferid M, Ezzaouia H and Bennaceur R 2003 *Appl Surf Sci* **210** 240
[25] Ben Jaballah A, Saadoun M, Hajji M, Ezzaouia H and Bessaïs B 2004 *Appl Surf Sci* **238** 199
[26] Unagami T 1980 *J. Electrochem. Soc* **127** 476
[27] Kooij E S, K. Butter and Kelly J J 1999 *Electrochem. Solid St. Lett* **2** 178
[28] Mughal A, El-Demellawi J K and Sahraoui C 2014 *Phys. Chem. Chem. Phys* **16** 25273
[29] Li X and Bohna P W 2000 *Appl Phys Lett* **77** 2572
[30] El-Demellawi J K, Anjum D H and Chaieb S 2015 *Mater. Res. Soc. Symp. Proc.* **1748**
[31] Tsujino K and Matsumura M 2005 *Adv. Mater. Weinheim, Ger* **17** 1045
[32] Treacy M M J and Borisenko K B 2012 *Science* **335** 950
[33] Street R A 2000 *Springer Series in Materials Science* **37** *Springer-Verlag Berlin Heidelberg* p. 1
[34] Wolford D, Scott B, Reimer J and Bradley 1983 J *Physica B+C*
[35] Chaieb S and El-Demellawi J 2015 U.S. Patent 2015/0309226 A1

Acknowledgments
We thank King Abdullah University of Science and Technology for financial support.

Synthesis of lithium doped zinc oxide by sol gel

K MEZIANE[1], A ELHICHOU[1], A ELHAMIDI[1], A ALMAGGOUSSI[1] and M CHHIBA[2]

[1]Groupe d'Étude des Matériaux Optoélectroniques (G.E.M.O), Université Cadi Ayyad, Faculté des Sciences et Techniques, BP 549, 40000 Marrakech, Maroc

[2]Laboratoire Rayonnement Matière et Instrumentation, Université Hassan 1, Faculté des Sciences et Techniques, 26000 Settat, Maroc

E-mail: a.elhichou@uca.ma/aelhichou@yahoo.fr

Abstract. Li-doped ZnO thin films were prepared by sol-gel method and deposed on glass substrate using spin coating technique. The effects of Li on structural and optical properties were investigated. The X-ray diffraction (XRD) analysis reveals that Li incorporation leads to the great improvement of the crystalline quality of ZnO thin films. Scanning electron microscopy (SEM) images showed that nanowires are aligned nearly perpendicular to the substrate plane and are affected significantly by Li incorporation. The optical transmission of the films was higher than 80% in the visible region. It is found that the optical gap and the refractive index remain practically constant.

1. Introduction

ZnO is a very promising material for semiconductor device applications [1–2]. It has a direct and wide band gap in the near-UV spectral region [3–4], and a large free-exciton binding energy [3–4] so that excitonic emission processes can persist at, even above, room temperature [4, 5]. It is well known that ZnO crystallizes in the wurtzite structure, the same as for GaN [4]. Its properties have been studied since the early days of semiconductor electronics [6]. But the use of ZnO as a semiconductor in electronic devices has been hindered by the lack of control over its electrical conductivity. ZnO having n-type electrical conductivity, which is due to intrinsic defects such as oxygen vacancies and zinc interstitials.

There are several deposition techniques utilized to prepare ZnO thin films such as chemical vapor deposition [7], magnetron sputtering [8], electrodeposition [9-10], spray pyrolysis [11] and sol-gel [12- 13] method. Among these methods, the sol-gel method process an attractive technique for obtaining thin film, due to controllability of compositions, simple facilities, low cost, etc.

In order to synthesize ZnO for several applications, can be doped with different elements (Na and K) group I for Zn sites, or (N, P and Sb) group V for oxygen sites as substitutional dopants. W. Water et al affirmed that doped Li atoms in ZnO films were involved in their substitution for Zn atom [1].

[1] A. ELHICHOU

The aim of this work is to study the influence of Li doping concentration on the structural and optical properties of zinc oxide films prepared by sol gel method. The films were deposited on glass slides with different Lithium salt concentration in the sol.

2. Experimental

The solution was prepared by zinc acetate dehydrate ($Zn (CH_3COO)_2,2H_2O$) and the 2-methoxy ethanol, the solution was found cloudy which became clear on addition of monoethanol amine (MEA), the molar ratio of MEA to zinc was 1/1 and the concentration of zinc acetate was o,6mol/l, the concentration of nitrate of lithium ($LiNO_3$) as a dopant was of 8, 10, 12, 15 and 20 at. % Li with respect to Zn.
The solution was stirred by a magnetic stirrer at 80C° for 2h to get a clear, homogeneous and transparent sol. The solution was dropped on the glass substrate, which was rotated at 2500 rpm for 30s. After spin coating, the substrates were kept at 120°C for 20min in a tube furnace, to evaporate the solvent. This procedure was repeated 10 times to obtain the desired thickness. The thickness was measured by profilometry and ranged around 250 nm. Finally, all the films were annealed at 500 °C for 1h30min in order to eliminate the organic component in films.
X-ray diffraction (XRD) measurements were realized using a Bragg-Brentano geometry (Bruker D8 Advance, Cu Kαradiation). The morphology of the films was studied using a Field Emission Scanning Electron Microscope (FE-SEM) (JEOL J7600F). The SEM images were acquired with a fixed electron beam at low accelerating voltage (10 kV).
For optical studies, a computer-controlled UV–vis–NIR spectrometer was used in the spectral range 250–2200 nm. Blank glass substrate was used as reference for transmittance measurements.
The thickness of the electroplated LZO film was measured through a profilometer (Taylor Hobson precision Form Talysurf 50 with μltra software), with accuracy on the height of 16nm.

3. Results and discussion

3.1. Structural and morphological characterizations

Figure 1 presents XRD patterns of ZnO thin films obtained by annealing at 500C° for 90 min with different Li-doping concentration (8, 10, 15 and 20 at. % Li). All the films are polycrystalline and have three pronounced diffractive peaks (100), (002) and (101) corresponding to 2Θ values of 31.60°; 34.30° and 36.13° respectively.
For the concentration of 20 at. % Li, the intensity of (002) peak was stronger than of (100) and (101) indicating that the thin films had a high preferential c-axis orientation. No other peaks were detected for all samples; indicating that all precursors were completely decomposed and no other complex product were formed. Furthermore, the X-ray diffraction patterns confirmed that the thin films synthesized in this study are ZnO with typical diffraction peaks of hexagonal (wurtzite) ZnO.

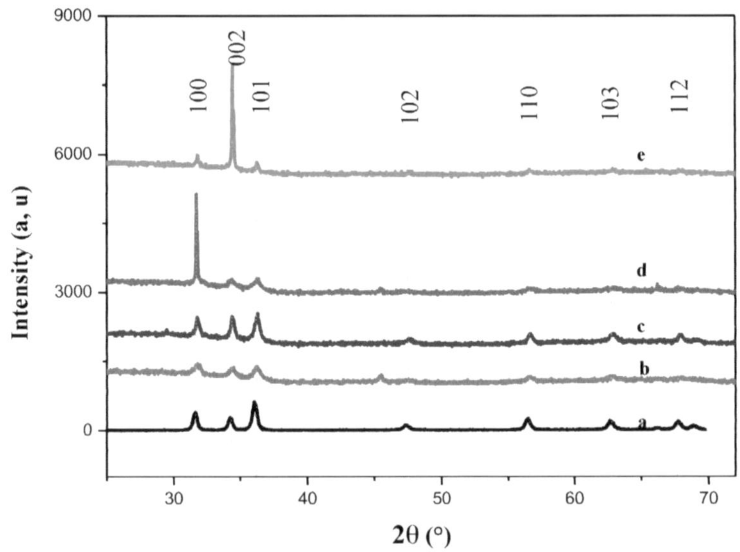

Figure 1. X-ray pattern of Li doped ZnO thin films deposited on glass substrates by sol gel at different Li concentration: (a) 0 at. % Li; (b) 8 at. % Li; (c) 10 at. % Li; (d) 15 at. %Li and (e) 20 at. % Li.

The lattice parameters a and c calculated using Equation (1) [14]:

$$\frac{1}{d^2} = \frac{4}{3}\left(\frac{h^2 + hk + k^2}{a^2}\right) + \frac{l^2}{c^2} \quad (1)$$

Where (hkl) are the plans reticular, and d distance between (hkl) taken from the X'pert Highscore.

It was found that, when the concentration was 15 at. % Li, the lattice parameters decrease with increasing of Li content (Figure 2). A decrease in the lattice parameters can be expected when Zn^{2+} ions are replaced by Li^{+} ions because of the smaller radius of Li^{+} ions (0.06 nm) than Zn^{2+} ions (0.074 nm). It is well known that the doping of semiconductors with electrically active atoms can cause in change of lattice constant could be attributed to Li incorporation and is the indicative of defect evolution in the lattice. This decrease in the lattice parameter for Li-doped ZnO was previously reported by water et al [1].

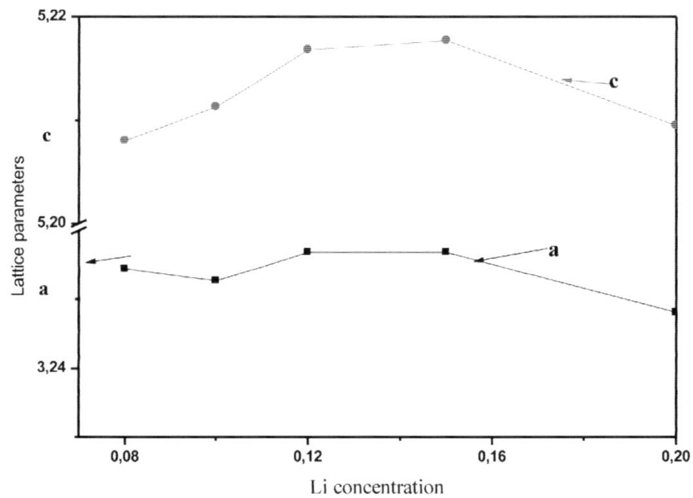

Figure 2. Related plots of the lattice parameters a (Å) and c (Å) for Li doped ZnO films deposited on glass substrates by sol gel at different Li content

The structural parameters such as D size of grains and the texture coefficient $T_{C\ (hkl)}$ were listed in Table1. The size of grains was determined first from XRD spectra, using Scherrer relation (2) [15]:

$$D = \frac{(0{,}9 * \lambda)}{(\beta * cos\theta)} \qquad (2)$$

Where λ is the wavelength, FWHM is the full width at half maximum of the peak and is the Bragg angle. Table 1 shows the variation of sol-grain size. Note that the Li increases is associated to an increase of the crystallite size.

We have studied the influence of Li on the film texture of the plane (002) by systematically calculating the texture coefficient (TC) for different pH according the following formula [16] (3):

$$Tc(002) = \frac{\frac{I(002)}{I_0(002)}}{N^{-1}\sum_n \frac{I(hkl)}{I_0(hkl)}} \qquad (3)$$

Where I(hkl) is the measured relative intensity of a plane (hkl), I_0(hkl) is the standard intensity of the plane (hkl) taken from the JCPDS data (36-1451), n is the reflection number and N is the number of diffraction peaks.

Table 1 shows that the maximum intensity occurs at 20 at. % Li for (002) peak intensity. This is strengthened by the highest value of TC value obtained for the same sample.

Table 1. TC (hkl) values and D size of grain of Li doped ZnO films deposited on glass substrate by sol gel.

	8at.%Li	10at.%Li	15at.%Li	20at.%Li
$Tc_{(100)}$	1.48	1.42	3.99	0.41
$Tc_{(002)}$	1.41	1.64	1.86	4.05
$Tc_{(101)}$	1.88	1.92	0.33	1.95
D(nm)	52.84	62.22	60.39	64.24

Figure 3 shows, a SEM micrograph of the ZnO nanowires grown. It can be seen that the nanowires are aligned nearly perpendicular to the substrate plane. The nanowires are between 20 and 50 nm in diameter and 100 nm in length. According to the figure 1, the high intensity of the ZnO (002) peak at 2Θ=34.3 ° for Li doped ZnO at 20% is due to the strong preferred orientation of the nanorods.

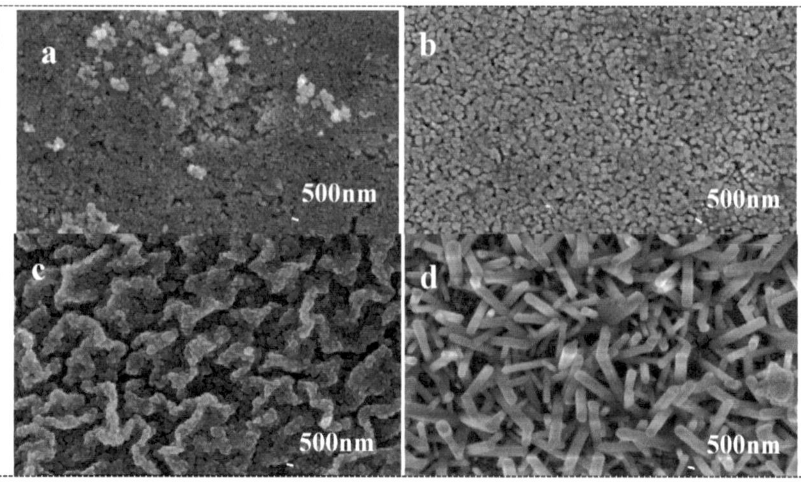

(3)

Figure 3. SEM images of Li-ZnO films deposited on glass substrate by sol gel at different Li concentration: (a) 8 at. % Li; (b) 10 at. % Li; (c) 15 at. % Li and (d) 20 at. % Li.

3.2. Optical properties

The optical transmittance is an important optical parameter for transparent conducting oxides. Optical transmittance spectra of ZnO thin films with different concentration of Li showed that all films are transparent in the visible regions (Figure 4). All films exhibit a high average transmittance (80%) and show a sharp fundamental absorption edge at about 380 nm. One can see that Li-ZnO prepared at 15 and 20 at. % Li show better transmittance than the other ones, in good agreement with what has preceded above.

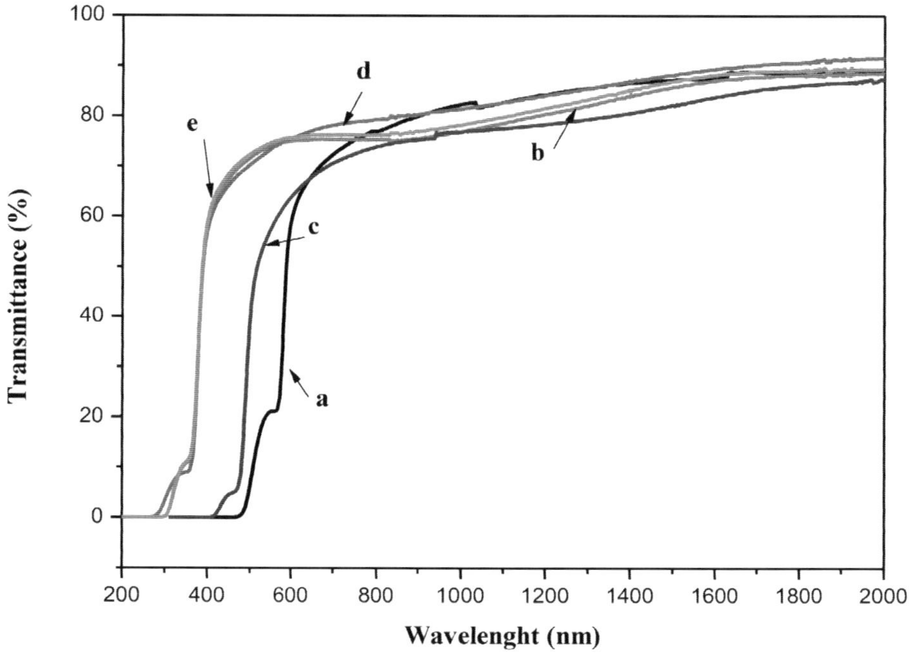

Figure 4. Transmission spectra vs. wavelength for Li doped ZnO films deposited on glass substrate by sol gel at different Li concentration: (a) 0 at. % Li; (b) 8 at. % Li; (c) 10 at. % Li; (d) 15 at. %Li and (e) 20 at. % Li.

The energy band gap (E_g) could be estimated by assuming a direct transition between the valence band (E_v) and the conduction band (E_c) by using the following equation [17]: (4)

$$\alpha . h\nu = A . (h\nu - E_g)^{1/2} \quad (4)$$

Where A is a constant. The optical band gaps were obtained by extrapolating the linear part of curves $(\alpha . h\nu)^2$ as a function of incident photon energy hν to intercept the energy axis.

The optical band gaps determined from these curves are listed in table 2. The founded values range around 3.20 eV which is in good agreement with literature [18]. The E_g values seem to be none affected by the Li values. This can be explained by the dependence of lattice parameter on the ZnO doped Li

Refractive indexes of the produced ZnO thin films were calculated using equation (5) [19].

$$n^2 = \frac{3}{\sqrt{Eg/20}} - 2 \quad (5)$$

Where n is the refractive index of the ZnO thin films and Eg is the band gap energies of the ZnO thin films. Refractive index practically constant values 2.3 table (2).

Table 2: Refractive index and band gap energies of Li doped ZnO films deposited on glass substrate by sol gel.

	8at.%Li	10at.%Li	15at.%Li	20at.%Li
Eg(eV)	3.182	3.271	3.218	3.220
n	2.35	2.32	2.34	2.34

4. Conclusion

ZnO doped Li thin films were prepared on glass substrates by the sol gel method, the effects of Li doped ZnO on the structural and optical properties were analysed: The structural and morphological studies revealed that the crystallinity was improved with increasing the Li, the high intensity exhibited strong (002) peak with preferred c-axis orientation, the nanowires are between 20 and 50nm in diameter. The transmittance in the visible is over 80% and the band gap energy is 3.2eV. The most ideal property is obtained at 20 at. %Li

Acknowledgement

The authors are grateful for the financial support from CNRST Morocco under grant number: 23UCA2015

5. References

[1] Water W, Chu S Y, Juang Y D and Wu S J 2002 *Mater. Lett.* **57** 998
[2] Abu El-Fadl A, El-Maghraby E M, Mohamad G A 2004 *J. Cryst. Res. Technol.* **39** **(2)** 143
[3] Thomas D G 1960 *J. Phys. Chem. Solids* **15** 86
[4] Chen Y, Bagnall D M, Koh H J, Park K T, Hiraga K, Zhu Z Q and Yao T 1998 *J. Appl. Phys.* **84** 3912
[5] Srikant V and Clarke D R 1998 *J. Appl. Phys.* **83** 5447
[6] Reynolds D C, Look D C and Jogai B 1996 *Solid State Commun.* **99** 873
[7] Wang X, Yang S,Wang J, Li M, Jiang X, Du G, Liu X and Chang R.P.H 2001, *J. Cryst. Growth*, **226** 123
[8] Jeong S H, Park B N, Lee S B and Boo J-H 2005 *J. Surf. Coat. Technol* **193** 340.
[9] El Hichou A, Stein N, Boulanger C and Johann L 2010 *Thin Solid Films*, **518** 4150
[10] Rousset J, Saucedo E and Lincot D 2009 *Chem. Mater.* **21** 534
[11] El Hichou A, Addou M, Bougrine A, Dounia R, Ebothé J, Troyon M and Amrani M 2004 *Mater. Chem. Phys.* **83** 43
[12] Schuler T, Krajewski T, Grobelsek I and Aegerter M A 2004 *J. Sol-Gel Science and Technology.* **31** 235
[13] Toyoda M, Watanabe J and Matsumiya T 1999 *J. Sol-Gel Science and Technology* **1-2** 93
[14] Hodgson J N 1970 *Optical Absorption and Dispersion in Solids, Chapman & Hall,London*
[15] Jung H J, Lee S, Choi H C and Choi M Y 2013 *Solid State Sciences* **21** 26
[16] Shinde S S, Shinde P S, Pawar S M, Moholkar A V, Bhosale C H, Rajpure K Y 2008 *Solid State Sci.* **10** 1209
[17] Addonizio M L, Aronne A, Daliento S, Tari O, Fanelli E, Pernice P 2014 *Appl. Surf. Sci.* **305** 194
[18] Srikant V, Clarke D R 1998 *J.Appl.Phys.***83** 5447
[19] Dimitrov V, Saka S 1996. *J. Appl. Phys.* **79** 1736

Journal of Physics: Conference Series **758** (2016) 012020 doi:10.1088/1742-6596/758/1/012020

Effect of Ionic Liquid (emim BF_4) on the Dispersion of Gold Nanoparticles

Mohamed R. Tchalala[1]*, Dalaver H. Anjum[2] and Sahraoui Chaieb[1]

[1]Physical Sciences and Engineering Division, *King Abdullah University of Science and Technology*, Thuwal 23955-6900, KSA
[2]Imaging and Characterization Laboratory, *King Abdullah University of Science and Technology*, Thuwal23955-6900, KSA

E-mail: mohammed.tchalala@kaust.edu.sa

Abstract. A simple method has been described for the design of hybrid materials made of 1-ethyl-3-methylimidazolium tetrafluoroborate (emim BF_4) stabilized gold (0) nanoparticles providing clear evidence about the coordination and stabilization behavior of imidazolium ionic liquids. The monodispersed gold nanoparticles were thoroughly verified by steady state spectroscopy and TEM microscopy, the hybrid material in comparison with the bare Au nanoparticles. This study revealed enhanced emission and a red-shifted absorption band which is attributed to charge transfer from IL to gold nanoparticles.TEM images of the nanoparticles corroborated the steady state results and showed similar particle sizes furthermore the nanoparticles appear often surrounded by IL layers. Based on the experiment results, which favor the interaction between IL anions instead of IL cations, we suggest the existence of an Au---F interaction and anionic Au-BF_4^- stabilization in the fluorine-containing ILs.

1. Introduction

In recent years, with advances in nanotechnology, various new nanomaterials have been investigated for application in the development of electrochemical sensors [1]. Composite of nanoparticles with ionic liquids represent a unique physical system where properties of the components wholly or partially mix and new behavior can emerge [2-4]. The combination of ionic liquids and nanoparticles are regarded as one of the very promising media whose function can be designed [4-10].

The study of the metal nanoparticles has been an incredibly active area in recent year because of their application in different domains, which will unlock new pathways in the field of nanotechnology [11]. As the name suggests, metallic nanoparticles are characterized by their small size and huge surface area to volume ratio, which allows for more points of contact than would be the case of their bulk equivalent compound, making them an excellent candidate for a variety of purpose in various fields [12, 16].

Among various metal nanoparticles, gold nanoparticles have tremendous interest in wide-ranging applications such as photonics, information storage, catalysis, electronic, therapeutics, and photovoltaics, [17, 18]. Bare metal nanoparticles are prepared by employing physical methods such as mechanic subdivision of metallic aggregates and evaporation of metal in vacuum by resistive heating or laser ablation and by chemical methods such as reduction of metal ions in a solution using appropriate reduction agents

[1] mohammed.tchalala@kaust.edu.sa

[19-22]. However, it is well known that metallic NPs are only kinetically stable, and they have high tendency to aggregate and make bigger structure due to large surface energy that drives thermodynamically favored coalescence [23]. Consequently, to obtain a stable dispersion of Au-NPs freely dispersed in solution, it is necessary to find means to prevent this phenomenon either by functionalization during synthesis or by collecting the particles in a compatible medium before they agglomerate or aggregate [24-26]. In this context, Ionic liquids have emerged as one of the most investigated classes of stabilizing agent for nanoparticles, because they allow stable dispersion without the need for additional stabilizers [27, 28]. They are structured solvents that can act as a template for the shape-selective synthesis of nanoparticles. Moreover, the unique combination of non-volatility and favorable solubilizing characteristics of ILs has motivated a growing interest as a replacement for organic solvents and as a new way to study nonaqueous solubility of metallic nanoparticles [27].

Ionic liquids are the generic term for a class of materials constituted entirely of ions and being liquid below 100 C. These unique substances composed entirely of ions offer a host of attractive properties, including ultra-low vapor pressure, high thermal stability, high ionic conductivity and a broad redox stability [29, 30]. Ionic liquids are characterized by weak interactions between a large cation and charge delocalized anion. This feature produces unusual liquid-like transport properties [31]. Moreover, owing to their extremely low vapor pressures, ionic liquids can be treated under high vacuum conditions and are regarded as fruitful media for vacuum sciences and technologies [32, 33].

Herein we represent and discuss the contribution of ionic liquid (emim BF_4) in the dispersion and stabilization of metallic nanoparticles (Au-NPs) for future application in catalysis and chemical sensors as well as in plasmonics.

2. Experimental:

All reactions involving gold nanoparticles and ionic liquids were carried out under an argon atmosphere.

1.1. Chemicals

All chemicals were used as received from the manufacturer without further purification. Gold nanoparticles (50 nm) were purchased from NanoCS.1-ethyl-3-methylimidazolium tetra fluoborate [emim] [BF_4] (99% purity) were purchased from Sigma-Aldrich.

1.2. Characterization

Samples for UV-Vis measurements were carried out using a Genysis 10S UV-Vis-Spectrometer (Thermo Scientific). Fluorescence spectra were determined using a Horiba Fluoromax-4 spectro fluorometer. For solution measurements, a conventional 10mm quartz cuvette was used. Room temperature time-resolved photoluminescence measurements were conducted using time-correlated single photon counting (TCSPC) [18] with a GaAs photocathode as a detector. Excitation was achieved by a Hamamatsu laser diode at λ= 373 nm.

2. Results and discussion:

Additionally to their small size and their high surface area, metallic nanoparticles have strong tendency to aggregate [24-26]. Since they are unstable from the standpoint of agglomeration, and in the absence of any extrinsic impediment, the unprotected particles aggregate primarily under the action of Van der Waals forces. To prevent aggregation process from occurring, the particles surface can be protected by electrostatic stabilization and steric stabilization [34]. Electrostatic stabilization is based on the Coulomb repulsion between individual particles promoted by adsorption of ions on metallic surface leading to the formation of multilayers [35-37]. Steric stabilization is builds on the steric hindrance achieved by involving the metal center, and a sterically bulky layer material, such polymer or surfactant, forming a protective layer promotes a steric barrier preventing the coagulation or aggregation [38, 39]. This type of stabilizing system can give rise to some type of nanocomposite material since the organic layers form a Nanometric scale second phase. These stabilization types are illustrated in Scheme 1.

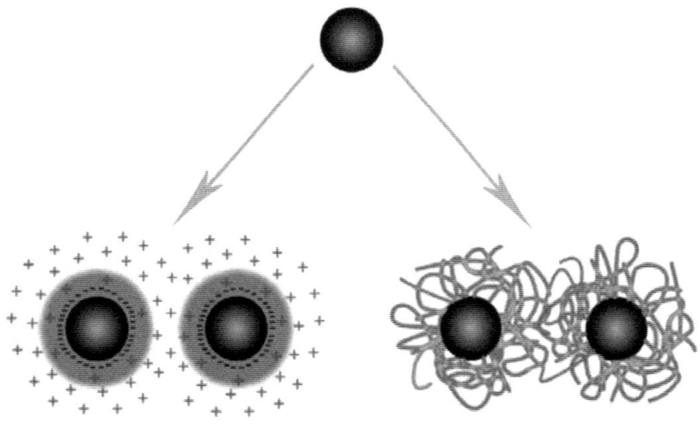

Scheme 1. Schematic representation of the electrostatic and steric stabilization in metal NPs.

Due to their properties cited above, ionic liquids can be an alternative to such ligand layers without adding any additional stabilizer, surfactants or capping ligands [40-42].

Most of the studies deal with a parallel and synergistic development of both nanoparticles and ionic liquids for materials chemistry [43], whereas other are devoted to the study of the dispersion of inorganic nanoparticles in ionic liquids [44]. In light of the various associations mentioned before, it will be shown how such hybrid gold nanoparticle/imidazolium compounds can be considered as new highly promising materials.

The optical properties of gold nanoparticles are dominated by surface plasmon resonance induced by light irradiation. The surface plasmons absorption originated from the interaction of external electromagnetic radiation with highly polarizable Au $5d^{10}$ of Au nanoparticles [45]. (Scheme3)

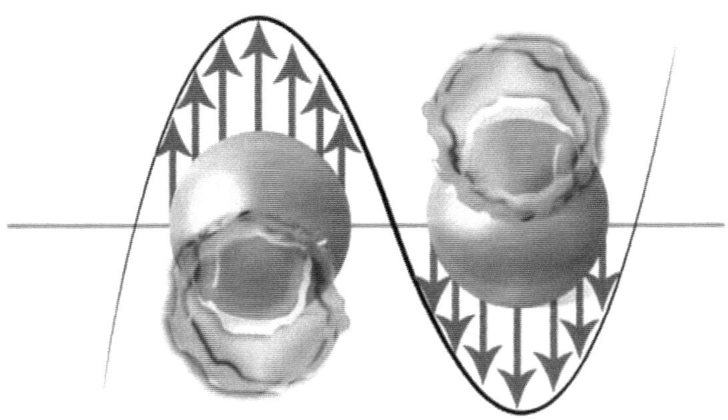

Scheme 2. Origin of surface Plasmon resonance due to the coherent interaction of the electrons in the conduction band with the electromagnetic field.

The optical absorption spectrum shows a surface plasmon resonance (SPR) band at 550nm Figure 1. indicating that it is in a metallic state.

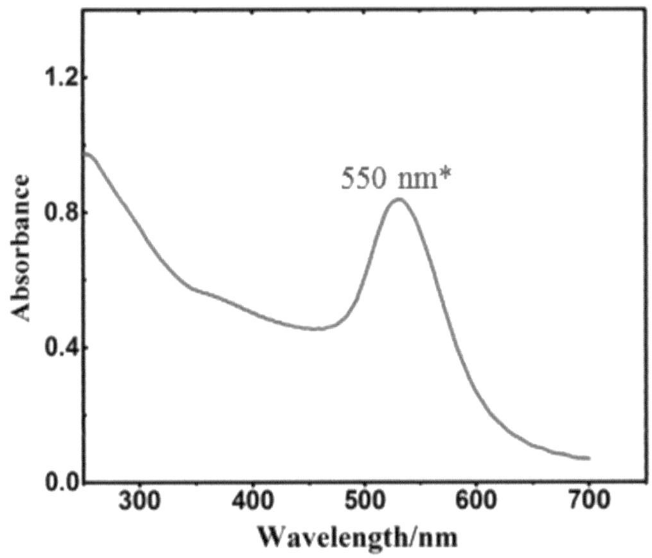

Figure 1. The UV-Visible absorption spectra of gold nanoparticles

Figure 2. shows the photoluminescence (PL) of gold metallic NPs, the emission spectra is characterized by the presence of one sharp peak in the red-band region showing the fundamental absorption of gold nanoparticles with the peak center at nm 426 nm attributed to quantum confinement of electrons from gold nanoparticles. The excitation spectrum is observed with a sharp and intense peak centered at 310nm.

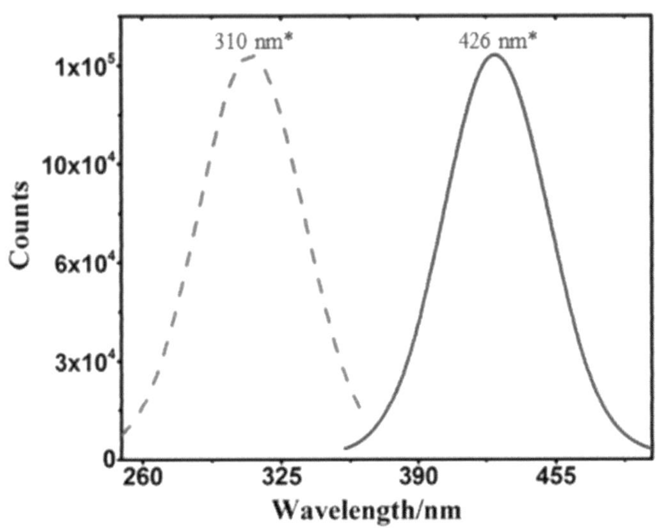

Figure 2. The fluorescence spectra of gold nanoparticles, Excitation spectra (dashed curve) and the emission spectra (solid curve)

The observed PL from this sample is due to electron-hole recombination that occurs between the excited electrons near the Fermi surface and holes in the d band [46]. The outermost d and s electrons of the

Journal of Physics: Conference Series **758** (2016) 012020 doi:10.1088/1742-6596/758/1/012020

constituent atoms must be treated together leadings to 6 bands. Five of them are relatively flat, lying a few eV below the Fermi level, they are usually denoted as d bands; the 6th one, which is almost free-electron like is known as the conduction band or sp band. A sketch of a possible transition occurring in the quantum dots related to the fluorescence is given in Scheme3.

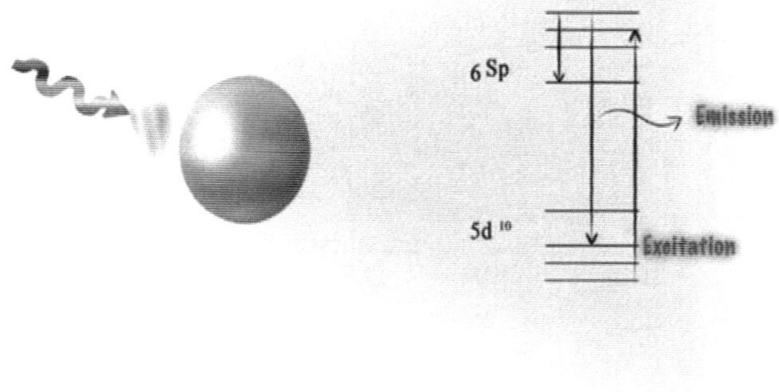

Scheme3. The possible transition is occurring in the Au quantum dots.

The gold nanoparticles, used as supplied are dispersed in ionic liquid emim BF_4. This procedure has no effect on the gold nanoparticles core shape or size. After adding the gold nanoparticles to the emim BF_4solution, the reaction pot was directly immersed in an ultrasonic bath and kept under sonication for 2 hours.

Characterization by TEM and UV-Vis spectroscopy of the Au-emim BF_4 was performed to evaluate the preservation of the morphology and optical properties of colloidal nanostructures after fusion with emim BF_4.

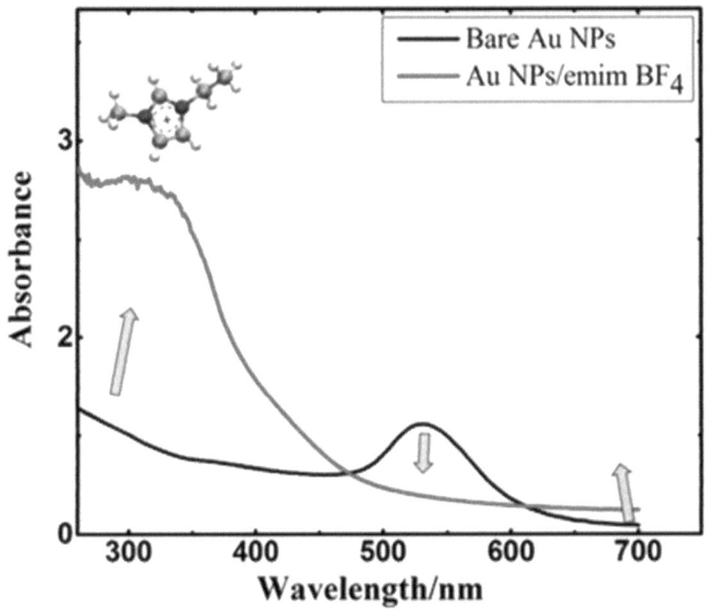

Figure 3. UV-Vis absorption spectra. Bare Au Nanoparticles (black), Au NPs + emim BF_4 (red)

Figure 3, shows the absorption spectrum of the Au-emim BF_4, The characteristic absorption spectrum of the gold nanoparticles dispersed in emim BF_4 attest to the presence of both inorganic gold nanoparticles and the emim BF_4 components in the UV spectral region. The absorption peaks appeared red shifted in the range of 438-560 due mainly to the influence of the π- π* transition of the imidazolium ring. As the plasmon absorption of gold nanoparticles is sensitive to the surrounding environment, the observed weak absorption and the absence of the usual bell-shaped band of the gold plasmon resonance might be attributed to the formation of bonded ionic liquid layer. The subsequent increase of the ionic liquid field on the surface of Au NPs, in other words is due to the weak surface plasmon resonance (SPR) of the gold nanoparticles embedded in ionic liquid medium. The tail which instead appear in the spectra of Au-emim BF_4 in the visible range, can be ascribed to the presence of energetically different associated structures, originating from an interaction among the π delocalized orbital of the imidazolium ring. Note that the intensity of the absorption was increased by a fold in the blue region. This tail probably due to the monomeric form of imidazolium ion free ion and has nothing to do with the composite [47, 48].

Useful information on the surface modification of gold nanoparticles was also obtained from photoluminescence spectra Figure 4.

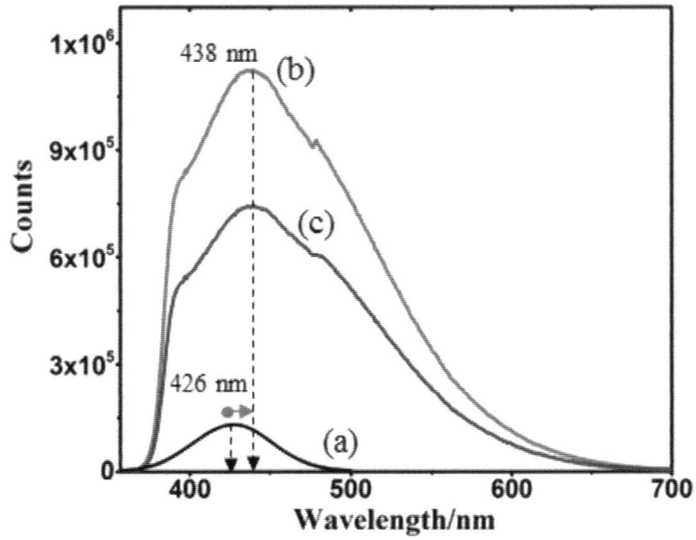

Figure 4. PL emission spectra. (a)Neatemim BF4 (b) Au nanoparticles (c) Au ''NPs'' +emim BF_4

By comparing to Au, emission spectra of emim BF_4-Au mixture exhibit a new shoulder peak at 438 nm, which may be attributed to the excimer emission of π-π interaction between the free imidazolium moieties. This is confirmed in the PL emission of blank emim BF_4, which shows the same shoulder at the same position meaning that is not coming from ionic liquid surrounding the gold nanoparticles. Moreover, the small shift of the emission maximum toward longer wavelength suggests the existence of string π-π stacking in the medium. The maximum of the band –edge PL emission of fusion material is weakly shifted, confirming that the capping exchanges does not modify the geometrical properties of the Au NPs. The increase in the absorption intensity is due to the contribution from the emim BF_4 as a consequence of the transition of the π* excited state to the π energy level in the aromatic ring [47, 48]. The broad emission of the nanocomposite is due to hierarchical stacking of the imidazolium structures around the particles which would influence the band.

The morphology of the dispersed gold nanoparticles was investigated by direct observation using in situ TEM (Figure 5.), which is possible in the ionic liquid medium despite the high vacuum condition and the high energy of the electron beam. TEM was performed on samples prepared using the Au-emim BF_4suspension ultrasonically dispersed and then deposited on holey carbon film supported on a copper grid; the grid was placed on filter paper to remove the excess material and was allowed to dry under vacuum. The solution contains particles with a uniform size of 50 nm. The micrographs in figure 4 also show that gold nanoparticles are well dispersed in the ionic liquid, but some of them aggregate into larger blobs. As can be seen from the illustration 4b, the ligne profile analysis shows a distinct surface layer of the ionic liquid emim BF_4 around the Au-NPs.

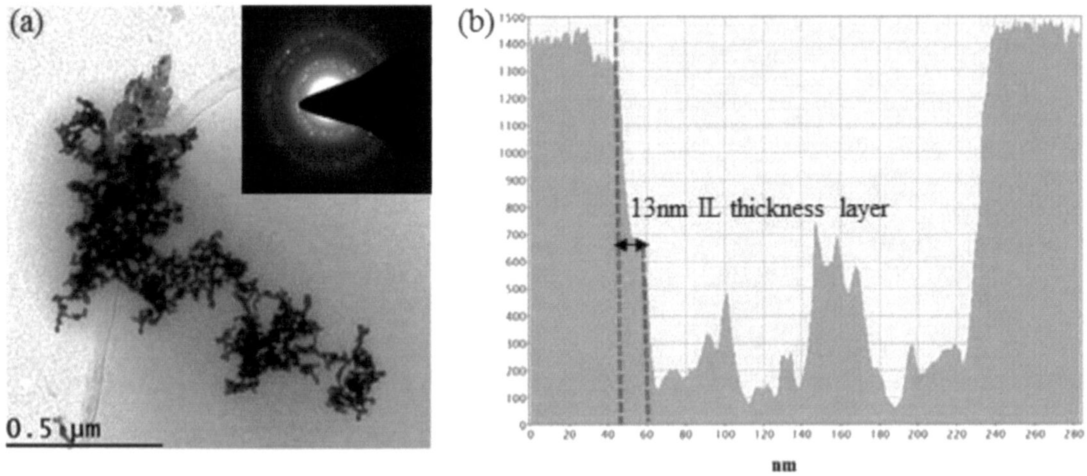

Figure 5. Typical TEM micrograph and ligne profile of Au nanoparticles

The presence of the IL around the NPs was confirmed by the time-resolved PL decay curves for Au and Au NPs in emim BF_4 at the emission wavelength peak of 310 nm. Emim BF_4 ionic liquid surrounding the nanoparticle surface act as effective stabilizers of Au nanoparticles which will reduce the average lifetime, in the Au-emim BF_4(2.74 ns) compared to pure IL(4.69 ns). This reduction in the transition lifetime indicates clearly that the charge transfer process is occurring from emim BF_4 to Au nanoparticles.

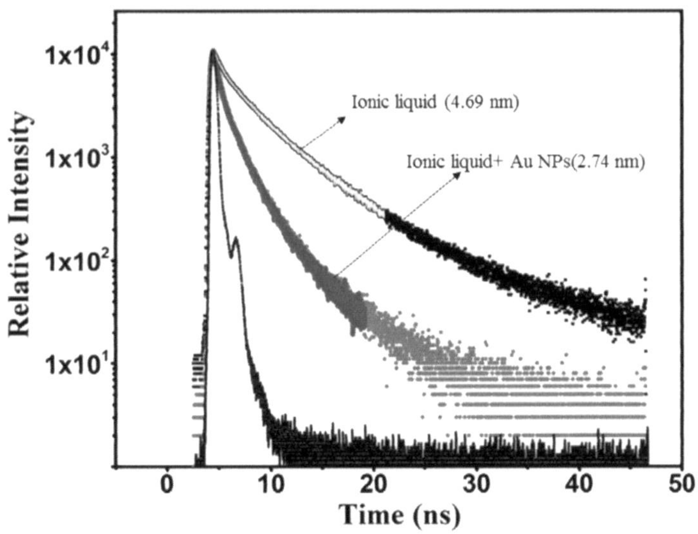

Figure 6. Lifetime decay of the IL and IL+ Au nanoparticles at 426 nm emission wavelength

The ionic liquid precursor emim BF_4 as shown below played a significant role in the stabilization of metallic nanoparticles. The stabilization of gold nanoparticles in the ionic liquid is better explained using the Derjaguin-Landau-Verwey-Overbeek (DLVO) theory [49, 50], which treat the IL ions as point charges, providing an electrostatic protection by the formation of '' protective'' anionic and cationic layers ''shells' around the NPs in core shell system. IL network should be considered as supramolecular polymeric structures with high degree of self-organization and weak interactions that bind weakly to the metal surface than to other anions or capping ligands [31]. When mixed with metallic nanoparticles IL becomes organized and nanostructured with two regions: polar and a nonpolar one [51-54]. Based on the above results and in accordance to DLVO theory gold nanoparticles are considered to be stabilized by the BF_4^- anion and large imidazolium cation, firstly by the formation of anionic BF_4^- shell as first inner shell, the Au/BF_4^-layer is expectedly surrounded by imidazolium cations (Scheme 4).

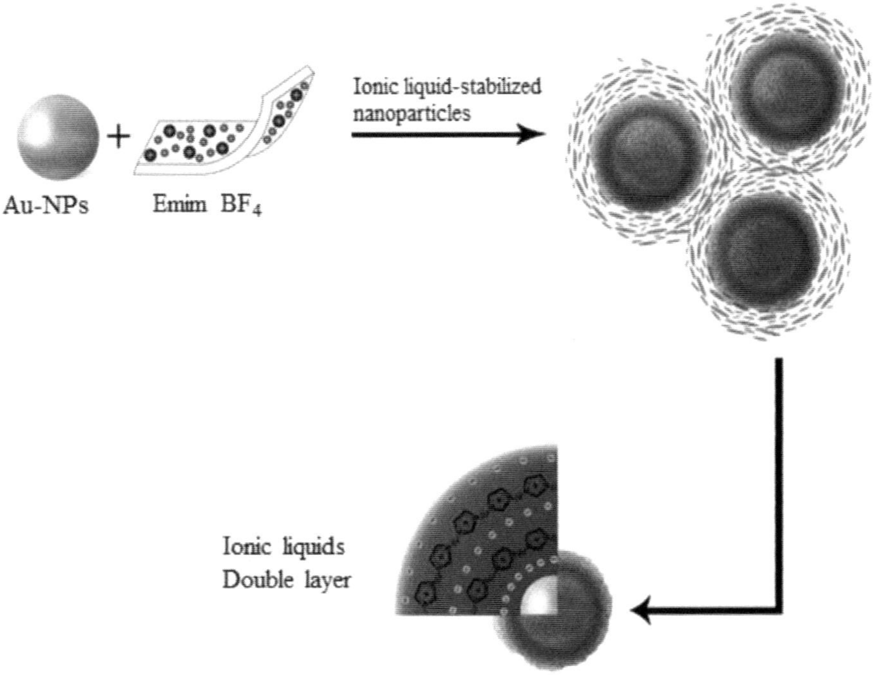

Scheme 4.Schématique représentation of structural organization of Au-emim BF_4 IL

3. Conclusion

The interaction between Au-NPs and imidazolium IL in a uniform hybrid material resulting from fusion of luminescent Au-NPs in emim BF_4 ionic liquid has been studied as model compound using spectroscopic and morphological techniques. From the result obtained by steady state spectroscopy as well as time-resolved spectroscopy and TEM micrographs show the existence of a layer of the IL on the Au-NPs surface via weak parallel coordination, due to an electrostatic stabilization. On the other hand Imidazolium moiety act as cationic barrier inducing steric stabilization avoiding further aggregation of Au NPs.

4. References:

[1] Shao Su, Wenhe Wu ac, Jimin Gao b, JianxinLu bb and Chunhai F 2012 *J. Mater. Chem.* **22** 18101
[2]Torimoto T, Tsuda T, Okazaki K and Kuwabata S 2010 *Adv. Mater* **22** 1196–1221
[3] Fukushima T, Kosaka A, Ishimura Y, Yamamoto T, Takigawa N, Ishii N and Aida T 2003 Science **300** 2072–2074
[4]Dupont J 2011*Acc. Chem. Res* **44** 1223–1231
[5] Kato T 2002 *Science* **295** 2414–2418
[6] Greaves T L and Drummond C J 2008 *Chem. Soc. Rev* **37**1 709–1726
[7]Kimizuka N and Nakashima T 2001 *Langmuir* **17** 6759–6761
[8] Nakashima T, ZhuJ, Qin M., Ho S S and Kotov N 2010 *Nanoscale* **2** 2084–2090
[9] Nakashima T and Kimizuka N 2011 *Langmuir* **27** 1281–1285
[10]Nakashima, T and Kimizuka, N 2012 *Polym. J* **44** 665–671
[11] El-Sayed M A 2001 Acc. Chem. Res **34** 257–264
[12]AndersonJ A and Fernandez-Garcıa M 2011 Imperial College Press, London, 2nd edn11

[13]Kamat P V 2002 *J. Phys. Chem. B* **106** 7729–7744
[14] XiaY, XiongY, Lim B and Skrabalak S E 2009 *Angew. ChemInt. Ed* **48** 60–103
[15] Chen H M andLiuRS 2011 *J. Phys. Chem. C* **115** 3513–3527
[16]Saha K, AgastiS S, KimC, Li X N and RotelloV M 2012 *Chem. Rev.***112** 2739–2779
[17] Rao R C, Kulkarni G U, Thomas J P, Edwards PP 2000 *ChemSoc Rev.* **29** 27–35
[18] Daniel M C, Astruc D 2004 *Chem Rev.***104** 293–346
[19]Turkevich J, Stevenson PC, Hillier J 1951 *Discuss Faraday Soc* **11** 55–75
[20]Kimling J, Maier M, Okenve B, Kotaidis V, Ballot H, Plech A 2006 *J Phys Chem B* **110** 15700–7
[21]Frens G 1972 *Colloid Polym Sci.* **250** 736–41
[22]Frens G 1973 *Na Phys Sci* **241** 20–2
[23]Luska KL, Moores A 2012 *ChemCatChem* **4** 1534-1546
[24]Sperling R A and Parak W J 2010 *Philosophical Transactions of the Royal Society A: Mathematical, Physical and Engineering Sciences* **368** no19151333–1383
[25]Seol S K, Kim D, Jung S, Chang W S, and Kim J T 2013 *Journal of Nanomaterials* **2013** 531760,6
[26]Hoppe C E, Lazzari M, Pardiñas-Blanco I and López-Quintela M A 2006 *Langmuir* **22** 167027–7034
[27]Bönnemann H, Brijoux W 1996 *Wiley-VCH, Weinheim* pp. 339
[28]Reetz M T, HelbigW, Quaiser S 1996 *Wiley-VCH, Weinheim* pp. 279
[29]Hapiot P and Lagrost C 2008 Chem Rev **108** 2238
[30]Banhote P, Dias A P, Papageorgiou N, Kalyansundaram K and Gratzel M, 1996 *InorgChem* **35** 1168
[31]Dupont, JBraz, J 2004 *Chem. Soc.* **15** 341. Consorti, C S, Suarez,P A Z, de Souza, R F, Burrow R A, Farrar, D H, Lough, A J, Loh W, Da Silva, L H M, Dupont J 2005 *J Phys. Chem. B* **109**,4341. Dupont J, Suarez P A Z,De Souza R F, Burrow R A, Kintzinger J P 2000 *Chem.sEur.J* **6** 2377
[32]Cremer T, Killian M, Gottfried J M,Paape N, Wasserscheid P, Maier F, Steinrück H P 2008 *ChemPhysChem* **9** 2185
[33]Waldmann T, Huang H H,.Hoster H E, Höfft O, Endres F, Behm R J 2011 *Chem Phys Chem* **12** 2565
[34]Kulkarni P, Sureshkumar R, Biswas P 2003 *J Colloid Interface Sci* **260** 36–48
[35]Turkevich J, Stevenson P C, Hillier J 1951 *Disc. Faraday Soc* **11** 55275
[36]Turkevich J, Kim G, 1970 *Science* **169** 873
[37]Turkevich J 1985 *Gold Bull* **18** 86291
[38] Heller, Wand PughT L 1954 *J. Chem. Phys***22** 1778
[39] Heller W 1966 *Pure Appl. Chem.***12** 249

[40] Fonseca G S, Umpierre A P, Fichtner P F P, Teixeira S Rand Dupont J 2003 *Chem.–Eur J* **9** 3263–3269
[41]Scheeren C W, Machado G, Dupont J, FichtnerP. F. P. And Texeira S R 2003 *Inorg. Chem* **42** 4738–4742
[42]Migowski P and Dupont J 2007 *Chem.–Eur. J* **13** 32–39
[43]Neouze MA 2010 *J. Mater. Chem* **20** 9593
[44] Rao C N R., Matte H S S,Voggu R and Govindaraj A 2012 *DaltonTrans* **41** 5089
[45] Jain P K, Huang X, El-Sayed I H, El-Sayed M A *2008 AccChem Res* **41** 1578–86
[46]Yen Hsun S, Sheng-Lung T, Shih-Wen T, Yun-Chorng C, Shih-Hui Cand Wei-Min Z 2010 *Nanoscale* **2** 2639
[47]Paul A, Mandal P K, Samanta A 2005 *J. Phys. Chem. B* **109** 9148-9153
[48]Paul A, Mandal P K, Samanta A, 2005 *Chem. Phys. Lett.* **402** 375-379
[49] Verwey E J W, Overbeek J T G 1999 *Dover Publications Mineola* pp. 1–218
[50] Ninham BW 1999 *Adv. Colloid Interface Sci* **83** 111 –136
[51] Gannon T J, Law G, Watson R P, Carmichael A J, Seddon K R 1999 *Langmuir* **15** 8429 –8434

[52] Law G, Watson R P, Carmichael A J, Seddon K R 2001 *Phys. Chem.Chem. Phys* **3** 2879 – 2885
[53] Canongia Lopes J N, Gomes MF C,dua AAH P 2006 *J. Phys.Chem. B* **110** 16816 – 16818
[54] Canongia Lopes J N,dua AAH P 2006 *J. Phys. Chem. B* **110** 3330 – 3335

Acknowledgments
SC and RT Thanks KAUST for financial support.

Journal of Physics: Conference Series **758** (2016) 012021 doi:10.1088/1742-6596/758/1/012021

The effect of sol aging time on Structural and Optical properties of sol gel ZnO doped Al

G El Hallani, N Fazouan*, A Liba, M Khuili
Materials Physics Laboratory, Sultan Moulay Slimane University, Faculty of Sciences and Technologies, BP 523, 23000, Beni Mellal, Morocco

* E-mail: fazouan@yahoo.fr

Abstract. Currently the doped or undoped ZnO semiconductor is of great importance in the field of electronic and optoelectronic devices such as transparent conductors and optical windows of solar cells based on silicon. ZnO thin films are produced by several techniques such as sol–gel method which is a chemical technique usually dependent on solution conditions. However, the sol gel aging time is an important parameter, which can have a significant impact on the properties of thin films. In this work we studied the effect of aging times (0h, 24h, 48h, 72h, 1 week) of the precursor solution on the structural and optical properties of ZnO doped Al (3 at.%). Thin films prepared by spin coating on glass substrates were investigated. The X-ray diffraction (XRD) analysis shows that the ZnO doped Al (3 at.%) exhibit the hexagonal würtzite structure with a preferential orientation along [002] direction. The shift of (002) peaks towards higher diffraction angles is observed with sol aging time and also, a variation of crystallite sizes and thickness of thin films are shown with increasing sol aging time. All films present an average optical transmittance around 90% in the visible range with some interference fringes indicating a relative smoothness of films. We note an increasing in transmittance level with sol aging time from 0h to 48h. We can conclude that the aging times of the precursor solution influences the structural and optical properties of studied thin films.

1. Introduction

In recent years, undoped and doped Zinc oxide (ZnO) thin films received extensive attention due to their excellent optical and electrical properties [1-2]. Doped ZnO thin films are investigated as converter of photons for conductive photovoltaic solar cells window layers [3] and as optoelectronic devices in the UV range [4.5]. This importance came from its high direct band gap (Eg= 3.37 eV) at room temperature and an exciton bending energy of 60 meV [6]. Several techniques were used to synthesize undoped and doped ZnO thin films, such as metalorganic chemical vapor deposition (MOCVD) [7], sputtering [8], pulsed laser deposition [9], spray pyrolysis [10,11] and sol-gel method [12,13,14].
In this work we have studied the effect of sol aging times (0h, 24h, 48h, 72h and 1 week) on structural and optical properties of Al doped ZnO at 3% deposited on glass substrates by the sol gel process using the spin coating method. The crystallinity of the undoped and AZO thin films was measured by

an X-ray diffractometer X'Pert Pro equipped with a CuKα source (λ= 1.5406 Å) and optical transmittance measurements were carried out using a LAMBDA 900 UV-VIS-IR spectrophotometer.

2. Experimental Method

The spin coating method was used to prepare ZnO and AZO thin films on glass substrates. The solution of undoped ZnO and AZO aged at different times were prepared. For synthesis, we have employed zinc acetate dihydrate [Zn(CH3CO2)2.2H2O], absolute ethanol [C2H6O], ethanolamine [MEA,C2H7NO] and aluminum nitrate nanohydrate [Al(NO3)3,9H2O], respectively as a precursor of zinc, solvent, stabilizer and source of dopant. Firstly, the zinc acetate dihydrate was dissolved in mixture of ethanol and ethanolamine MEA solution. The zinc acetate concentration in solution was kept to 0.5M and the molar ratio of MEA/Zinc was maintained to 1.0. Aluminum nitrate nanohydrate was added at the solution with Al/Zn concentration of 3%. The finale solution was refluxed in magnetic stirrer for 1h at 80°C until clear and homogeneous solution was obtained. The procedure to fabricate the AZO thin films is similar to the process used for elaboration of undoped ZnO thin films as described in our previous work [15]. The flow chart of deposition process is shown in Figure 1.

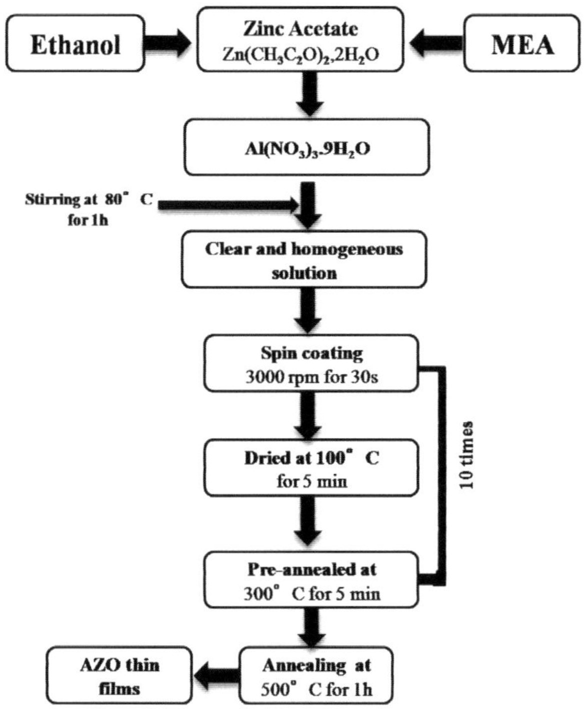

Figure 1. Procedure for preparing spin coated AZO thin films.

3. Results and Discussion

3.1. Structural properties

Figure 2 shows X-ray diffraction patterns of ZnO doped Al thin films with different aging times (0h, 24h, 48h, 72h and 1 week) of the precursor solution. The analysis of the peaks intensity shows that all films are polycrystalline with a hexagonal wurtzite structure with preferential growth along the c-axis direction. We note an increasing in the (002) peak intensity when the precursor solution aging rised of 0h to 48h and then decreased for 72h and 1 week. This means a smoothness growth at 48h sol aging and indicating an optimum sol aging time condition. A gradual shift in (002) peak was observed toward higher angles with increasing sol aging time, 2θ= 34.31° for 0 h and reached at 2θ= 34.46° for

1 week aged sol. The shift of the peak (002) to the wide angle is due to the incorporation of Al in the matrix of ZnO as reported by Shahzad et al. [16].

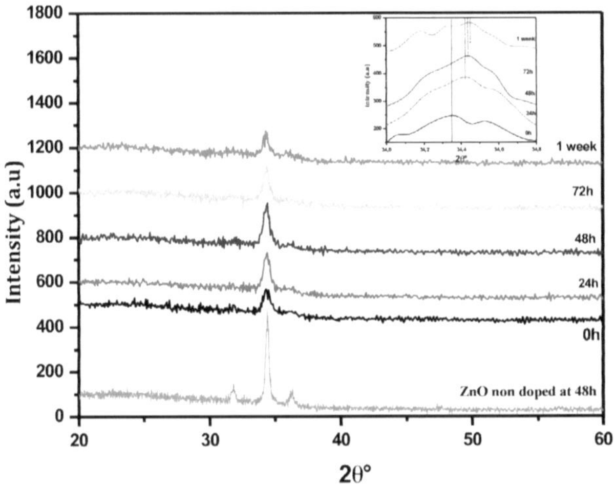

Figure 2. XRD spectrum of undoped ZnO and AZO thin films at different aging time

The crystallite size (D) was estimated using the (002) diffraction peak from the XRD data in accordance to the Debye-Scherer's formula [17].

$$D = 0.9\lambda / \beta\cos(\theta) \quad (1)$$

where β is the full width half maximum (FWHM) of the diffraction peak measured in radians, λ (1.5406 Å) is the wavelength of the X-ray, and θ is the Bragg's diffraction angle of (002) plane. An average crystallite size value corresponding to ZnO and AZO films deposited at different aging time is presented in Table 1. For the aging time 48h the crystallite size for ZnO and AZO decreases of 35.214nm to 17.235nm respectively, we explain this change by substitution of Zn sites by Al dopant of weak ionic ray and low structure degradation as shown by DRX spectra and observed by other authors [16]. For AZO thin films, the crystallite size increases with time aging due to the viscosity ascent of precursor solution as illustrated in figure 3.

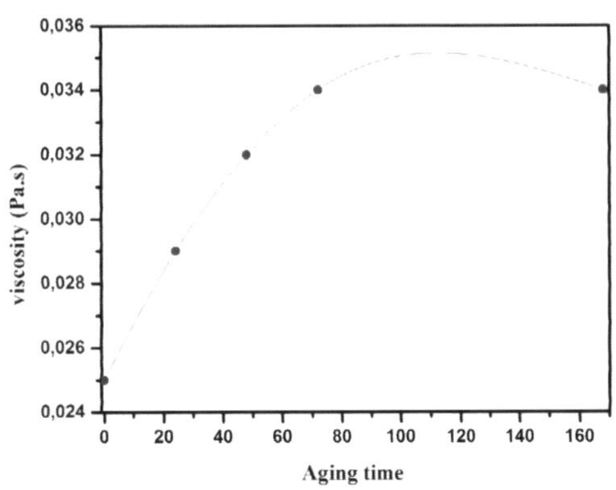

Figure 3. Viscosity of AZO thin films at different aging times

The average uniform strain (ε) in the AZO films along c-axis was calculated by using biaxial strain model [18] :

$$\varepsilon = C_{film} - C_{bulk} / C_{bulk} \tag{2}$$

where C_{bulk} (5.20661 as JCPDS , 36–1451) is the lattice constant of undoped bulk ZnO for hexagonal structure and C_{film} is the lattice one of AZO elaborated films. The C_{film} was calculated using the DRX data by the relation:

$$C_{film} = \lambda / \sin(\theta_{002}) \tag{3}$$

For the calculated strain values we note that the aged sample at 48h presents the minimum strain witch indicates a good crystallinity as identified by XRD.

Table 1. Various structural parameters of AZO thin films, and the ε strain

Sol aging time (h)	Crystallite size (nm)	(002) peak position 2Θ	FWHM (deg)	Lattice constant c (Å)	Strain ε
Undoped film (48h)	35,214	34,4567	0,1181	5,208	-1,0564E-4
0	14,767	34.3152	0.5628	5.223	0,00315
24	14,768	34.3300	0.5628	5.220	0,00257
48	17,235	34.4630	0.4824	5.200	-0,00127
72	20,6815	34.4453	0.4020	5.203	-6,934E-4
168	21,016	34.4508	0.3956	5.206	-1,172E-4

3.2. Optical Properties

Figure 4 shows the plots of optical transmittance and absorption coefficient spectra of the AZO thin films at different sol aging times. It was showed that the films present an averaged optical transmittance around 85% in the wavelength range of 400–800 nm with some interference fringes. No variation of optical transmittance is observed with prolonged sol aging time.

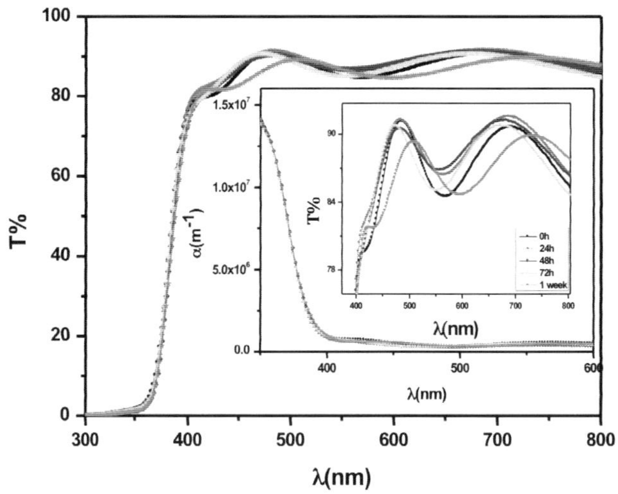

Figure 4. Optical Transmittance and Absorption coefficient plots of Al doped ZnO at different sol aging times

It was observed in figure 5 that at 48h aging time, the transmittance decreased from 95% to about 87% from un-doped ZnO film to doped ZnO at 3%. This effect can be due to the degradation of the cristallinity of doped films as observed in XRD spectra (there is more scattering of light and this decreases the transmittance).

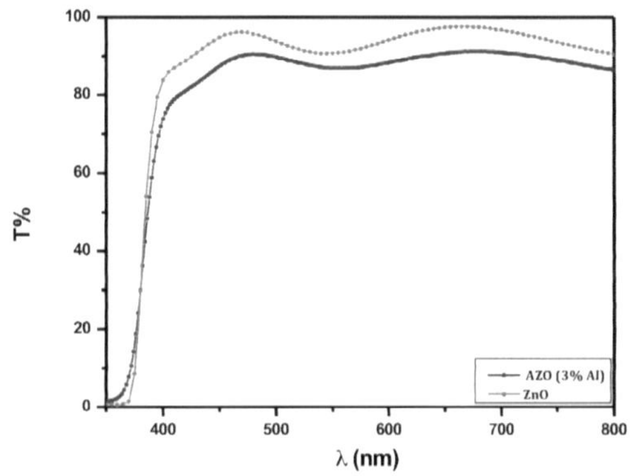

Figure 5. Optical Transmittance for ZnO and AZO thin films at 48h

The optical band gap Eg of the films are calculated using the Tauc relationship [19]:

$$(\alpha h\nu)^2 = B(h\nu - Eg) \tag{4}$$

where B is an independent energy constant, hν is the photon energy and α is the absorption coefficient. The Tauc's plot $(\alpha h\nu)^2$ versus hν is shown in Figure 6. The optical band gap Eg for AZO thin films was estimated by extrapolating the linear portion of the each curve with the (hν)-axis. In the lower energy range α varies exponentially with photon energy and follows the Urbach formula [20]:

$$\alpha(h\nu) = \alpha_0 \exp(h\nu / E_u) \tag{5}$$

where α_o is a constant and E_u is the Urbach energy that is the width of the tailing in the bandgap. Inside Figure 6, we have presented, the plot of lnα versus the photon energy hν for Al doped ZnO thin films at different sol aging times.

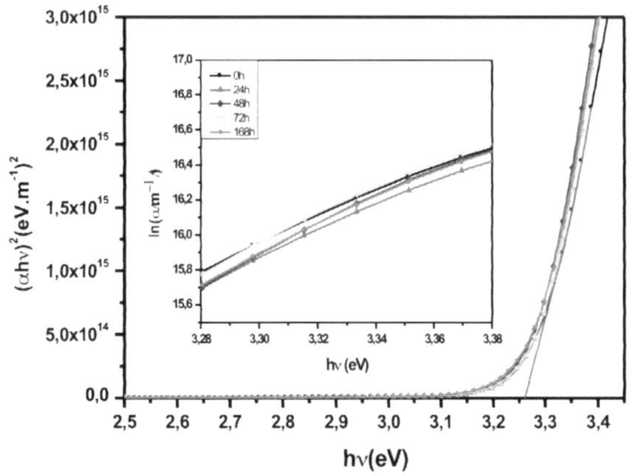

Figure 6. The Plots of $(\alpha h\nu)^2$ and Ln(α) vs photon energy of AZO thin films at different sol aging time

Table 2 gives the band gap Eg, the Urbach E_u energies and the estimated thickness of the films which was estimated using the transmittance spectra fringes.

Table 2. The optical parameters of undoped and doped Al(3%at) ZnO thin films.

Sol aging time (h)	*Thickness (nm)*	*Eg (eV)*	*Eu(ev)*
Undope film (48h)	204,526	3.2837	0.065
0	198,862	3.2609	0.103
24	202,614	3.2663	0.099
48	200,774	3.2636	0.105
72	193,696	3.2645	0.103
168	208,323	3.2600	0.114

Table 2 shows decreasing of the optical band gap Eg for doped AZO films compared to undoped ZnO thin films aged at 48h. Also it can be seen that AZO at 48h produce an Urbach band tail width around 0.105 eV higher value of about 0.065 eV observed for undoped ZnO films, which indicate an increase of defects level in the band gap. No significantly variation of optical band gap, thickness and Urbach energy with sol aging time was observed, as showed by [12]

4. Conclusion

The effects of sol aging time on structural and optical properties of sol gel ZnO doped Al thin films deposited by sping coating on glass substrate were investigated. The results show that the undoped and Al doped ZnO thin films were polycrystalline and present a hexagonal wurtzite structure with a preferential orientation (002) along c-axis. The results indicated a smoothness growth at 48h sol aging exhibiting an optimum sol aging time condition. The films were highly transparent in the visible range. We note no significantly effect on optical properties with sol aging time.

References

[1] Wolf S A, Awschalom D D, Buhrman R A, Daughton J M, Molnár S Von, Roukes M L, Chtchelkanova A Y, and Treger D M 2001 *Science* **294** 1488.
[2] Fortunato E, Nunes P, Marques A, Costa D, Águas H, Ferreira I, Costa M E V, Godinho M H, Almeida P L, Borges J P, and Martins R 2002 *Adv. Eng. Mater.***4** 610
[3] Trupke T, Green M A, and Würfel P 2002, *J. Appl. Phys.* **92** 1668
[4] Breivik T H, Diplas S, Ulyashin A G, Gunnæs A E, Olaisen B R, Wright D N, Holt A, and Olsen A 2007 *Thin Solid Films* **515** 8479.
[5] Ryu Y R, Lee T S, Lubguban J A, White H W, Park Y S, and Youn C J 2005 *Appl. Phys. Lett.* **87** 153504.
[6] Ozgur U, Alivov Y I, Liu C, Teke A, Reshchikov M A, Dogan S, Avrutin V, Cho S J, Morkoc H (2005) *J. Appl. Phys.* **98** 041301.
[7] Fragala M E, Malandrino G, Giangregorio M M, Losurdo M, Bruno G, Lettieri S, Amato L S, and Maddalena P 2009 *Chem. Vap. Deposition* **15** 327–333
[8] K C Park , D Y Ma , K H Kim 1997 *Thin Solid Films* **305** 201-209
[9] Agura H, Suzuki A, Matsushita T, Aoki T, Okuda M 2003 *Thin Solid Films* **445** 263–267.
[10] Douayar A, Bihri H, Mzerd A, Belayach A and Abd-Lefdil M 2014 *Sensors & Transducers* **27** 156-160
[11] Kaid M A, Ashour A 2007 *Appl. Surf. Sci.* **253** 3029–3033.
[12] Shahzad B, Qi Y 2011 *Adv. Mater. Res.* **216** 271-277.
[13] Krunks M, Mellikov E 1995 *Thin solid films* **270** 33.
[14] Smirnov M, Baban C, Rusu G I (2010) *Appl. Surf. Sci.* **256** 2405.
[15] El Hallani G, Fazouan N, Liba A, Laanab L 2014 IRSEC conference IEEE 82-84.
[16] Shahzad M B, Qi Y, Lu H, Wang X 2013 *Thin Solid Films* **534** 242–248.
[17] Scherrer P, 1918 *Goettinger Nachrichten Math. Phys.* **2** 98-100.
[18] Cebulla R, Wendt R, Ellmer K 1998 *J. Appl. Phys.* **83** 1087
[19] Tauc J, Grigorovici R, Vancu A 1996 *Phys. Status Solidi (b)* **15** 627.
[20] Pankove J I 1971 *Prentice-Hall Inc Englewood Cliffs, NJ* USA.

Journal of Physics: Conference Series **758** (2016) 012022 doi:10.1088/1742-6596/758/1/012022

Overall challenges in incorporating micro-mechanical models into materials design process

M BENNOURA [1] and A ABOUTAJEDDINE [2]
Mechanical engineering laboratory, Faculty of Science and Technology, University of Sidi Mohamed Ben Abdellah, Fes, Morocco

E-mail: [1]mounia.bennoura@usmba.ac.ma
[2]ahmed.aboutajeddine@usmba.ac.ma

Abstract. Using materials in engineering design has historically been handled using the paradigm of selecting appropriate materials from the finite set of available material databases. Recent trends, however, have moved toward the tailoring of materials that meet the overall system performance requirements, based on a process called material design. An important building block of this process is micromechanical models that relate microstructure to proprieties. Unfortunately, these models remain short and include a lot of uncertainties from assumptions and idealizations, which, unavoidably, impacts material design strategy. In this work, candidate methods to deal with micromechanical models uncertainties and their drawbacks in material design are investigated. Robust design methods for quantifying uncertainty and managing or mitigating its impact on design performances are reviewed first. These methods include principles for classifying uncertainty, mathematical techniques for evaluating its level degree, and design methods for performing and generating design alternatives, that are relatively insensitive to sources of uncertainty and flexible for admitting design changes or variations. The last section of this paper addresses the limits of the existing approaches from material modelling perspective and identifies the research opportunities to overcome the impediment of incorporating micromechanical models in material design process.

1. Introduction

Our innovation capabilities are tied strongly to our ability to engineer materials that meet requirements of novel product concepts. Indeed, the acceleration of material discovery is crucial for speeding up the advance of technology necessary to the design innovative products. The ultimate way to discover new materials is by the possibility of designing materials following desired functional requirement. A main goal of the field of materials design is the potential for tailoring materials from atomic scale to the system scale.

Historically, Engineering design has been taught using the paradigm of selecting materials on the basis of tabulated databases of properties [1] .Consequently, the performance of many engineering products and systems is limited fundamentally by the properties of available constituent materials. Obviously, to innovate one can no longer limit the performance of systems to available materials. Thus, the recent trends have moved toward concurrent design of material composition and microstructure together with the system level, to obtain materials with desired properties for particular applications and satisfying performance requirements on the system level. The goal is to tailor materials to meet specified ranges of performance requirements of the overall system. These performances are often in conflict when the microstructure is chosen. It is important to note, that these desired material properties and performance characteristics usually depend on phenomena that operate at different length and time scales. Material design is then built on a multi-scale journey from the smallest scale to the system level. One of the important multi-scale passages is the bridging of micro-scale to meso-scale through micromechanical models. These last models play a crucial role in the achievement of materials design initiative. These micromechanical models aim to predict accurately the response of a material under uniform load conditions from the properties of their constituents. The primary objective of this work is to emphasize a number of challenges related to integrating micromechanical models in material design process. These challenges are mainly associated to uncertainty due to idealization resulting from the lack of complete knowledge of microstructure and its description. In this respect, the key methods and approaches for dealing with this uncertainty were reviewed and critically evaluated from material modelling point view; first to identify their drawbacks and second to draw the research opportunities in this field.

2. Issues in integrating micromechanical model in material design process

Micromechanical models aim to predict accurately the response of a structure or device under service load conditions. Unfortunately, virtually all micromechanical models are embryonic from material design point view. It is important to emphasize that micro-mechanical models inevitably incorporate assumptions and approximations that impact the precision and accuracy of predictions. When added to the material design process, they bring their drawbacks and the uncertainty may be magnified when a model is utilized near the limits of its intended domain of applicability or when information propagates through a series of models [2]. Furthermore, microstructure itself has random character, such as sizes and distributions of grains, phases, and so forth. These sources of uncertainty give rise to the need to consider sensitivity of properties and responses of interest to variation of microstructure at various scales, as well as propagation of model uncertainty through multiscale model chains. So, one of the most important and challenging problems is the uncertainty in these models associated to various sources, including [3]:

- Insufficient knowledge about microstructure.
- Natural variability and randomness in materials (phases introduced by processing conditions) that are not taken into account.
- Ad-hoc parameters used by these models to presumably describe the microstructure.

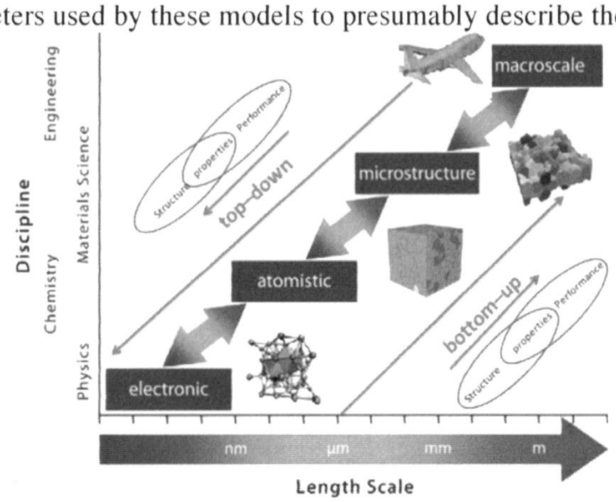

Figure 1. Hierarchy of levels from atomic scale to system level in concurrent materials design [1].

Uncertainty is endemic in most facets of micro-mechanical models, the uncertainty of microstructure models and model parameters impact significantly material design process. Consequently, a combined strategy of bottom–up and top–down modelling 'figure 1' to target

Journal of Physics: Conference Series **758** (2016) 012022 doi:10.1088/1742-6596/758/1/012022

performance requirements will be very difficult [2,3]. In this respect, decreasing the uncertainty effect and partially managing its propagation through multi-scale models is of paramount importance. The only way to gain this goal is by having a good understanding of microstructure (approximations and simplifications), in order to let models be more realistic and transmit information without loss; by combining the approach with simulation tools, and finally by easing the decision design making.

A practical approach for accounting uncertainty involves many aspects [2,4]: quantify uncertainty to the extent possible and then mitigate uncertainty by seeking the robust solutions that are less sensitive to variation of microstructure and various other sources of uncertainty. Uncertainty quantification, propagation and management will be addressed in later sections of this paper.

3. Overview of the incumbent uncertainty managing approaches

In this section, a review of existing methods and approaches of the design process under uncertainty is presented; these methods comprise uncertainty quantification in engineering systems and uncertainty management methods.

3.1. Uncertainty quantification

3.1.1. Uncertainty classification.

The strategy and methods for quantifying uncertainty in engineering systems depend on the types of uncertainty, for that, it is necessary to categorize the types of uncertainty in a model. An understanding of these classifications is necessary to properly position the methods for design under uncertainty. There are two approaches to classify the types of uncertainty. One is classification according to the nature of uncertainty by Ayyunb and Klir [5, 6], and the other is by the source of uncertainty by Der Kiruregian and others [7, 8, 9, 10]. The classification by source of uncertainty is the most widely adopted and in this case the uncertainty is split as either Aleatory (irreducible) or Epistemic (reducible) based on the causes of the uncertainty. Aleatory uncertainty is inherent in the physical system and can only be quantified in a statistical sense. Epistemic uncertainty, on the other hand, due to the lack of knowledge, which may results from idealizations (simplified relationships between inputs and outputs), approximations, or numerical errors. Examples include size of the representative volume element (RVE) of microstructure used to predict properties or responses. It can be diminished by improvements in measurements and model formulation or by increasing the accuracy or sample size of data.

3.1.2. Quantification of uncertainty.

In this section, several approaches for quantifying uncertainty will be reviewed; the uncertainty quantification is the process of identifying different sources of uncertainty and developing corresponding mathematical representations, by estimating output uncertainty (probability distribution of system response) with variability in input parameters [2]. One must mentioned that quantification of aleatory uncertainty is relatively easier than quantification of epistemic uncertainty if measuring system behaviour is feasible.

Depending on the types of uncertainty analysis methods, the types have been classified as non-probabilistic and probabilistic methods. The probabilistic methods are further devised on statistical and non-statistical approaches. Non-probabilistic methods include Interval Analysis and Fuzzy Logic methods, which are particularly useful when a system model is deterministic (there are no random errors in a system response) and input parameters are uncertain (without probability density functions due to lack of information). If a system model (computational model) is deterministic and input parameters are known as probabilistic density functions, then there are lots of techniques available in literature. If a system model is non-deterministic and input parameters are known as probability density functions then statistical methods for uncertainty analysis are also useful for uncertainty quantification. However, if a system model is non-deterministic and input parameters are uncertain, and then no existing method is applicable to uncertainty analysis. Existing uncertainty analysis methods depending on the characteristics of a system are presented in Table 1.

As discussed above, there are many available uncertainty analysis methods for deterministic system models; however, as shown in Table 1 uncertainty analysis methods for non-deterministic system models have not been developed in literature. Monte Carlo simulation is the only method for quantifying non-deterministic system response variation with the probability distributions of input parameters. If system behaviour is non-deterministic due to non-parametric input variation, it is impossible to quantify uncertainty by using statistical or non-statistical efficient uncertainty analysis methods. Uncertainty analysis for nondeterministic simulation is indispensable for multiscale materials design since the simulation models for heterogeneous material behaviour tend to be stochastic [4].

Table 1. Uncertainty analysis methods versus the characteristics of a system.

System model	Deterministic		Non-deterministic	
Input	Uncertain	Probability distribution	Uncertain	Probability distribution
Available uncertainty analysis methods	Non-probabilistic methods ▪ Interval analysis ▪ Fuzzy logic	Probabilistic methods - Statistical methods ▪ Monte carlo simulation ▪ Latin Hypercube sampling -Non- statistical method ▪ First and second order moment methods ▪ Polynomial chaos expansion ▪ Stochastic response surface methods	N/A	Probabilistic methods Statistical methods ▪ Monte Carlo

As mentioned the Aleatory system uncertainty is quantifiable; however, epistemic uncertainty that arises from incomplete knowledge of a system is virtually impossible to quantify [2]. Therefore, reducing uncertainty (instead of being quantified) is feasible when designer has large amounts of data and by increasing the knowledge or getting more information about system. Instead, researchers [11, 12, 13, 14] have focused on procedures to reduce uncertainty and validate a model's accuracy. However, the restriction of these methods is that designers must have some true results. In some cases, we cannot obtain any better data than current data due to severe uncertainty in a system or restrictions of computational or experimental expenses, which is the case of multiscale materials design specially micromechanical model.

3.2. Uncertainty management

3.2.1. Robust Design.

Uncertainty management refers to reducing the uncertainty effect on systems design. Several techniques have been performed to achieve target system performance in the presence of uncertainty such as robust design and reliability based design and optimization (RBDO). Robust design techniques aimed to minimize the effects of uncertainty on system performance [15], while reliability-based design optimization techniques are based on finding optimized designs characterized by low probability of failure [16]. The robust design is approach for managing uncertainties; it's allowed to design a system to be insensitive to uncertainty without eliminating or reducing its sources in the system.

There are several categories of robust design that deal with different types of uncertainty. Type I robust design, originally proposed by Taguchi [17], focuses on achieving insensitivity of performance with respect to noise factors—parameters that designers cannot control in a system. Type II robust design [18] relates to insensitivity of a design to variability or uncertainty associated with design variables-parameters that a designer can control in a system. A method for types I and II has been proposed, namely the Robust Concept Exploration Method [19]. These types of robust design have been extended to include Type III [4], which considers sensitivity to uncertainty embedded within a model (model parameter/structure uncertainty).

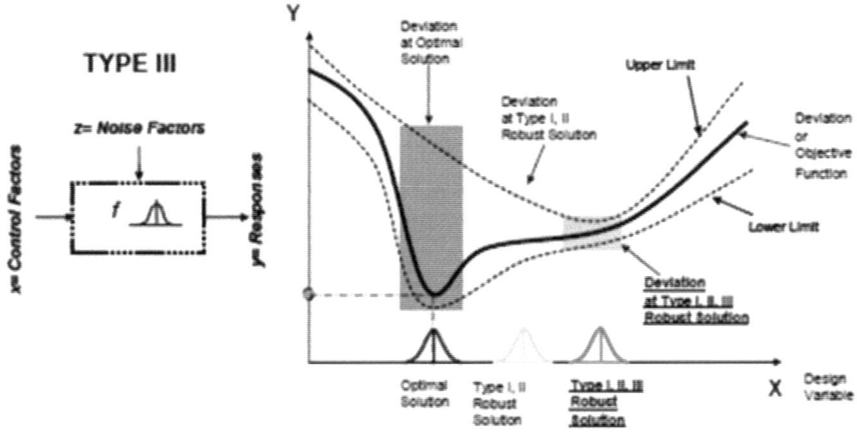

Figure 2. Type III Robust Design [2,4]

As illustrated in 'Figure 2', type III Robust Design Considers not only the objective function but also the two uncertainty limits due to the non-parametric variability, un-configured variability, and model parameter uncertainty, the optimal and Type I and II robust solution have larger performance deviations than the Type I, II, and III robust solution. The most accurate way to incorporate the embedded uncertainties as well as the uncertainty in control and noise factors during design exploration is to perform actual simulation using statistical techniques. Monte Carlo Simulation is a popular method to measure variations of performance by simulating input variations (uncertainty analysis).Even though this approach could produce accurate results in design exploration; it requires a large number of experiments for uncertainty analysis even in a single evaluation during a design exploration process. It is nearly impossible to employ this approach in materials design exploration even if a sampling technique have been applied, such as Latin Hypercube sampling, to reduce number of experiments. So the computationally inexpensive uncertainty analysis method is necessary to solve this problem.

3.2.2. Robust design of multidisciplinary systems

There are many techniques for multi-level deterministic optimization which taken account the uncertainty. An example; the Analytical Target Cascading (ATC) approach [20], ATC is an approach for cascading overall system design targets to sub-system specifications based on an hierarchical multi-level decomposition [20]. The subsystem design problems are formulated as minimum deviation optimization problems and the outputs of the subsystem problems are used as inputs to the system-level problem. This approach eliminates bottom-up uncertainty propagation though a multi-level chain since the problem is decomposed and design exploration is executed only at each level, also the super-level iteration must wait for the results of sub-level optimization; therefore, multi-level optimizations are highly interdependent and cannot be parallelized.

To have the possibility to parallelize the bottom–up simulations and to identify candidate ranged solutions that must be selected in the context of a top-down strategy in the design system, an iterative approach is essential for bottom-up information flow (simulations, experiments), combined with top down guidance from applications performance requirements. Choi et al. [21, 22] have developed an

approach called the Inductive Design Exploration Method (IDEM), schematically shown in Figure 3. IDEM [22,23] has two major objectives: (1) to explore top–down, requirements driven design, guiding bottom–up modeling and simulation, and (2) to manage uncertainty in model chains.
It's based on three steps:

- Step 1: It is necessary to define the rough design space (x space in 'Figure 3'), the interdependent space (y space), and the performance space (z spaces). Discrete points are generated in each of these spaces (Discretizing).
- Step 2: The discrete points which are generated are evaluated based on the mapping models (models f and g in 'Figure 3') and the evaluated data sets which are composed of a discrete input point and output range are stored in a database.
- Step 3: Feasible regions in y and x spaces are sequentially identified, along with a given final performance range in z space. We call this step Inductive Discrete Constraints Evaluation (IDCE).

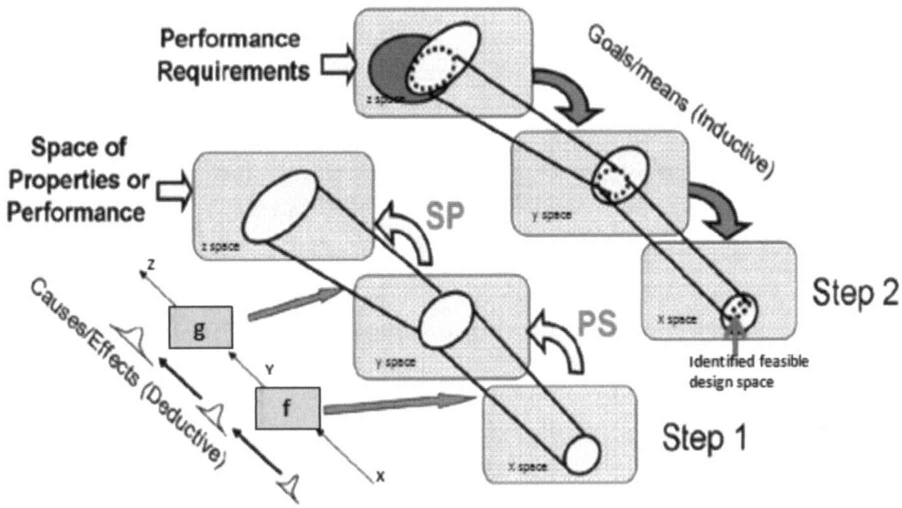

Figure 3. A schematic of steps 1 and 2 in IDEM [1].

The two first steps of IDEM process is as follows:

- Discretizing: all possible combinations of discrete input of associated input variables are created
- Grouping: The discretized points created by the discretization process are grouped as input sets for mapping models.
- Mapping: The points in each group are evaluated in parallel by each mapping model and the evaluation results are stored with the input points.
- Merging: Sets of input points and corresponding outputs obtained in the mapping step are merged to form a set of evaluated original discrete points.

The third step (IDCE) of IDEM aims to find feasible regions in y and x spaces, by using the Hyper Dimensional Error Margin Indices (HD_EMIs) as metrics
Despite of the advantages offered by IDEM, it's still short, first because it cannot be overcome the uncertainty in material models due to idealization associated with a lack of complete knowledge of a given phenomenon and its description. For example, there is uncertainty in the constitutive models, which are used to represent behavior of individual phases. Second IDEM may be computationally intensive if the number of design variables is large or the simulation model is expensive.

Recently, IDEM has been extended to account for model parameter uncertainty [23]. Further work needs to be carried out to account for other representations of epistemic uncertainty such as intervals and fuzzy sets.

4. Critical evaluation of uncertainty managing approaches from material modelling perspective

After presenting an overview of available methods and approaches of the design under uncertainty process, the capabilities of those methods and approaches are critically evaluated from material multi-scale modelling perspective. Then, the research opportunities in this field are discussed, in order to develop new perspective for overcoming the shortness of present approaches.

The table 2 summarizes all the gaps among the management taxonomy of uncertainty including quantification and mitigation uncertainty discussed above.

Table 2. Summary of critical evaluation of the reviewed approaches and methods.

Issues	Approach	Limits
Classification of uncertainty	Classification by the nature of uncertainty	The classification is difficult to incorporate with mathematical modelling in systems design
	Classification by the sources of uncertainty	There are disagreements among the researchers for defining types of uncertainty. None of the reviewed approaches studied explicitly the type of uncertainty in computational models that predict material behaviours.
Analysis and quantification of uncertainty	Existing methods for uncertainty analysis (Interval analysis, Monte Carlo method, etc.)	Uncertainty analysis methods for nondeterministic system models have not been developed in literature except for Monte Carlo (MC) method. The MC method is computationally intensive to apply into materials design
	Methods for reducing uncertainty	Designers must have some true results. In some cases, we cannot obtain any better data than current data due to severe uncertainty in a system or restrictions of computational or experimental expenses, which is the case of multiscale materials design
Robust design under uncertainty	Type I robust design (Taguchi method), Suh's Axiomatic design	Type I robust design is not valid If noise factors cannot be quantified as numeric parameters or quantitative levels. However, the types of uncertainty in materials design problem tend to be unparameterizable.
	Type II robust design (RCEM)	No study has been done for establishing robust design methods for designing a system to be insensitive to uncertainty and unquantifiable variability (such as morphology changes in materials) in a system model.
	Type III robust design	Any computationally inexpensive uncertainty analysis method have been developed for this approach
Robust design of multidisciplinary systems	Analytical Target Cascading (ATC)	Eliminates bottom-up uncertainty propagation though a multi-level chain since the problem is decomposed and design exploration is executed only at each level. Multi-level optimizations are highly interdependent and cannot be parallelized (the super-level iteration must wait for the results of sub-level optimization).
	Inductive design exploration method (IDEM)	It cannot be overcome the uncertainty in material models due to idealization associated with a lack of complete knowledge of a given microstructure and its description. May be computationally intensive if the number of design variables is large or the simulation model is

expensive.

The problem of uncertainty in material design process has been widely studied, but The main conclusion to be drawn is that none of the outlined approaches for mitigation uncertainty had satisfactory results to deal with this endemic uncertainty in constitutive models, which are used to represent behaviour especially in micromechanical modelling, that why the improvement of the accuracy of this models and facilitate their incorporating in material design process is our primary proposes. In this perspective, new approach for modelling micromechanical models has to be carried out, which will be able to gather between homogenization requirements and between material design exigencies.

5. Closure

The main issues in integrating micromechanical models in material design process were investigated. This investigation was mainly based on the uncertainty characterization and quantification in these models and the robust design under uncertainty. Each of these issues has been studied rigorously. We, also, reviewed the literature related to these issues and we critically evaluate existing methods dealing with uncertainty; first to identify their drawbacks and second to clarify the research opportunities in this field. The principal conclusion is that, up to now, there is no approach to deal with uncertainty in micromechanical models, due to idealization resulting from the lack of complete knowledge of microstructure and its description. As a perspective, we aim to develop a robust material design perspective framework for developing efficient micromechanical models.

References

[1] McDowell D L and Story T L 1998 New directions in materials design science and engineering *Report of a Workshop Sponsored by the US National Science Foundation* Georgia Institute of Technology and Morehouse College Atlanta GA.

[2] Panchal J H, Surya R Kalidindi and David L McDowell 2013 Key computational modeling issues in Integrated Computational Materials Engineering *Computer-Aided Design* **45** 4–25

[3] McDowell D L and G B Olson 2008 Concurrent design of hierarchical materials and structures *Springer* 10.1007/s10820-008-9100-6

[4] Choi H J 2005 *A robust design method for model and propagated uncertainty* Ph.D. Dissertation, The GW Woodruff School of Mechanical Engineering, Georgia Institute of Technology, Atlanta, GA.

[5] Ayyub Bilal M 1997 *Uncertainty modelling and analysis in civil engineering.* CRC Press New York, 3-32.

[6] Klir, George, and Bo Yuan 1995 *Fuzzy sets and fuzzy logic* New Jersey: Prentice hall **4.**

[7] Isukapalli S S, Roy A and Georgopoulos P G 1998 Stochastic Response Surface Methods (SRSMs) for Uncertainty Propagation: Application to Environmental and Biological Systems *Risk Analysis* **18** 351-63.

[8] Der Kiureghian A 1989 Measures of Structural Safety under Imperfect State of Knowledge J. *Structural.Eng* **115** 1119-39.

[9] Haukass T 2003 Types of Uncertainties, Elementary Data Analysis, Set Theory, Reliability and Structural Safety *Lecture Notes*, University of British Columbia.

[10] Nikolaidis E 2005 Types of Uncertainty in Design Decision Making *Engineering Design Reliability Handbook* CRC Press New York, pp. 8-1.

[11] Jin R, X Du and Chen W 2003 The Use of Metamodeling Techniques for Optimization under Uncertainty *Journal of Structural & Multidisciplinary Optimization* **25** 99-116.

[12] Sargent R G 2003 Verification and Validation of Simulation Models *Winter Simulation Conferences* New Orleans LA 37- 48.

[13] Kennedy M C and O'Hagan A 2000 Predicting the output from a complex computer code when fast approximations are available *Biometrika* **87** 1-13.

[14] Qian Z , Seepersad C C , Joseph V R, Wu C F J and Allen J K 2004 Building Surrogate Models based on Detailed and Approximate Simulations *ASME DETC04* Salt Lake City UT DETC2004/DAC57486.

[15] Allen JK, Seepersad CC, Choi H J and Mistree F 2006 Robust design for multidisciplinary and multiscale applications *ASME Journal of Mechanical Design* **128** 832–43.

[16] GANO, Shawn E, RENAUD, John E, AGARWAL, Harish, *et al* 2006 Reliability-based design using variable-fidelity optimization. *Structures and Infrastructure Engineering* **2** 247-60.

[17] Taguchi G 1993 Taguchi on Robust Technology Development Bringing Quality Engineering Upstream *ASME Press* New York.

[18] Chen W, Simpson T W, Allen J K and Mistree F 1996 Using Design Capability Indices to Satisfy a Ranged Set of Design Requirements *Advances in Design Automation* Irvine CA 96-DETC/DAC-1090.

[19] Chen W 1995 *A robust concept exploration method for configuring complex systems* Ph.D Dissertation, GW Woodruff School of Mechanical Engineering, Georgia Institute of Technology Atlanta Georgia.

[20] Kim, Hyung Min, Michelena, Nestor F, Papalambros, Panos Y, *Et Al* 2003 Target cascading in optimal system design *Journal of mechanical design* **125** 474-80.

[21] Panchal J H, Choi H J, Shepherd J, Allen J K, McDowell D L and Mistree F 2005 A strategy for simulation-based multiscale,multifunctional design of products and design processes *ASME Design Automation Conference* Long Beach CA Paper Number DETC2005-85316.

[22] Choi H J, Allen JK, Rosen D, McDowell DL, Mistree F 2005 An inductive design exploration method for the integrated design of multi-scale materials and products in: *Design engineering technical conf – design automation conf*, Long Beach, CA, Paper Number DETC2005 85335.

[23] Sinha A, Panchal JH, Allen JK and Mistree F 2011 Managing uncertainty in multiscale systems via simulation model refinement *ASME 2011 international design engineering technical conf and computers and information in engineering conf American Society of Mechanical Engineers* 471-83

Some characteristic behaviours of a spin-1/2 Ising nanoparticle

M El Hamri[1], S Bouhou[1], I Essaoudi[1], A Ainane[1,2,3,4], R Ahuja[3] and F Dujardin[4]

[1]Laboratoire de Physique des Matériaux et Modélisation des Systèmes, (LP2MS), Unité Associée au CNRST-URAC 08, University of Moulay Ismail, Physics Department, Faculty of Sciences, B.P. 11201, Meknes, Morocco.
[2]Max-Planck-Institut für Physik Complexer Systeme, Nöthnitzer Str. 38 D-01187 Dresden, Germany.
[3]Condensed Matter Theory Group, Department of Physics and Astronomy, Uppsala University, 75120 Uppsala, Sweden.
[4]Université de Lorraine, LCP-A2MC, Institut de Chimie, Physique et Matériaux, 1 Bd. Arago, 57070 Metz, France.

E-mail: ainane@pks.mpg.de

Abstract. By using the effective-field theory with correlations based on the probability distribution technique; the magnetization, susceptibility, internal energy, specific heat and the free energy expressions for a ferromagnetic spin-1/2 Ising nanoparticle have been developed and calculated numerically for different surface shell exchange coupling parameter. A number of interesting phenomena have been observed, depending on the surface shell exchange coupling term.

1. Introduction

Magnetic nanoparticle (NP) systems have been receiving considerable attention in the last 15 years from both experimental and theoretical researchers [1, 2]. This is motivated by numerous possibilities of their applications in permanent magnets, recording media, microwave absorption, and in biomedical applications [3]. Furthermore, these NPs show excellent new properties which are totally different from those observed in the bulk counterparts, such as superparamagnetism, high field irreversibility, high saturation field, extra anisotropy contributions or shifted loops after field cooling [4, 5], because the NPs are greatly affected by the particle size [6].

From the theoretical point of view, these systems have been studied by a variety of techniques such as mean-field theory (MFT) [7], effective-field theory (EFT) with correlations [8, 9], Green Functions (GFs) formalism [10], and Monte Carlo (MC) simulations [11, 12]. By using the EFT with correlations, the magnetic and the thermodynamic properties of several magnetic NPs have been studied by different theoretical researchers. For instance, Canko *et al.* [13] have investigated the effect of crystal field on the magnetization, magnetic susceptibility, specific heat, internal energy, specific heat and the free energy of a Blume-Capel Ising nanotube. The authors found that the system undergoes first- and second- order phase transitions. The same results were found by Kantar *et al.* [14], by studying the thermodynamic and magnetic properties of ternary mixed Ising NPs with core/shell structure. Similarly, Şarlı [15] has investigated the temperature dependencies of the magnetic properties of the both ferromagnetic and the antiferromagnetic cylindrical mixed spin-1/2 core and spin-1 shell Ising nanotube system. The author found that the nanotube system undergoes a first-order phase transition for the weak surface/shell exchange interaction. However, it undergoes only a second-order phase transition for the strong interaction. Moreover, Kantar *et al.* study [16] on the thermal variation of magnetization of the mixed spin-1/2 core and spin-3/2 shell Ising NPs, with a crystal field interaction show that this system exhibits just a second-order phase transition as it is proved by the free energy of the system, and they have also found one or two compensation points in the magnetization curves. However, only one compensation point has been found by Kocakaplan *et al.* [17] in the thermal variations of the total magnetization of the spin-1/2 hexagonal Ising nanowire system with core/shell structure. Recently, in our work [18, 19], we have investigated the effect of the exchange interaction between core and surface shell spins on the thermodynamic and the magnetic properties of a transverse antiferromagnetic Ising NPs. We have found a number of interesting phenomena, such as the existence of the compensation points. In addition, two and even three compensation points have been found in the magnetization curves of a double-wall cubic metal nanotube by Liang *et al.* [20], which are not predicted in the Néel theory [21]. They have also observed a discontinuity at transition temperature in all the initial susceptibility curves.

The purpose of this work is to investigate the effect of the surface shell exchange coupling on the magnetic and thermodynamic properties of a ferromagnetic Ising nanocube. For this aim, we have organized this paper as follows: In Section 2, we outline the formalism. In Section 3, we present the results and discussion, and finally we give a brief conclusion in Section 4.

2. Model and formalism

We consider a 5 × 5 × 5 ferromagnetic cubic NP composed of a ferromagnetic spin-1/2 core which is interacting with a ferromagnetic spin-1/2 surface shell as depicted in the figure below (figure 1).

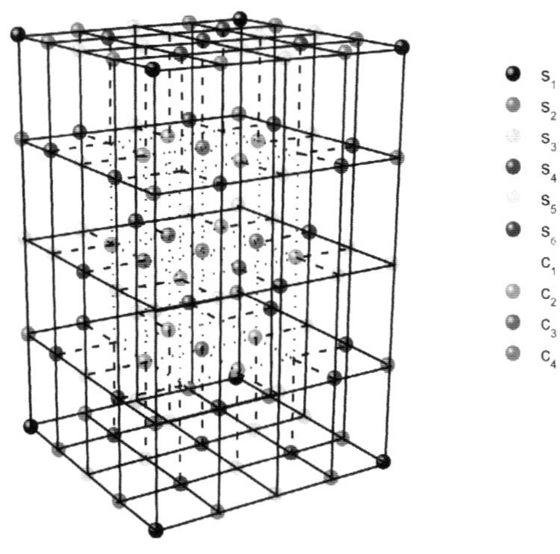

Figure 1: Schematic representation of magnetic spins in a cubic nanoparticle. Solid, dotted, and dashed lines represent the exchange interactions of the surface shell, core, and interface, respectively. The spin holders at the surface sites are denoted by S_1, S_2, S_3, S_4, S_5 and S_6, whereas C_1, C_2, C_3 and C_4 define the spin holders in the core site.

The inner spins are called the core (c) region which is surrounded by the outer spins that are known as the surface shell (s) of the particle. The number of core spins is $N_c = 27$ and the number of surface shell spins is $N_s = 98$. The Hamiltonian of the system is expressed as follows:

$$H = -J_s \sum_{<ij>} \sigma_i^z \sigma_j^z - J_c \sum_{<nm>} \sigma_n^z \sigma_m^z - J_{cs} \sum_{<im>} \sigma_i^z \sigma_m^z, \tag{1}$$

where σ_i^z denotes the z component of a quantum spin operator $\vec{\sigma}$ of magnitude $\sigma = 1/2$ at the site i. J_s, J_c and J_{cs} are the exchange interactions between the nearest-neighbour magnetic spins in the surface shell, the core and between the core and the surface shell interface ($J_{cs} > 0$) of the nanocube, respectively.

The theory to be used is the EFT in which the attention is focused on the cluster comprising just a single selected spin and the neighbouring spins with which it directly interacts. To this end, the Hamiltonian is split into two parts

$$H = H' + H_i. \tag{2}$$

The first term denoted by H' does not depend on the site i, while the second term (H_i) includes all contributions associated with the site i:

$$H_i = -(\sum_j J_{ij} \sigma_j^z) \sigma_i^z \tag{3}$$

The term J_{ij} represents the strength of the exchange interaction between the spins at nearest-neighbour sites i and j, where i is the single selected spin and j represents the neighbouring spins with which it directly interacts.

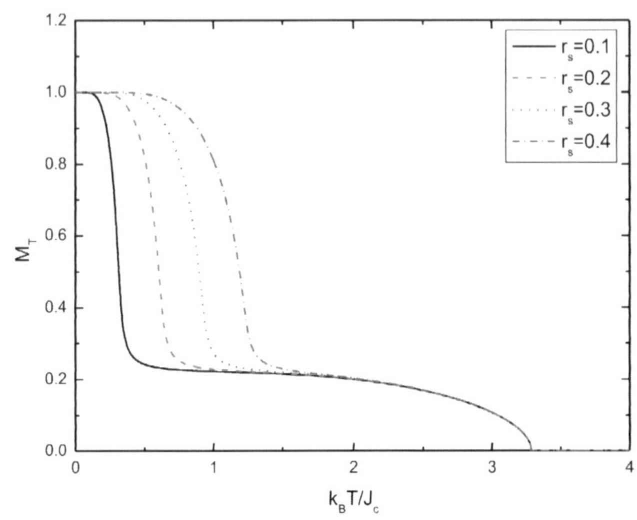

Figure 2: Reduced temperature dependence of the total magnetization of the system for some selected values of the surface shell exchange coupling r_s ($r_s = 0.1, 0.2, 0.3$ and 0.4), with $r_{cs} = 0.01$.

Using the approximation introduced by Sá Barreto *et al.* [22] we obtain the identity

$$\langle \sigma_i^z \rangle = \left\langle \frac{Tr\left[\sigma_i^z e^{-\beta H_i}\right]}{Tr\left[e^{-\beta H_i}\right]} \right\rangle, \tag{4}$$

where the angular bracket <...> denotes the canonical thermal average, $\beta = (k_B T)^{-1}$ with k_B stands for the Boltzmann constant and T is the temperature. If the exchange interactions were restricted only to the nearest-neighbours, and using the EFT with a probability distribution technique [23], the longitudinal magnetization of the system would be given by

$$m_i^z = \langle \sigma_i^z \rangle = \left\langle f_z\left(\sum_j J_{ij}\sigma_j^z \right) \right\rangle, \tag{5}$$

with

$$f_z(\sum_i J_{ij}\sigma_i^z) = \frac{1}{2}\tanh(\frac{\beta}{2}\sum_i J_{ij}\sigma_i^z). \tag{6}$$

To perform the thermal averaging on the right-hand side of equation (5), we follow the general approach described in reference [23]. First of all, in the spirit of the EFT, multispin-correlation functions are approximated by products of single spin averages. We then take advantage of the integral representation of the Dirac's delta distribution in order to write equation (5) in the following form

$$m_i^z = \int dy f_z(y) \frac{1}{2\pi}\int \left[d\lambda \exp(iy\lambda) \prod_j \langle \exp(i\lambda J_{ij}\sigma_j^z) \rangle \right] \tag{7}$$

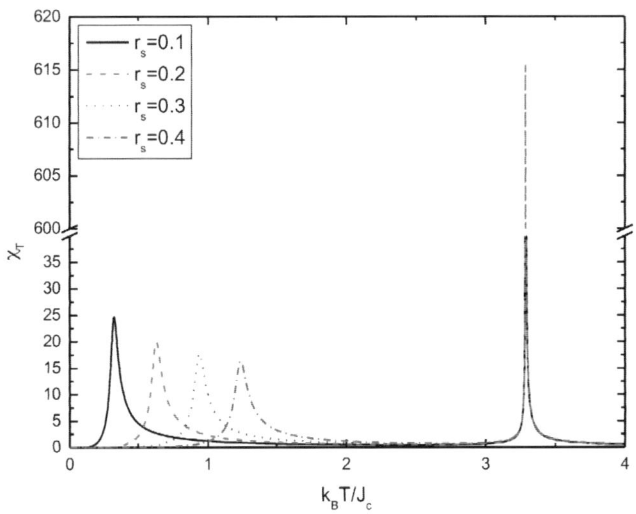

Figure 3: Reduced temperature dependence of the total susceptibility of the system for some selected values of the surface shell exchange coupling r_s (r_s = 0.1, 0.2, 0.3 and 0.4), with r_{cs} = 0.01.

In the calculation of equation (7), the commonly used approximation has been made according to which the multi-spin correlation functions are decoupled into products of the spin average. To make progress, we introduce the probability distribution of the spin variable σ_j^z:

$$\mathrm{P}(\sigma_j^z) = \frac{1}{2}[(1-2\mathrm{m}_i^z)\delta(\sigma_j^z + \frac{1}{2}) + (1+2\mathrm{m}_i^z)\delta(\sigma_j^z - \frac{1}{2})] \tag{8}$$

The explicit formulation of magnetizations has been given in reference [32]. The total longitudinal magnetization per site is given by

$$M_T = \frac{1}{125}(98M_s + 27M_c) \tag{9}$$

where M_s and M_c represent respectively the longitudinal magnetization of surface shell and core of the nanocube, which are given by

$$M_c = \frac{1}{27}(8m_{c_1}^{z} + 12m_{c_2}^{z} + 6m_{c_3}^{z} + m_{c_4}^{z}), \tag{10}$$

and

$$M_s = \frac{1}{98}\left(8m_{s_1}^{z} + 24m_{s_2}^{z} + 12m_{s_3}^{z} + 24m_{s_4}^{z} + 24m_{s_5}^{z} + 6m_{s_6}^{z}\right) \tag{11}$$

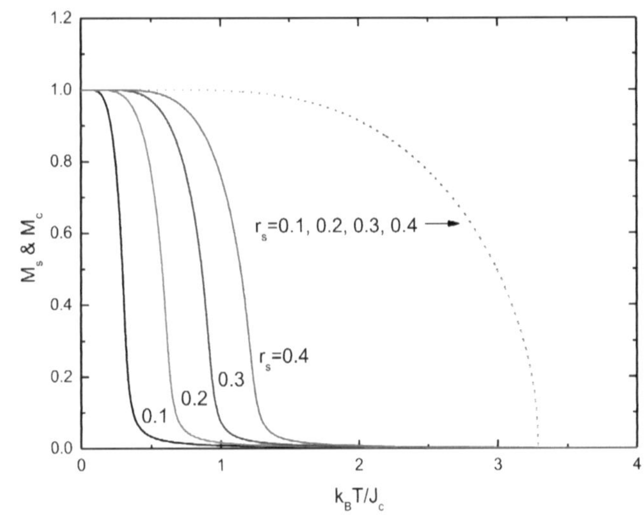

Figure 4. Reduced temperature dependence of the core and surface shell magnetizations of the system for some selected values of the surface shell exchange coupling r_s (r_s=0.1, 0.2, 0.3 and 0.4), with r_{cs}=0.01.

It is also interesting to study the behaviour of the longitudinal susceptibility of each site of the nanocube, which is defined by

$$\chi_T = \frac{1}{125}(98\chi_s + 27\chi_c) \tag{12}$$

where the core and the surface shell susceptibilities are given by

$$\chi_c = \frac{1}{27}\left(8\chi_{c_1}^{z} + 12\chi_{c_2}^{z} + 6\chi_{c_3}^{z} + \chi_{c_4}^{z}\right) \tag{13}$$

and

$$\chi_s = \frac{1}{98}\left(8\chi_{s_1}^{z} + 24\chi_{s_2}^{z} + 12\chi_{s_3}^{z} + 24\chi_{s_4}^{z} + 24\chi_{s_5}^{z} + 6\chi_{s_6}^{z}\right) \tag{14}$$

Journal of Physics: Conference Series **758** (2016) 012023 doi:10.1088/1742-6596/758/1/012023

The details of calculus of the each site of the longitudinal susceptibility have been given in reference [32].
By using the approximated spin correlation identities introduced by Sá Barreto *et al.* [24]

$$\langle f_i \sigma_i^z \rangle = \left\langle f_i \frac{Tr\left[\sigma_i^z e^{-\beta H_i}\right]}{Tr\left[e^{-\beta H_i}\right]} \right\rangle, \tag{15}$$

we can easily obtain the internal energy U of the system from the thermodynamic average of the Hamiltonian, as it has done by Kaneyoshi *et al.* [25] in the mixed-spin system

$$U = \langle H \rangle = -\frac{1}{2}(\frac{1}{125}(8\langle u_{c_1}^z \rangle + 12\langle u_{c_2}^z \rangle + 6\langle u_{c_3}^z \rangle + \langle u_{c_4}^z \rangle \\ + 8\langle u_{s_1}^z \rangle + 24\langle u_{s_2}^z \rangle + 12\langle u_{s_3}^z \rangle + 24\langle u_{s_4}^z \rangle + 24\langle u_{s_5}^z \rangle + 6\langle u_{s_6}^z \rangle)) \tag{16}$$

where

$$\langle u_i^z \rangle = \left\langle \sum_j J_{ij} \sigma_j^z f_z \left(\sum_j J_{ij} \sigma_j^z \right) \right\rangle. \tag{17}$$

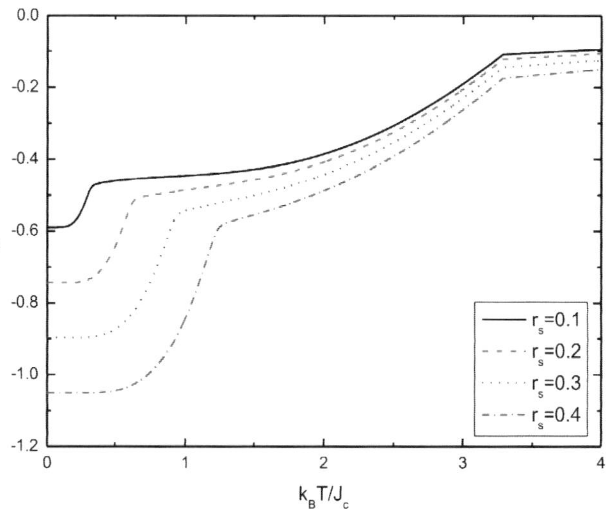

Figure 5. Reduced temperature dependence of the internal energy of the system for some selected values of the surface shell exchange coupling r_s ($r_s = 0.1, 0.2, 0.3$ and 0.4), with $r_{cs} = 0.01$.

The specific heat of the system is obtained from the relation:

$$C_h = \frac{\partial U}{\partial T} \tag{18}$$

The free energy of the system is defined as:

$$F = U - T\int_0^T \frac{C_h}{T'} dT' \tag{19}$$

By solving all these equations numerically, we can easily obtain the magnetic and the thermodynamic properties of the surface/shell nanocube system.

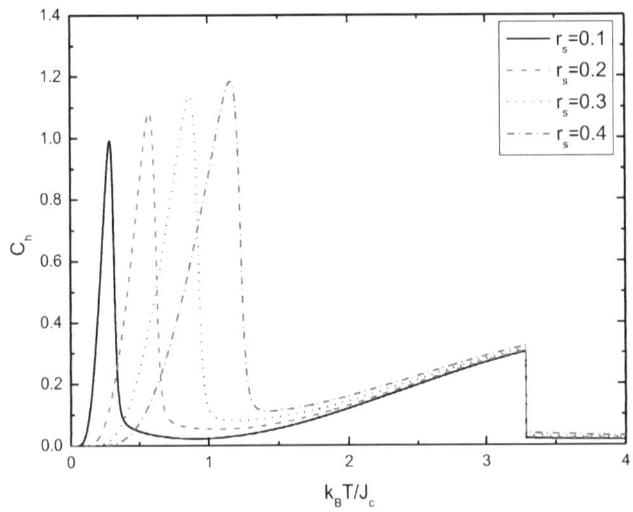

Figure 6. Reduced temperature dependence of the specific heat of the system for some selected values of the surface shell exchange coupling r_s (r_s = 0.1, 0.2, 0.3 and 0.4), with r_{cs} = 0.01.

3. Numerical results and discussion

In this section, we investigate the effect of the surface shell exchange coupling r_s on the magnetic and thermodynamic properties of a ferromagnetic spin-1/2 Ising nanocube. In the following discussion, we take J_c as a unit of the energy and we define the reduced exchange interactions as ($r_{cs}=J_{cs}/J_c$ and $r_s = J_s/J_c$).

The effect of the surface shell exchange coupling on the magnetic and thermodynamic properties of the particle is examined in figures 2, 3, 4, 5, 6 and 7 for some selected values of r_s (r_s = 0.1,0.2,0.3 and 0.4), and with a fixed value of r_{cs} = 0.01. Figure 2 shows the total magnetization versus reduced temperature (k_BT/J_c). As seen from this figure, all the magnetizations are 1.0 at zero temperature, then they suddenly decrease at the reduced temperatures: k_BT/J_c = 0.322, 0.629, 0.935 and 1.24 when r_s = 0.1, 0.2, 0.3 and 0.4, respectively and become zero at the same critical temperature: k_BT_c/J_c = 3.286. Therefore, a second-order phase transition occurs at this critical temperature and the phase transition is from the ferromagnetic phase to the paramagnetic one. The shape of the magnetization curves is S-type behaviour which is not predicted in the Néel nomenclature [21]. This shape is also found theoretically in references [16, 17; 26-31]. Figure 3 shows the temperature dependence of the total susceptibility χ_T, for some selected values of r_s. Two distinct peaks appear in these curves for each value of r_s as seen in this figure. The first one which has a sharp shape corresponds to the fluctuation of the magnetization curves, whereas the second one emerges at the critical temperature. Figure 4 illustrates the thermal variation of the core and surface shell magnetizations for the same parameters as those in the above figures. In this figure, $M_s = M_c$ = 1.0 at zero temperature. In addition, with increasing temperature the M_c curves show a monotonic decrease exhibiting the Q-type behaviour [21] which indicate that r_s does not have any considerable effect on the core magnetization, whereas M_s curves take several magnitudes in depending on r_s values which indicate that r_s has an apparent effect on the thermal variation of the surface shell magnetization. These results indicate that the main contribution of the total magnetization behaviour is the surface shell magnetization of the particle as it is found in our previous work of an antiferromagnetic transverse spin-1/2 Ising nanocube [32]. Figure 5 shows the effect of r_s on the internal energy. One can see that, the internal energy curves have a fluctuation at the reduced temperatures: 0.322, 0.629, 0.935 and 1.24 when r_s =0.1, 0.2, 0.3 and 0.4, respectively, and represent a discontinuity of the curvature at the critical temperatures. In addition, for any fixed value of k_BT/J_c, the stronger surface shell exchange coupling makes internal energy U decreasing. The specific heat-reduced temperature is plotted in figure 6. It is clearly seen that the C_h profile undergoes a sharp drop at the critical temperature after the change of the curvature at the fluctuation of the magnetization curves. The same profile is also observed in the specific heat curve of the spin-$\frac{3}{2}$ Blume-Capel model [33]. In figure 7 we plot the numerical results of free energy of the system for the same parameters in previous figures. We note from this figure that the system presents only a second order phase transition because the free energy curves do not exhibit a discontinuous behaviour at the critical temperature. We can also notice that the free energy is equal to the internal energy at the ground state.

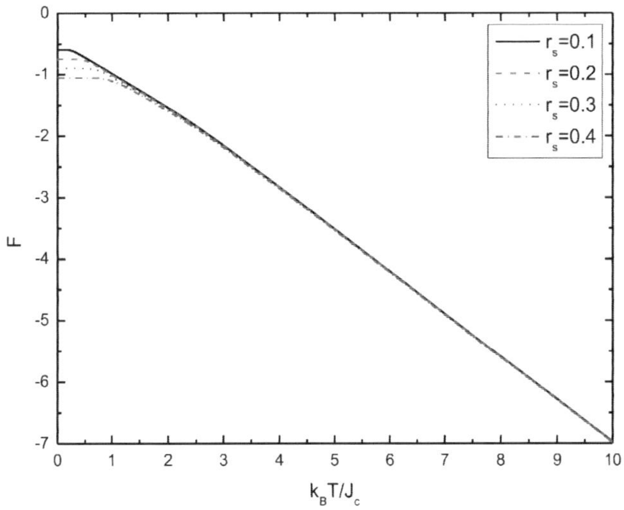

Figure 7. Reduced temperature dependence of the free energy of the system for some selected values of the surface shell exchange coupling r_s (r_s= 0.1, 0.2, 0.3 and 0.4), with r_{cs}= 0.01.

In order to investigate the effect of r_s on the critical temperature of the system, we have plotted the phase diagram (k_BT_c/J_c- r_s) in figure 8. It is clearly seen from this figure that, when r_s is less than 1.07 the critical temperature remains almost constant for all value of r_s, whereas, for r_s > 1.07 it increases linearly. The same behaviour is observed in a ferromagnetic cylindrical Ising nanowire [34].

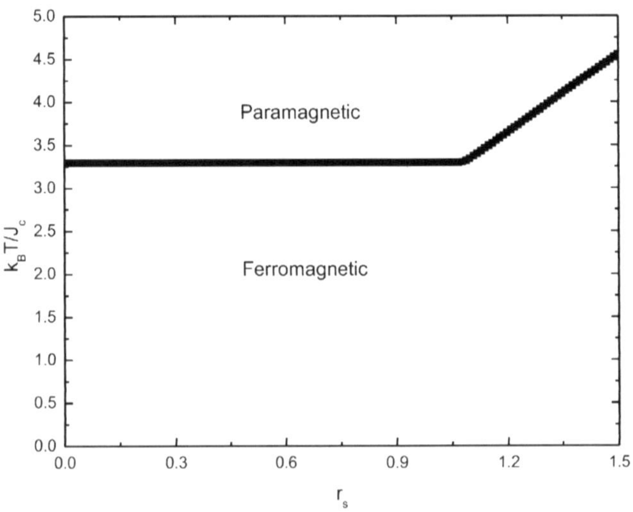

Figure 8. Phase diagram of the system in (k_BT_c/J_c-r_s) plane for $r_{cs} = 0.01$.

Conclusion

In this work, we have studied the thermodynamic and magnetic properties of a ferromagnetic nanocubic particle by employing the EFT based on the probability distribution technique with correlations. It is found that the surface shell exchange coupling r_s has a strong effect on the behaviour of the magnetization, internal energy, specific heat and free energy of the system.

Acknowledgement

This work has been initiated with the support of URAC: 08, the project PPR:(MESRSFC-CNRST) and the Swedish Research Links programme dnr-348-2011-7264 and completed during a visit of A. A. at the Max Planck Institut für Physik Komplexer Systeme Dresden, Germany. The authors would like to thank all the organizations.

4. References

[1] Kodama R H 1999 Magnetic nanoparticles *J. Magn. Magn. Mater.* **200** 359-372
[2] Vasilakaki M and Trohidou K N 2009 Numerical study of the exchange-bias effect in nanoparticles with ferromagnetic core/ferromagnetic disordered shell morphology *Phys. Rev. B* **79** 144402-144409
[3] López-Ortega A, Estrader M, Salazar-Alvarez G, Roca A G and Nogués J 2014 Applications of exchange coupled bi-magnetic hard/soft and soft/hard magnetic core/shell nanoparticles *Phys. Rep.* **553** 1-32
[4] Skumryev V, Stoyanov S, Zhang Y, Hadjipanayis G, Givord D and Nogués J 2003 Beating the superparamagnetic limit with exchange bias *Nature* **423** 850-853
[5] Nogués J, Sort J, Langlais V, Skumryev V, Suriñach S, Muñoz J S and Baró M D 2005 Exchange bias in nanostructures *Phys. Rep.* **422** 65-117
[6] Batlle X and Labarta A 2002 Finite-size effects in fine particles: magnetic and transport properties *J. Phys. D: Apply. Phys.* **35** R15-R42
[7] Kaneyoshi T 2009 Magnetizations of a nanoparticle described by the transverse Ising model *J. Magn. Magn. Mater.* **321** 3430-3435

[8] Bouhou S, Essaoudi I, Ainane A, Dujardin F, Ahuja R and Saber M 2013 Magnetic properties of diluted nanowire *J. Supercond. Nov. Magn.* **26** 201-211
[9] Kocakaplan Y and Keskin M 2014 Hysteresis and compensation behaviors of spin-3/2 cylindrical Ising nanotube system *J. Appl. Phys.* **116** 093904-093912
[10] Garanin D A and Kachkachi H 2003 Surface contribution to the anisotropy of magnetic nanoparticles *Phys. Rev. Lett.* **90** 65504-65508
[11] Iglasias O, Batlle X and Labarta A 2005 Microscopic origin of exchange bias in core/shell nanoparticles *Phys. Rev. B* **72** 212401-212404
[12] Masrour R, Bahmad L, Hamedoun M, Benyoussef A and Hlil E K 2013 The magnetic properties of a decorated Ising nanotube examined by the use of the Monte Carlo simulations *Solid State Commun.* **126** 53-56
[13] Canko O, Erdinc A, Taskin F and Atis M 2011 Some characteristic behavior of spin-1 Ising nanotube *Phys. Lett. A* **375** 3547-3551
[14] Kantar E and Keskin M 2014 Thermal and magnetic properties of ternary mixed Ising nanoparticles with core-shell structure: effective-field theory approach *J. Magn. Magn. Mater.* **349** 165-172
[15] Şarlı N 2013 Band structure of the susceptibility, internal energy and specific heat in a mixed core/shell Ising nanotube *Physica B* **411** 12-25
[16] Kantar E, Deviren B and Keskin M 2013 Magnetic properties of mixed Ising nanoparticles with core-shell structure *Eur. Phys. J. B* **86** 253-265
[17] Kocakaplan Y and Kantar E 2014 An effective-field theory study of hexagonal Ising nanowire: thermal and magnetic properties *Chin. Phys. B* **23** 046801-046808
[18] Bouhou S, El Hamri M, Essaoudi I, Ainane A and Ahuja R 2015 Magnetic properties of a single transverse Ising ferrimagnetic nanoparticle *Physica B* **456** 142-150
[19] El Hamri M, Bouhou S, Essaoudi I, Ainane A, Ahuja R and Dujardin F 2015 Thermodynamic properties of the core/shell antiferromagnetic Ising nanocube *J. Supercond. Nov. Magn.* **28** 3127-3133
[20] Liang J Y, Zou C L, Jiang W and Li X X 2015 Ferrimagnetic behaviors in a double-wall cubic metal nanotube *Physica E* **69** 81-88
[21] Néel L 1948 Propriétés magnétiques des ferrites: ferrimagnétisme et antiferromagnétisme *Annales de Physique* **3** 137-148
[22] Sá Barreto F C and Fittipaldy I P 1985 Thermodynamical properties of the transverse Ising model *Physica A* **129** 360-373
[23] Saber M 1997 A simple approximation method for dilute Ising systems *Chin. J. Phys.* **35** 577-583
[24] Sá Barreto F C, Fittipaldi I P and Zeks B 1981 New effective field theory for the transverse Ising model *Ferroelectrics* **39** 1103-1106
[25] Kaneyoshi T, Jaščur M and Fittipaldi I P 1993 Transverse Ising model with arbitrary spin *Phys. Rev. B* **48** 250-255
[26] Deviren B, Ertaş M and Keskin M 2012 Dynamic magnetizations and dynamic phase transitions in a transverse cylindrical Ising nanowire *Phys. Scr.* **85** 055001-055010
[27] Deviren B and Keskin M 2012 Thermal behavior of dynamic magnetizations, hysteresis loop areas and correlations of a cylindrical Ising nanotube in an oscillating magnetic field within the effective-field theory and the Glauber-type stochastic dynamics approach *Phys. Lett. A* **376** 1011-1019
[28] Deviren B, Kantar E and Keskin M 2012 Dynamic phase transitions in a cylindrical Ising nanowire under a time-dependent oscillating magnetic field *J. Magn. Magn. Mater.* **324** 2163-2170
[29] Kocakaplan Y, Kantar E and Keskin M 2013 Hysteresis loops and compensation behavior of cylindrical transverse spin-1 Ising nanowire with the crystal field within effective-field theory based on a probability distribution technique *Eur. Phys. J. B* **86** 420-428

[30] Kantar E and Ertaş M 2014 Magnetic hysteresis and compensation behaviors in spin-1 bilayer Ising model *Solid state commun.* **188** 71-76
[31] Ertaş M and Kantar E 2015 Cylindrical Ising nanowire with crystal field: existence of a dynamic compensation temperatures *Phase transit.* **88** 567-581
[32] El Hamri M, Bouhou S, Essaoudi I, Ainane A and Ahuja R 2015 Investigation of the surface shell effects on the magnetic properties of a transverse antiferromagnetic Ising nanocube *Superlattices Microstruct.* **80** 151-168
[33] Kaneyoshi T and Jaščur M 1992 Specific heat of the spin-3/2 Blume-Capel model *Phys. Rev. B* **46** 3374-3379
[34] Yüksel Y, Akıncı Ü and Polat H 2013 Investigation of bond dilution effects on the magnetic properties of a cylindrical Ising nanowire *Phys. Status Solidi B* **250** 196-206

First principle study of structural, electronic, optical and electrical properties of Ga doped ZnO with GGA and mBJ approximations

M Khuili[1], N Fazouan[1], H Abou El Makarim[2], G El Hallani[1] and E H Atmani[3]
[1]Laboratory of Materials Physics, Faculty of Sciences and Technologies, B.P 523, 23000 Beni Mellal, Morocco.
[2]Laboratory LS3ME, Team Theoretical Chemistry and Molecular Modeling, University of Mohammed V, Faculty of Sciences, Department of Chemistry, BP1014 Rabat Morocco.
[3]Laboratory of condensed matters and renewables energies, Faculty of Sciences and Technologies, B.P 146, 20650 Mohammedia, Morocco

E-mail: fazouan@yahoo.fr

Abstract. Physical properties of Gallium doped ZnO in wurtzite phase has been studied. We have used the Generalized Gradient Approximation (GGA) to determine the structural parameters of each concentration of Ga. The modified Becke–Johnson potential TB-mBJ has been used to calculate the partial and total density of states. The dielectric function, refractive index, optical reflectivity, absorption coefficient and transmittance spectra were predicted using the same potential. The calculated energies of bands structure has been used with the Boltzmann transport equation to calculate the electrical conductivity of Ga doped ZnO. We have observed that the donor concentration, optical band gap and electrical conductivity can be widely tunable with the doping levels of Ga. The results of pure and doped ZnO were in agreement with experimental and other theoretical studies. The obtained results confirm that Ga doped ZnO is a transparent conductor dedicated for applications in photovoltaic devices.

1. Introduction

Transparent conductive oxides (TCO) thin films are materials which have good optical transparency in the visible range and high electrical conductivity. These two characteristics of TCOs depend on the nature, number and atomic arrangement in the crystalline structure, the morphology of the layer and the presence of intrinsic (oxygen vacancies and interstitials of the metal) or extrinsic (dopants) defects. Zinc oxide is one of the TCO, frequently used in several areas such as solar energy and optoelectronics [1],due to its thermal stability [2]. Its wide band gap 3.37eV which allows it to be transparent to visible and infrared radiation, and its large exciton binding energy (60meV) [3].
To enhance its electrical conductivity, Zinc Oxide is usually doped with group III elements such as Ga, B, Al and In as effective donors [4]. Gallium doped zinc oxide has comparable structural and optical properties as pure ZnO seen its ionic and covalent radiuses, which are similar to those of Zn, and is a desirable material for many applications. Moreover, there are studies, which classify it better

conduction stability than Al doped ZnO for the reason that Ga has low reactivity and better resistivity to oxidation than Al [5].
Many theoretical studies have used ab-initio methods, to determine structural, electronic, optical and electrical properties of pure and doped ZnO, and also the percentage of doping to be used to get a material with clearly defined properties. However, density functional theory (DFT) underestimates the band gap of many transition metal oxides as ZnO. For example, the band gap calculated of pure ZnO is about 0.77 eV by the Generalized Gradient Approximation (GGA) [6] which is smaller than experimental results 3.4eV [7]. To get a correct band gap, many theoretical studies was introduced such as GGA+U [8], the Heyd–Scuseria–Ernzerh (HSE) of screened hybrid density functional [9] and GW implementations [10]. The last two methods lead to very expensive time calculations. Recently, Tran and Blaha introduce exchange–correlation potential TB-mBJ [11] which allows to have band gap values similar to those found by the hybrid functional in reasonable time calculation.
In this study, we report the structural, electronic, optical and electrical properties of Ga doped ZnO with different Ga concentrations. To calculate these properties, we employ density functional theory (DFT) employing the Full Potential Linearized Augmented Plane Wave (FL-LAPW) and using TB-mBJ in order to have a good estimate of the optical gap approaching the experimental value.

2. Computational methods

We have employed FP-LAPW approach as implemented in the WIEN2k code [12].A revised Perdew-Burke-Ernzerhof exchange-correlation functional PBEsol[13] is used to study the structural properties of our material. The electronic, optical and electrical properties are determined using TB-mBJ approximation. The values of the smallest muffin tin radius of atoms RMT were chosen as 1.93a.u. for zinc, 1.66a.u. for oxygen and 1.87a.u. for gallium, and an energy cutoff $K_{max} \times R_{MT}=7.5$ was used. The complex dielectric in this code is expressed by the well-known relation,

$$\varepsilon(\omega) = \varepsilon_1(\omega) + i\varepsilon_2(\omega) \tag{1}$$

the optical constants, absorption coefficient $\alpha(\omega)$, reflectivity $R(\omega)$, and transmittance $T(\omega)$ can be derived from $\varepsilon_1(\omega)$ and $\varepsilon_2(\omega)$:
The imaginary part of the $\varepsilon_2(\omega)$ in the long wavelength limit has been obtained directly from the electronic structure calculation [14].

$$\varepsilon_2(\omega) = (4\pi e^2/\omega^2 m^2) \sum_{i,j} \int \langle i|M_{cv}|j\rangle^2 f_i(1-f_i) * \delta(E_f - E_i - \omega) d^3k \tag{2}$$

$M_{cv}(k)$ The transition moments elements, i the initial states and j the final states, f_i is the Fermi distribution.
The real part of the dielectric function follows from the Kramers-Kronig relation [15]

$$\varepsilon_1(\omega) = Re(\varepsilon(\omega)) = 1 + (2/\pi) P \int_0^\infty \omega' \varepsilon_2(\omega')/(\omega'^2 - \omega^2) d\omega' \tag{3}$$

Where P implies the principal value of the integral
The absorption coefficient $\alpha(\omega)$and the reflectivity $R(\omega)$ follows from Fresnels formula[16]:

$$\alpha(\omega) = \omega\sqrt{2}\left[\sqrt{\varepsilon_1^2(\omega) + \varepsilon_2^2(\omega)} - \varepsilon_1(\omega)\right]^{1/2} \tag{4}$$

$$R(\omega) = \left|\frac{(\varepsilon(\omega)-1)^{1/2}}{(\varepsilon(\omega)+1)^{1/2}}\right|^2 \tag{5}$$

To obtain the electrical conductivity we have used the BoltzTrap package [17] based on semi-classic Boltzmann theory which use the calculated band structure data obtained from wien2k.

3. Results and discussion:

3.1. Structural parameters

Zinc oxide often crystallizes in a hexagonal wurtzite structure belonging to space group P63mc [18]. This structure is known by three parameters: the parameter of the cell a, the ratio c/a and the internal parameter u, which must be optimized. The ground state lattice constants of Ga doped ZnO are obtained by using 2D-optimize package available in the Wien2k code [19]. It helps to have a good volume and c/a simultaneously for solid better than those obtained by the 1D based on the minimization of energy by volume. First, we get value of c/a, therefore the energy for each volume, then for those volumes and energies the equation of state (EOS) will be obtained, and finally the optimum value of c/a is calculated from optimal volume. From these results of calculated parameters a and c, we found that they increase with increasing concentration of the incorporation of Ga into ZnO. In the Figure 1 we plotted the calculated lattice constants a and c versus doping concentration of Gallium. Results show that the a and c parameters increase slightly with increasing Ga concentration. This change in lattice constant caused by Ga doping ZnO is due to difference between effective ionic radii between Ga^{3+} (0.57 Å) and Zn^{2+} (0.74 Å) and by Mulliken charge of Zn, Ga and O leading to a weak attractive coulomb interaction which extends the bond lengths of Zn–O and Ga–O when the Ga concentration increases as reported by H.C Wu et al. [20].

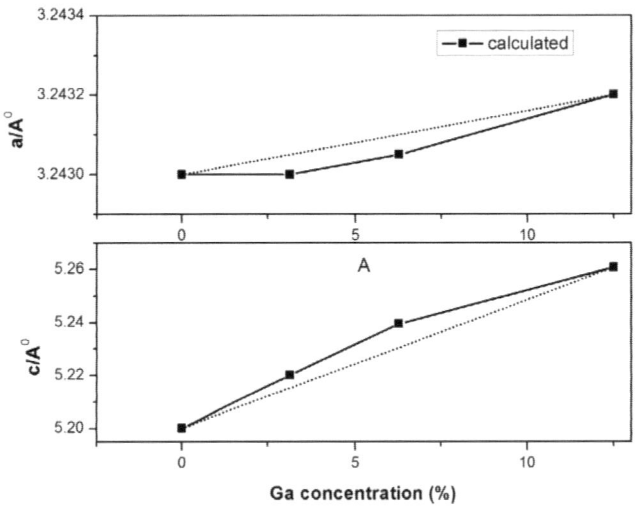

Figure 1. The lattice constants as a function of Ga concentration for (x=0, 3.125%, 6.25% and 12.5%).

3.2. Electronic structure

In this section, Figure 2 shows the band structure of pure ZnO and doped ZnO with 6.25% of Gallium. The band structure is calculated at the theoretical equilibrium configuration wurtzite, using our optimized lattice constants. Zinc oxide displays a direct band gap of about 2.8eV which is in a good agreement with experimental values [21] calculated between the minimum of the conduction band and the maximum of the valence band at Γ-point. In the case of Ga doped ZnO at 6.25%, the Fermi level shifts upward into the conduction band, the band gap is measured between the valence band maximum and the Fermi level in the conduction band [20,22]. The same we have calculated the band gap for the others concentrations of Ga, which are presented in the table 1, and compared with other theoretical and experimental values. We found that increasing the Ga concentration shift up the band gap value, which can be explained by producing a degenerate n-type semiconductor related to a pronounced Burstein–Mott-effect [23, 24].

Table1: Calculated band gap for various concentration of Ga

Concentration of Ga	Calculated band gap (eV)	Theoretical band gap (eV) [20]
0%	2.79	3.37
3.125%	3.34	-
6.25%	3.95	4.37
12.5%	4.32	5.5

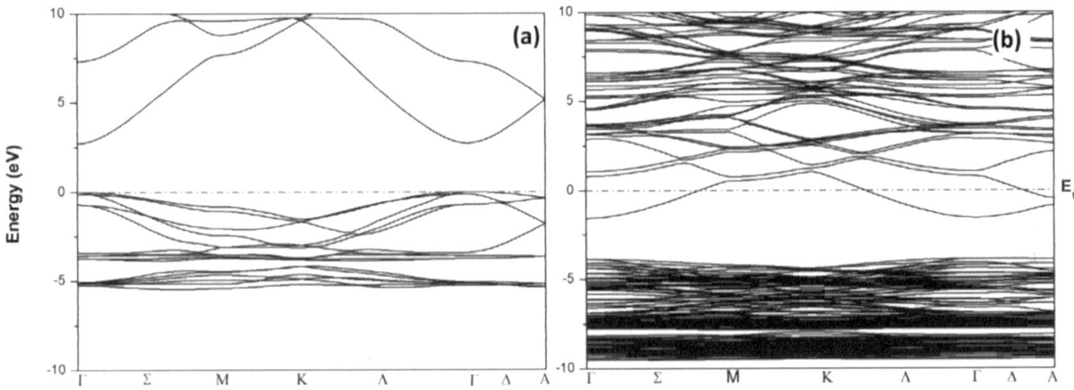

Figure 2. Band structure of (a) pure and (b) Ga doped ZnO in a 2x2x2 supercell

The calculated density of states (DOS) and partial density of states (PDOS) are shown for pure and Ga doped ZnO in Figure 3. The valence band essentially consists of the 2p and 2s states of O and 3d states of Zn. In the uppermost valence band, O 2p states are found between -4 and 0 eV. While the O 2s states appear in the range from -18 to-16eV. The position of peak of Zn 3d plays an important role in the electronic structure of ZnO system. From PDOS of Zn, we can see that the upper valence band -6eV to 0eV is dominated by Zn 3d with high intensity. In addition, the lowest conduction band is dominated by Zn 4s states, and experimental results also indicate that it is primarily derived from O 2p and Zn 4s states. The calculated band gap with GGA is found in our case around 0.8eV which is far from the experimental band gap 3.4eV due to the hybridization between the Zn 3d and O 2p levels. The introducing of the potential TB-mBJ can correct the energy level of Zn 3d states to -6.2eV and the gap value improve, to 2.8eV. The hybrid functional, such as PBE0 and HSE, can shift the Zn 3d states to -5.8, -5.9 and -6.4, and the gap is 3.24–3.26, that is very close to the experimental value of 3.4eV, but it is computationally expansive [25].

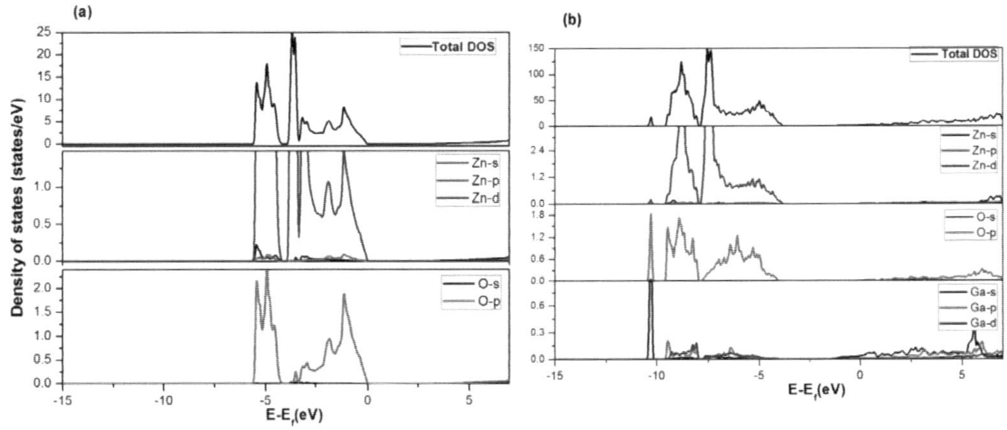

Figure 3. Total and partial density of states of: (a) pure ZnO and (b) Ga doped ZnO in a 2 x2x2 supercell, respectively

The total and partials densities of states (TDOS, PDOS) of pure and Ga doped ZnO in 2x2x2 supercell (x = 6.25%) are plotted in the same figure 3. Here we find that the Ga-3s orbital contributes to the occupied states around Fermi level. These donor states around the Fermi level could be considered as the origin of the conductivity increasing in Ga doped ZnO sample.

3.3. Optical properties

The optical band gap (Eg) of pure ZnO and Ga doped ZnO was estimated by Tauc relation [26]:

$$\alpha h\nu = A\left(h\nu - E_g\right)^n \tag{6}$$

In this equation, α is absorption coefficient, hν is photon energy, and A is the proportionality constant for a direct interband transition. The power n is equal to the ½ since ZnO is a direct band gap material. We have plotted the $(\alpha h\nu)^2$ vs hν for various concentrations, and fitted for the straight region. The extrapolation of this straight line to hν axis gives the value of the band gap. Figure 4 shows that the value of band gap Eg nincreases with increasing Ga doping concentration. This is attributed to Moss-Burstein effect, caused by an increase in free electron concentration due to Ga doping.

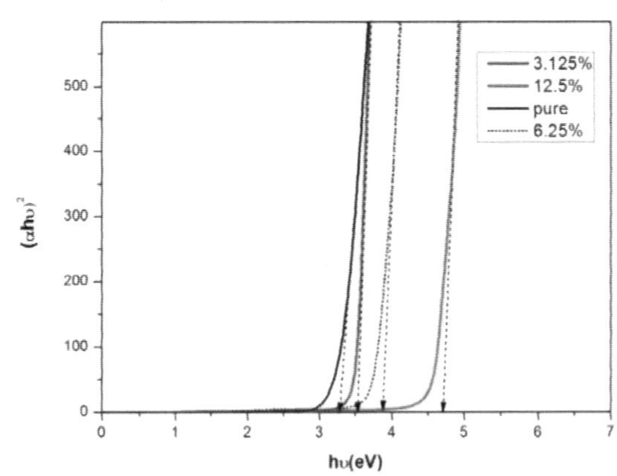

Figure 4. Plot of $(\alpha h\nu)^2$ versus (hν)

To determine the other optical properties of the Ga doped ZnO, it is necessary to calculate the imaginary part, and the real one of the dielectric function ε_2 and ε_1 respectively. From which the optical constants, such as refractive index $n(\omega)$, absorption coefficient $\alpha(\omega)$, reflectivity $R(\omega)$, and energy-loss spectrum $L(\omega)$ are derived [27].
Reflectivity as a function of wavelength in the range of 250–900 nm at various Ga concentrations is displayed in the Figure 5a. As the Ga concentration increases from 0 to 12.5 at. %, the reflectivity decreases and rest less than 7%.
Figures 5b and 5c show the absorption coefficient and optical transmittance spectra respectively at various Ga concentrations. The average transparency in the visible range was around~80-90%. It appears that doping ZnO with Ga results in an improvement of transmission in the wider range of the solar spectrum. Except at higher doping concentrations (12.5%) showing a reduction in the transmittance, which is explained experimentally by the increased scattering of photons by crystal defects created by doping [28]. The absorption edge is blue shifted with increasing Ga doping concentration, which means broadening of the optical band gap, related to Burstein-Moss effect.

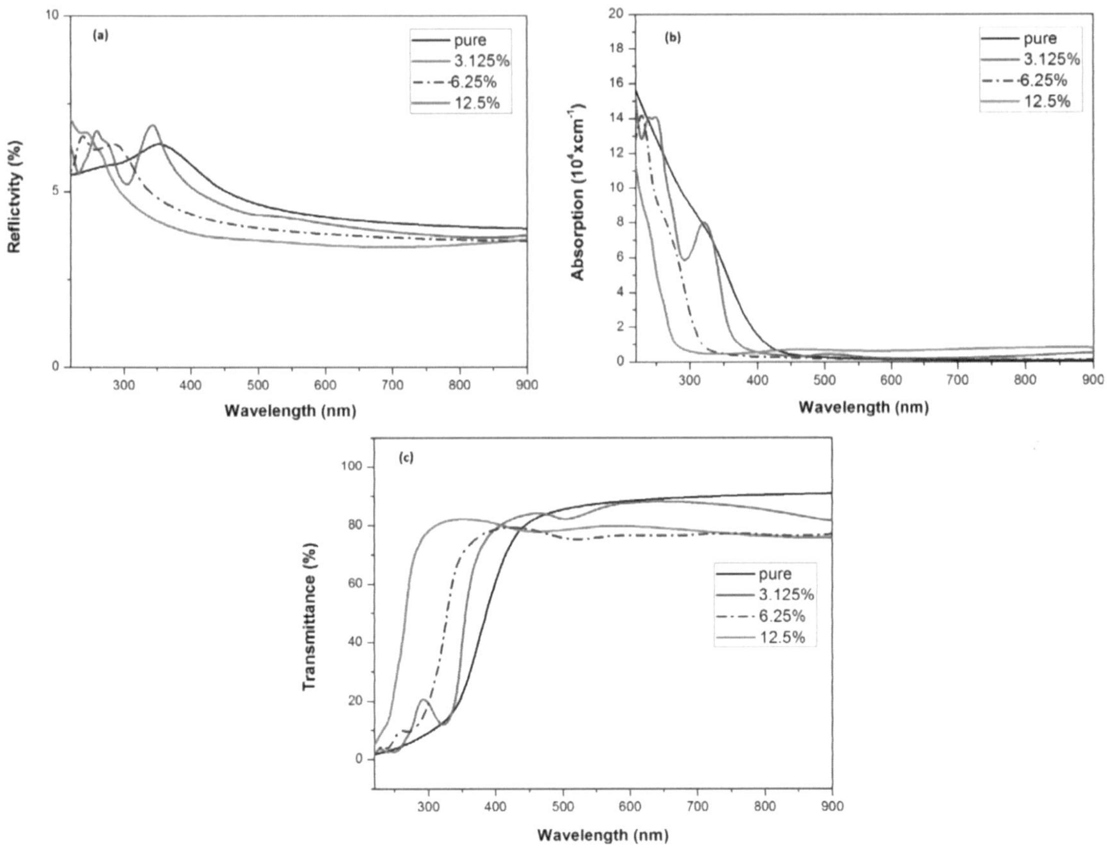

Figure 5. The variation of (a) reflectivity, (b) absorption coefficient, and (c) transmittance for doped ZnO with various Ga concentrations.

3.4. Electrical properties

Based on the calculated electronic band structures and the semi-classical Boltzmann equations with the relaxation time approximation were used to evaluate the electric transport coefficients for pure and Ga doped ZnO for various concentrations.

Using the electronic structure it is possible to calculate σ/τ as a function of n and T, where τ is the relaxation time in second, T the temperature in Kelvin, n the carriers concentration in $cm^{-3}$and σ the conductivity in $(\Omega cm)^{-1}$. But it is not possible to calculate σ itself without knowledge of the scattering rate τ^{-1}. In order to proceed, we use relationship written by Ong et al.[29]:

$$\tau = 2.53\times10^{-5}T^{-1}n^{-1/3} \quad (7)$$

then we calculate σ as σ/τ × τ knowing that:

$$\tau = 2.53\times10^{-5}T^{-1}n^{-1/3} \quad (8)$$

Here in the above equations, $f\mu$ is the Fermi distribution function, μ= chemical potential (temperature dependent) and Ω the volume. The indices α and β denote the components of tensor conductivity.

Figure 6 shows electrical conductivity as function of concentration of Ga. From this Figure, we can see at low concentration ≤ 6.25% of Ga the electrical conductivity increases with increasing concentration of Ga, explained by the replacing Zn^{2+} by Ga^{3+} atom which can release one electron to the ZnO material, and increases the mobility and the carrier concentration in the low concentration range. However, the electrical conductivity drastically decreased at the high doping concentration over 6.25% due to the saturation for high concentration of Ga. The difference of ionic radii of Zn^{2+}(0.74 Å) and Ga^{3+}(0.62Å), there should be a limit of solid solution of Ga into ZnO crystal and so does the increase of carrier concentration.

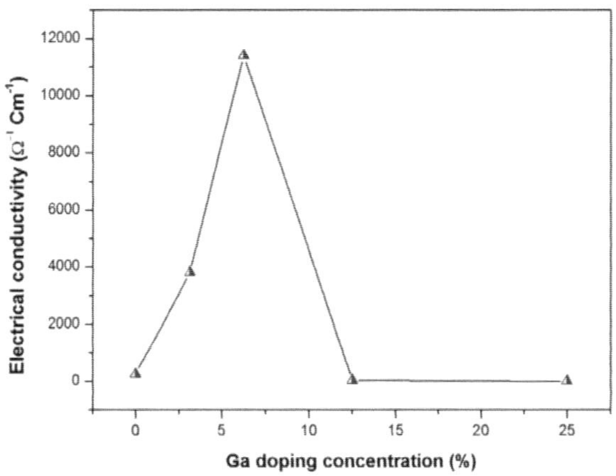

Figure 6. The electrical conductivity of pure and Ga doped for various Ga concentrations

4. Conclusion

In this paper, we have studied the effect of incorporation of Gallium in the Wurtzite Zinc oxide using First Principles and semi-classic Boltzmann theory. We calculate the structural, electronic, optical and electrical properties of Ga doped ZnO at various concentrations of Ga, compared to undoped ZnO. For low concentration of Ga the reflectivity and absorption are reduced, the transmittance and optical band gap is enhanced. Also, the electrical conductivity is increased due to increasing of the donor concentration, but at high concentration of Ga the electrical properties is decreased. The structural parameters and the band gap found are in a good agreement with theoretical and experimental results. These results make this doped material with low concentration of Ga desirable in applications of photovoltaic field.

References

[1] Chen M, Pei Z L, Sun C, Wen L S and Wang X 2000 *J. Cryst. Growth* ***220*** *254*-62
[2] Fernandez S, Martinez-Steele A, Gandia J J and Naranjo F B 2009, *Thin Solid Films* **517** 3152-56
[3] Hvam J M 1978 *Solid State Commun* **26** 987–990
[4] Gong L, Ye Z, Lu J, Zhu L, Huang J, Gu X, Zhao B 2010 *Vaccum* **84** 947
[5] Nomoto J, Konagai M, Okada K, Ito T, Miyata T and Minami T 2010 *Thin Solid Films* **518** 2937
[6] Usuda M and Hamada N 2002 *Phys. Rev. B* **66** 125101
[7] Hummer K, 1973. *Phys. Stat. Sol* **56**: 249-60
[8] Xinguo Ma ,Ying Wu, Yanhui Lv and Yongfa Zhu 2013 *J. Phys. Chem. C* **117** (49) 26029–39
[9] Wrobel J, Kurzydłowski K J, Hummer K, Kresse G and Piechota *2009* J, *J. Phys. Rev. B* **80** 155124–131
[10] Dixit H, Saniz R, Lamoen D and Partoens B 2010 *J. Phys.: Condens. Matter* **22** 125505–11
[11] Tran F, Blaha P 2009, *Phys. Rev. Lett.* **102** 226401
[12] Blaha P, Schwarz K, Madsen G, Kvasnicka D and Luitz J 2001 An augmented plane wave plus local orbitals program for calculating crystal properties, Vienna University of Technology,Austria
[13] John P Perdew, and all, *Phys. Rev. Lett.* ***100*** 136406
[14] Khenata R, Bouhemadou A, Sahnoun M, Ali H Reshak, Baltache H, Rabah M 2006 *Comp.Mater.Sci* **38** 29–38
[15] Amin B, Ahmad M I, Maqbool S, Said G and Ahmad R 2011 *J. Appl. Phys.* **109** 023109
[16] Sun J, Wang H, He J, Tian Y 2005 *Phys. Rev. B* **71** 123132
[17] Madsen G K H and Singh D J 2006, *Comput. Phys. Commun.* **175** 67
[18] Kisi E H, Elcombe M M, Acta *Crystallogr* 1989. *C* ***45*** 1867–70
[19] Ali H Reshak and Morteza Jamal 2013 *J. Alloys Compd.* **555** 362–6
[20] Hsuan Chung Wu, Yen Chun Peng and Chieh Cheng Chen 2013 *Opt. Mater.* **35** 509–15
[21] Ozgur U, Alivov Ya I, Liu C , Teke A, Reshchikov M A, Dogan S, Avrutin V, Cho S J and Morkoç H 2005, *J. Appl. Phys.* **98** 041301
[22] Zhou X H, Hu Q H and Fu Y 2008, *J. Appl. Phys.* **104** 063703
[23] Burstein E 1954 *Phys. Rev.* **93** 632
[24] Moss T S 1954 *Proc. Phys. Soc. B* **67**
[25] Oba F, Togo A, Tanaka I, Paier J and Kresse G 2008, *Phys. Rev. B* **77** 245202
[26] David A and Mott N F 1970, Philos. Mag. 22 **903**
[27] Gajdosc M, Hummer K, Kresse G, Furthmuller J and Bechstedt F 2006, *Phys. Rev. B* **73** 045112
[28] Babar A R, Deshamukh P R, Deokate R J, Haranath D, Bhosale C H and Rajpure K Y 2008, J. Phys. D: Appl. Phys. **41** 135404
[29] Khuong P Ong, David J. Singh, and Ping Wu 2011, *physical review B* **83** 115110

Journal of Physics: Conference Series **758** (2016) 012025 doi:10.1088/1742-6596/758/1/012025

Ab initio study of structural and electronic properties of (GaN)n/(AlN)n superlattices

L. Djoudi [a,b*], M. Merabet [a,b], F. Dahmane [b], M. Boucharef [a], S. Benalia [a,b], D. Rached [a]

[a]Laboratoire des Matériaux Magnétiques, Faculté des Sciences, Université Djillali Liabès de Sidi Bel-Abbès, Sidi Bel-Abbès (22000), Algeria

[b]University of Tissemsilt, Institute of Science and Technology 38000, Algeria

E-mail: djoudilakhdar@yahoo.fr

Abstract. Structural and electronic properties of binary ***GaN*** and ***AlN*** compounds and their superlattices (SLs) $(GaN)_n/(AlN)_n$ are investigated using the first-principles full potential linear muffin-tin orbitals method (FP-LMTO) . The exchange-correlation potential is treated with the local density approximation of Perdew and Wang (LDA-PW). The ground-state properties are determined for the bulk materials ***GaN***, ***AlN***, and their superlattices ***(GaN)n/(AlN)n)*** in cubic phase. The calculated structural properties of ***GaN*** and ***AlN*** compounds are in good agreement with available experimental and theoretical data. It is found that ***AlN*** exhibit an indirect fundamental band gap while that ***GaN*** and the superlattices (SLs) exhibit a direct fundamental band gap, which might make the superlattices $(GaN)_n/(AlN)_n$ materials promising and useful for optoelectronic applications. The fundamental band gap decreases with increasing the number of monolayer.

1. Introduction

The study of semiconductor heterostructures, like quantum wells and superlattices, has resulted in an ever expanding and dynamic area of research over the past three decades. Semiconductor heterostructures are ultra-thin layered materials whose thicknesses are in the range of a few atomic layers. The presence of these ultrathin layers leads to the "quantum size effect" when the physical dimensions of the layers are comparable to the De Broglie wavelength of the charge carrier [1].This new effect is of great interest both in fundamental physics and in devices applications [2]. Semiconductor superlattices consist of thin layers of alternating semiconductors typically 10-100 layers ranging from 10 to 500 Å in thickness. The constituent materials have different band gap and the conduction band edges for the two materials are not aligned; consequently, carriers experience a periodic potential along the growth axis, essentially forming a two-dimensional artificial crystal.

A superlattice is a periodic heterostructures formed by two types of semiconducting materials, one type that acts as a quantum well and the other acting a quantum barrier. The structure are grown by alternating the well and the barrier material such that the layers of well material are all of width L_w and the barrier layers are all of width L_b. The superlattice is thus a periodic structure with period $d = L_w + L_b$. The artificial periodicity created in superlattices can lead to the tailoring of certain material properties for applications in fields such as optoelectronics [3, 4]. The ***III–V*** superlattices which have wide band gaps having physical properties where that are very promising for the development of the optoelectronic devices in the visible and the ultraviolet regions of the electromagnetic spectrum with photon energies spread in a wide energy range [5]**.** Noteworthy, progress has been made during the past few years in the growth of ***III–V*** nitride materiel and devices. Due to the wide range of direct band gaps provided by nitride system. Band gap tuning is possible by forming alloys of the compounds with gaps of desired ranges [6]. However, understanding the basic properties of wide band-gap ***GaN*** and ***AlN*** and their superlattices, are currently of great academic and technological interest from the designing of novel materials to the fabrication of nanoelectronic and photonic devices [7].

The relevance of the materiel combination ***(GaN/AlN)*** is in combination of different band gap type and combination of very different direct band gap values (4.1 eV for ***AlN*** and 1.64 eV for ***GaN***) [8, 9], special attention has to be paid to the ***GaN/AlN*** compounds, because the two binary compounds which are involved have common atom which is the nitride atom. Superlattices come predominately in two types. In the so-called "type-I" superlattices, where electrons and holes are confined to the same layers, is exemplified by the $(GaP)_n/(AlP)_n$ and $(CdTe)_n/(ZnTe)_n$ [10, 11]. In "type-II" superlattices, the electron and holes are confined to different layers, example of "type-II" superlattices composition is $(BP)_n/(BAs)_n$ and $(BeTe)_n/(CdS)_n$ [12, 13].

Knowing that, the ***III-V*** nitride system and their compounds have attracted much interest for years for this very reason. Having a much higher breakdown field, saturation velocity and thermal conductivity than ***GaAs***, ***GaN*** offers great potential for high power devices [14]. Gallium nitride ***(GaN)***, aluminum nitride ***(AlN)*** and related materials are of considerable current interest because of their applications in light emitting devices operating in the visible and deep ultraviolet (UV) spectral regions [15-20]. In addition, ***AlN*** and ***GaN*** have a high melting point, a high thermal conductivity, and a large bulk modulus [21]. These properties, as well as the wide band gaps, are closely related to their strong (ionic and covalent) bonding. These materials can therefore be used for short wavelength light-emitting diodes (LED's) laser diodes, and optical detectors, as well as for high-temperature, high power, and high-frequency devices [22].

In this paper, we report first-principles study of the structural and electronic properties of bulk ***GaN***, ***AlN*** and superlattices $(GaN)_n/(AlN)_n$ (n number of monolayers; n = 1, 2 and 3) in the zinc-blend structure (B3), using the full-potential linear muffin-tin orbital (FP-LMTO) method, in the framework of the density functional theory (DFT) within the LDA. The organization of the article is as follows. The computational method we have adopted for the calculations is described in section 2. We present our results in section 3. Finally, conclusions are given in section 4

2. Calculation Method

In the present study, the structural and electronic properties of binary ***GaN*** and ***AlN*** compounds and their $(GaN)_n/(AlN)_n$ superlattices are investigated using the first-principles full potential linear muffin-tin orbitals method (FP-LMTO) simulation program based on the density functional theory (DFT) within the local density approximation of Perdew and Wang (LDA-PW) [23]. In this theory, the quantum many body problems of interacting electrons and nuclei can be mapped onto a system of one electron equations called Kohn–Sham equations [24, 25]. The space in the (FP-LMTO) method is divided into non-over-lapping muffin-tin (MT) spheres centered at the atomic sites separated by an interstitial region (IR). In the IR regions, the basis set consists of plane waves. Inside the (MT) spheres, the basis sets is described by radial solutions of the one particle Schrodinger equation (at fixed energy) and their energy derivatives multiplied by spherical harmonics. The exchange and correlation potential is treated by the local density approximation (LDA) [23]. The details of calculations are as follows: the charge density and the potential are represented inside the muffin-tin (MT) spheres by spherical harmonics up to $l_{max} = 6$. The self-consistent calculations are considered to be converged when total energy of the system in stable within 10^{-5} Ry. Accurate Brillouin integrations zone is performed using the tetrahedron method [26]. The corresponding integrating points over the irreducible Brillouin (IBZ) are 22 k- points for the bulk binary compounds (***GaN*** and ***AlN***) , 64 k-points for SL(1,1) and 32 k-points for SL(2,2) and 98 k-points for SL(3,3).
In order to achieve energy eigenvalues convergence, the wave functions in the interstitial regions were expanded in plane waves up to 270.9592 and 291.5794 Ry for ***GaN*** and ***AlN***, respectively. For superlattices, the wave functions are expanded in plane waves up to approximately 222.6561, 222.6027 and 240.6854 Ry for SL (1,1), SL (2,2) and SL (3,3) respectively. The number of 12050 planes waves is used for both compounds ***GaN*** and ***AlN***. For the superlattices, the number of plane waves is 16242, 32458 and 48690 for SL (1,1), SL (2,2) and SL (3,3) respectively. To avoid the overlap of atomic spheres the (MT) radius for each atomic position is taken to be different for each case.

3. Results and Discussions

3.1. Structural properties

The ground-state structural parameters have been obtained by minimizing the total energy with respect to the volume, by fitting this total energy versus the volume data on the non-linear Murnaghan equation of state [27]. Hence, first we computed the lattice parameter of the GaN and AlN in their cubic structure with the space group F-43m (no. 216). Geometry of (GaN)n/(AlN)n superlattice are shown in Fig. 1. Then, we studied the (GaN)n/(AlN)n superlattices where n is the number of monolayers the selected n: 1, 2 and 3. The lattice mismatch for GaN/AlN superlattice is a 2.89 % between its two constituent bulk semiconductors, so the lattice mismatch problem does not arise. In Table 1, we summarize the calculated lattice constants and the bulk module and its pressure derivative of binaries (GaN and AlN) and the superlattices (GaN)n/(AlN)n together with the available experimental and theoretical data.
As can be seen, the obtained lattice constants for GaN and AlN are in reasonable agreement with the experimental data, for GaN is 0.88% greater and for AlN is 0.8% greater, than the experimental value, which ensures the reliability of the present first-principles computations. It is found out that the calculated bulk modulus for GaN is 12.86% less than the experimental value. The lattice parameter of (GaN)n/(AlN)n superlattices (n: 1, 2 and 3) has been found that a1,1 is equivalent to [a0(GaN)+ a0(AlN)]/2, a2,2 is about twice larger than a1,1 and a3,3 is threefold larger than a1,1. The lattice constant of (GaN)n/(AlN)n superlattices depending on n (the number of monolayers) is shown in Fig. 2. It is very clear from Fig. 2 that the lattice constant of (GaN)n/(AlN)n superlattices increase linearly with the number of monolayers increasing.

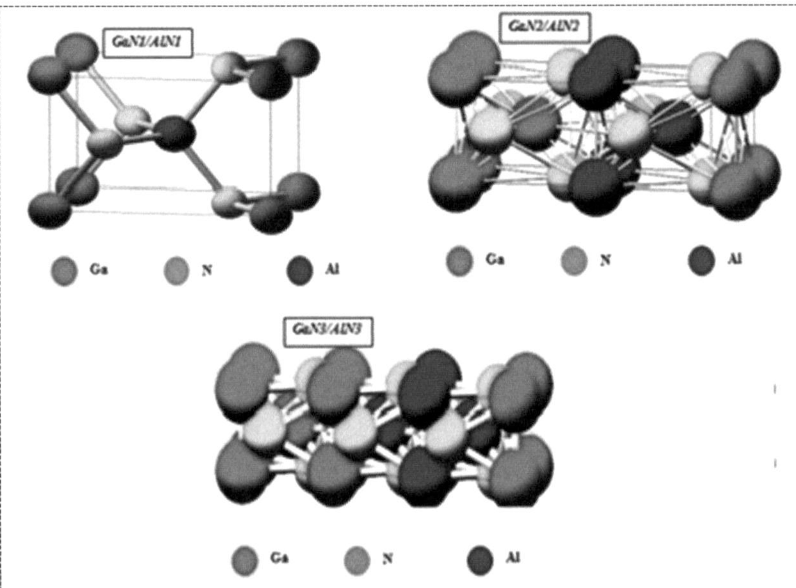

Figure 1. Geometry of (GaN)n/(AlN)n superlattice (n = 1, 2 and 3).

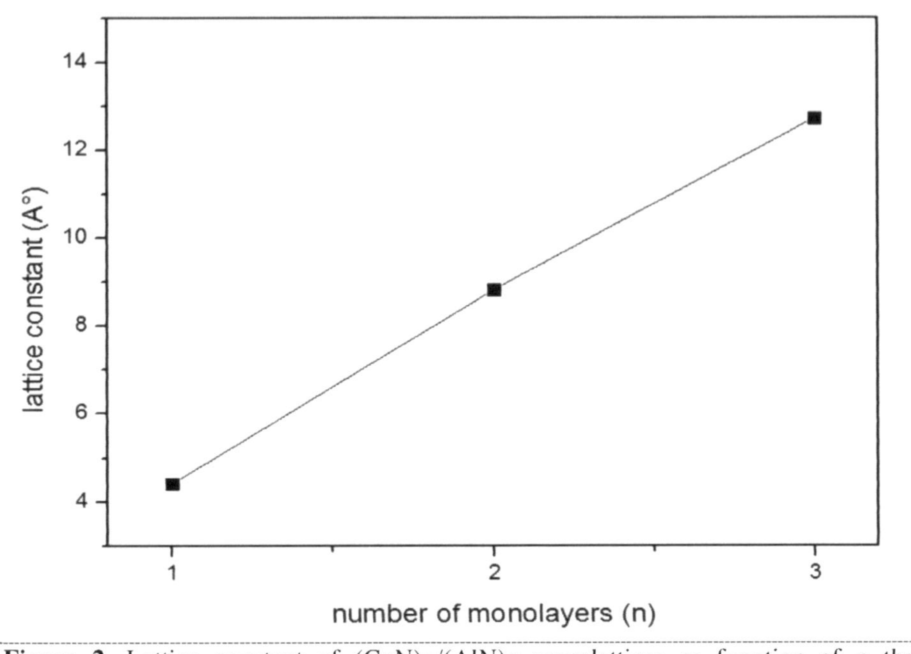

Figure 2. Lattice constant of (GaN)n/(AlN)n superlattices as function of n the number of monolayers.

Table 1. Calculated lattice parameter a_0, the bulk modulus B_0 and its pressure derivative B'_0 for zinc-blende GaN, AlN and the superlattices (GaN)n/(AlN)n (n = 1, 2 and 3) compounds at equilibrium volume compared to the available theoretical and experimental data .

	Present work			Experimental works			Other theoretical works		
	a_0 (Å)	B_0 (GPa)	B_0' (GPa)	a_0 (Å)	B_0 (GPa	B_0' (GPa)	a_0 (Å)	B_0 (GPa)	B_0' (GPa)
GaN	4.570	165.562	3.8507	4.49 [28] 4.53[29] 4.50 [30]	190 [31]	·	4.46[32], 4.51[33] 4.494 [8, 9]	202 [32] ,191[33] 200 [8, 9]	4.43 [32] ,4.14 [33] 4.75 [8, 9]
AlN	4.405	182.03	3.835	4.37 [28]	-	-	4.346[34] 4.377 [8, 9]	206 [35] ,212 [34] 201 [8, 9]	3.97 [35] ,3.77 [34] 3.66 [8, 9]
(GaN)1/(AlN)1	4.411	194.528	3.977	-	-	-	-	-	-
(GaN)2/(AlN)2	8.823	200.851	4.176	-	-	-	-	-	-
(GaN)3/(AlN)3	12.728	201.880	3.967	-	-	-	-	-	-

From Table 1, we noticed that the calculated bulk moduli for the (GaN)n/(AlN)n superlattices increase non-linearly with the number of monolayers increasing and a non-linear variation of the derivative of the bulk moduli for the (GaN)n/(AlN)n superlattices has been also observed.

3.2. Electronic properties

We have calculated the electronic band structure properties of GaN, AlN and (GaN)n/(AlN)n superlattices where n is the number of monolayers the selected n: 1, 2 and 3 in the zinc-blende structure at this equilibrium lattice constants. We note that the good knowledge of electronic band structure in semiconductors provides valuable information as regarding their potential utility in fabricating electronic and optoelectronic devices. In table 2, we have listed the obtained direct and indirect band-gaps, at the high-symmetry point Γ in the Brillouin zone, for the systems investigated and compared the results with available experimental and theoretical data. The results clearly show that the band gap are, on the whole, underestimated by about 54.69 % compared with experimental data for GaN, but are in good agreement with other theoretical data. The large difference in the calculated values of the band gaps as compared to the experimental values can be explained by the well-known fact that, in the electronic band structure calculations within DFT, LDA underestimates the energy gaps in semiconductors, we can see this in several papers that have already been published. The results show that GaN has a direct-gap with the minimum of conduction band at the Γ point (Fig. 3. a); AlN has an indirect-gap with the minimum energy gap at the X point (Fig. 3. b).

Table 2. Calculated direct and indirect band gaps for zinc-blende GaN, AlN and the superlattices (GaN)n/(AlN)n (n = 1, 2, and 3) compounds at equilibrium volume compared to the available theoretical and experimental data.

	Present work		Experimental works		Other theoretical works	
	Eg (Direct) (eV)	Eg (Indirect) (eV)	Eg (Direct) (eV))	Eg (Indirect) (eV)	Eg (Direct) (eV))	Eg (Indirect) (eV)
GaN	1.45	4.46	3.2[36]	-	1.52[32] 2.1[37] 1.64[8, 9]	3.22[32] 3.19[8, 9]
AlN	4.09	3.3752			4.13[35] 4.10[8, 9]	3.20[8, 9]
(GaN)1/(AlN)1	3.1213	3.3302	-	-	-	-
(GaN)2/(AlN)2	3.0454	3.1209	-	-	-	-
(GaN)3/(AlN)3	3.0477	3.0067	-	-	-	-

The calculated band gap energies of (GaN)1/(AlN)1, (GaN)2/(AlN)2 and (GaN)3/(AlN)3 along the higher symmetry directions of Brillouin zone are shown in (Fig. 4. a), (Fig. 4. b) and (Fig. 4. c) respectively. From the results of the calculated band gap energies, we find that these materials have direct band gap for n =1 and 2, which is of interest for optoelectronic devices, but for n = 3 this superlattice become have indirect band gap. It is important to note that the results found are automatically small, because we must not forget that the LDA method greatly underestimates the band gap in semiconductors, so we can predict that the direct band gap values for (GaN)n/(AlN)n superlattices for n = 1, 2 and 3 are 6.8887, 6.7212 and 6.7262 eV, respectively. It is very clear from table 2 that the direct and indirect band-gaps decrease with the increasing of the n number of monolayers. We note that the difference in the band gaps of the (GaN)n/(AlN)n superlattices and the end compounds is usually attributed to charge transfer, reflected in terms of the electronegativity difference of the mixing cations and anions.

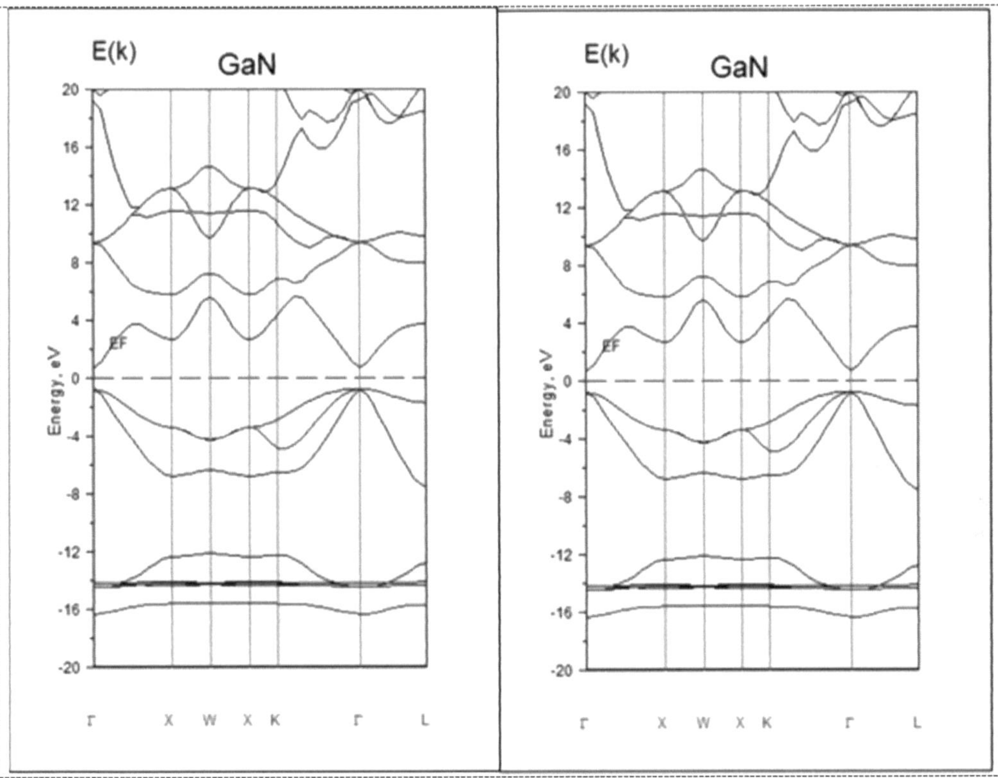

Figure 3 (a and b). Band structure along the symmetry lines of the Brillouin zone at the equilibrium lattice constant for GaN and AlN compounds. The position of the Fermi level is shown by horizontal solid line.

To obtain a deeper insight into the electronic structure, we have also displayed in (Fig. 5. a) and (Fig. 5. b), the total and the partial atomic site-decomposed density of states (TDOS and PDOS) for GaN and AlN compounds, respectively and in (Fig. 6. a), (Fig. 6. b) and (Fig. 6. c) their (GaN)n/(AlN)n superlattices (n =1, 2 and 3) in order to analyze the contributions from the different states of Ga, Al and N atoms. The results shown in (Fig. 5. a) and (Fig. 5. b) shows that the upper valence bands from - 19.4272 to 0.1727 eV and from - 19.4272 to -1.2217 eV for GaN and AlN, respectively, are governed by hybridization N-2s, Ga-4s and N-2p states for GaN and N-2s, N-2p and Al-3s states for AlN, in the two intervals (- 19.4272 to 0.1727 eV for GaN) and (- 19.4272 to -1.2217 eV for AlN) can be observed two structures for GaN and two structures for AlN. The first structure (from - 19.4272 to - 11.0272 eV for GaN) and (- 19.4272 to - 13.8217 eV for AlN) is due to the N-2s and with a small

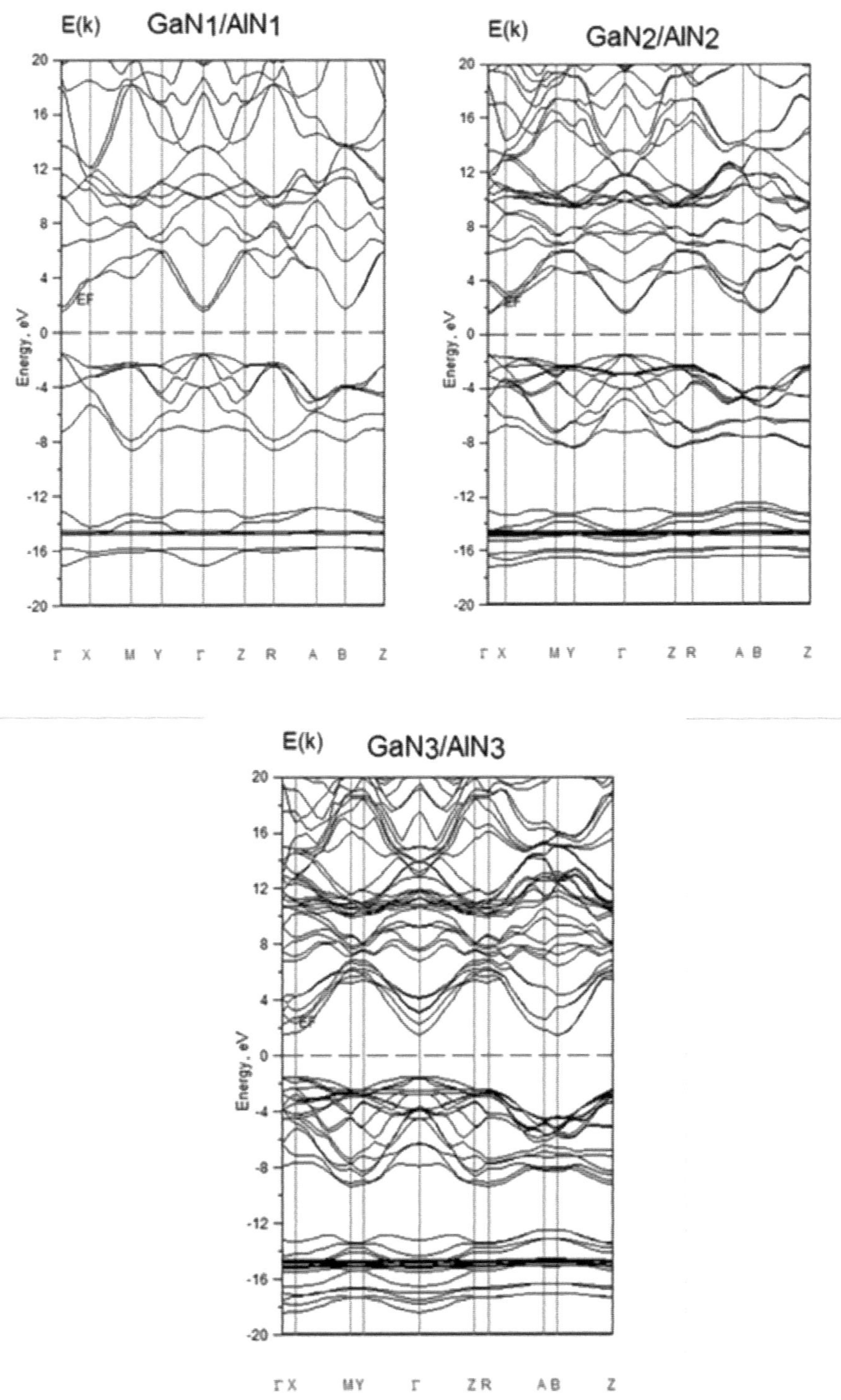

Figure 4 (a, b and c). Band structure along the symmetry lines of the Brillouin zone at the equilibrium lattice constant for (GaN)n/(AlN)n superlattices (n = 1, 2 and 3). The position of the Fermi level is shown by horizontal solid lineshown by horizontal solid line.

contribution from Ga-4s for GaN and is due to the N-2s and with a small contribution from Al-3s for AlN, the second structure (from - 8.2272 to 0.1727 eV for GaN) and (from -8.2217 to - 1.2217 eV for AlN) is due to the N-2p and with a small contribution from Ga-4s for GaN and to the N-2p and with a small contribution from Al-3s for AlN .The structure above the Fermi level is broad and extends up to 20 eV for both compounds. As in the two binary, the calculated total and partial densities of states of these superlattices show a resemblance with some small differences in details. We noticed that the contribution of different orbitals is proportional to the number of monolayer of the two components of the superlattice. The densities of states of the valence and conduction bands of the superlattice are characterized by three regions separated by gaps. Following these figures Fig. 6. a), Fig. 6. b) and Fig. 6. b), the first region, in the valence bands, for the three configurations is mainly due to the contribution of Ga-3d and N-2s states with a small contribution from Al-3s states. The second structure is due to the Al-3p and N-2p with a very small contribution from Al-3s. The third region in the conduction band the three superlattices have the same contribution in the total density of states. These contributions are formed by a hybridization of s, p and d states of the three atoms.

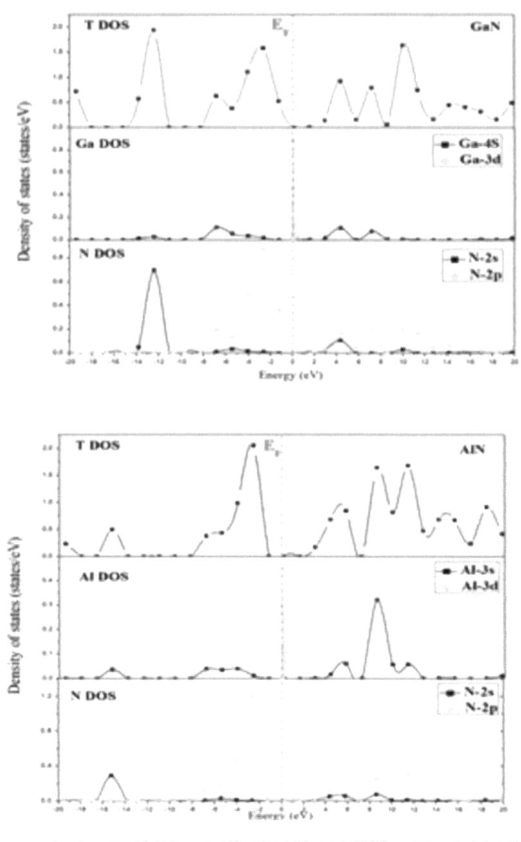

Figure 5 (a and b). Total and partial density of states (DOS) for GaN and AlN compounds.

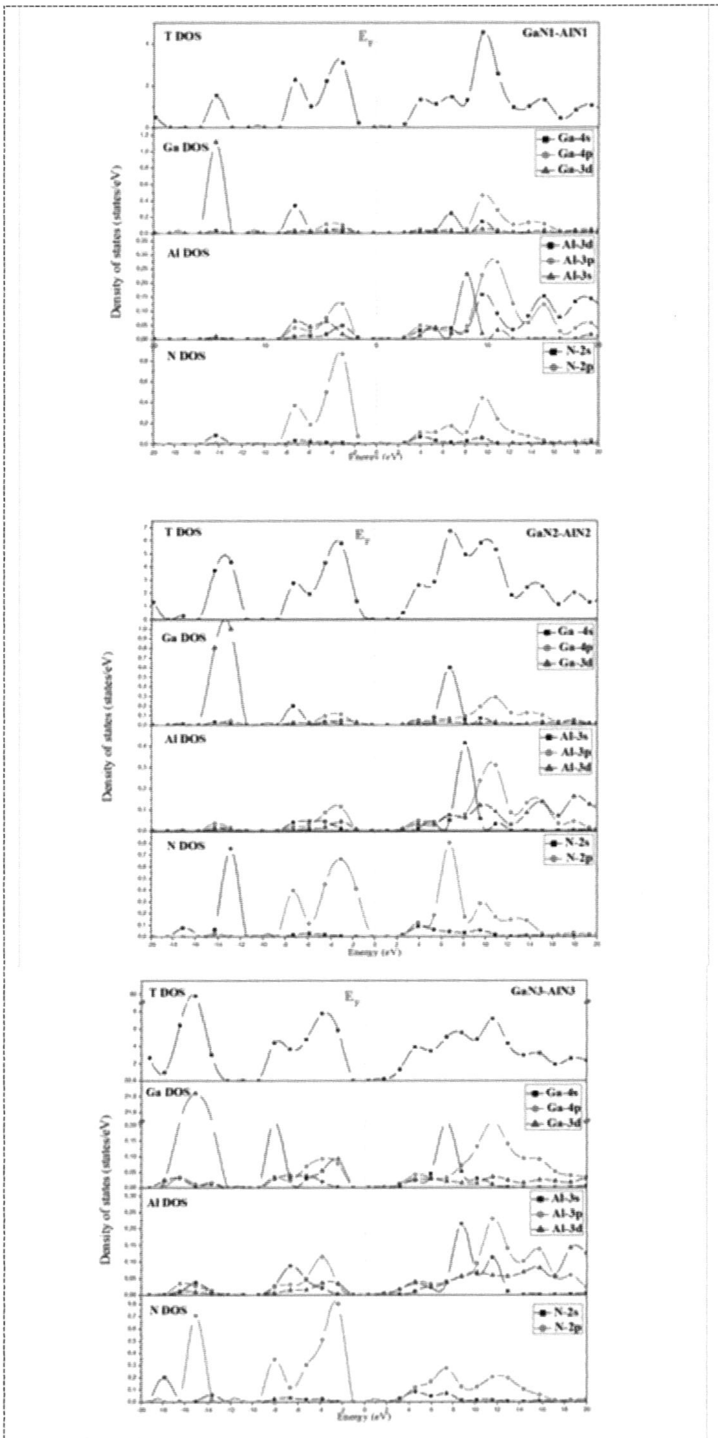

Figure 6 (a, b and c). Total and partial density of states (DOS) for (GaN)n/(AlN)n superlattices (n = 1, 2 and 3).

4. Conclusion

The full potential linear muffin-tin orbitals method (FP-LMTO) approach based on the density functional theory (DFT), within the local density approximation of Perdew and Wang (LDA-PW) is utilized to investigate the structural and electronic properties of GaN and AlN compounds and their (GaN)n/(AlN)n superlattices. The calculated lattice parameter and bulk modulus of each binary compounds are in reasonable agreement with experimental values. From the calculated results, it is found that the lattice constant of (GaN)n/(AlN)n superlattices is depending on n the number of monolayers. The calculated band structure show that GaN has a direct-gap, AlN has an indirect-gap and the (GaN)n/(AlN)n superlattices has a direct gap and decreases with increasing the number of monolayers also we noticed that, from n equals three the superlattice become have indirect band gap. Finally, one can point out that the direct gap character that the (GaN)n/(AlN)n superlattices exhibit is of great importance for the optical transitions and might be useful for the design of quantum well lasers and solar cell heterostructures. To the best of our knowledge, there is no publications available on the structural (a_0, B_0 and B_0') and electronic (band gaps) properties of for the studied superlattices are available to be compared with our theoretical results.

5. References

[1] G. Bauer, F. Kuchar and H. Heinrich 1984 Springer Series in Solid-State Sciences Vol. **53** 2
[2] L. Esaki 1986 IEEEJ. Quantum Electron. **22** 1611
[3] C. Weisbuch, B. Vinter 1991 Quantum Semiconductor Structure, Academic Press, SanDiego
[4] A. Yariv 1989 Quantum Electronics, 3rd ed., John Wiley & Sons, New York
[5] N.A. Noor, W. Tahir, Fatima Aslamc and A. Shaukat 2012 Physica B **407** 943
[6] Bal K. Agrawal, Savitri Agrawal and Rekha Srivastava 1999 Surface Science **431** 84
[7] K. Yamamoto, M. Uchida, A. Yamamoto, A. Masuda and A. Hashimoto 2002 Phys. Stat. Sol (b), **234** 915
[8] L. Djoudi, A. Lachebi, B. Merabet and H. Abid 2012 Acta Physica Polonica A, No. **4**, Vol. **122**
[9] L. Djoudi, A. Lachebi, B. Merabet and H. Abid 2012 Modern Physics Letters B, Vol. **26**, 24 1250159.
[10] M. Merabet, S. Benalia, D. Racheda, R. Khenata, A. Bouhemadoud, S. Bin Omranc, Ali H. Reshake and M. Rabah 2011 Superlattices And Microstructures, **49** 132–143.
[11] M. Boucharef, S. Benalia, D. Rached, M. Merabet, L. Djoudi, B. Abidri and N. Benkhettou 2014 Superlattices and Microstructures, Vol. **75** 818-830.
[12] M. Merabet, D. Racheda, R. Khenata, S. Benalia, B. Abidri, N. Bettahar and S. Bin Omran 2011 Physica B **406** 3247–3255.
[13] L. Djoudi , M. Merabet , M. BouchareF, S. Benalia and D. Rached 2014 superlattices and microstructures, Vol. **75** 233-244.
[14] S.C. Binari, "GaN FETs for microwave and high-temperature applications. Electrochemical Society Proceedings, 1996. 95-21.
[15] S. Strite and H. Morkoç, J. Vac 1992 Sci. Technol. B **10**, 1237
[16] C.G. Van de Walle 1993 Physica B **185**, ix-x
[17] H. Morkoç, S. Strite, G.B. Gao, M.E. Lin, B. Sverdlov and M. Burns 1994 J. Appl. Phys. **76**, 1363
[18] F.A. Ponce and D.P. Bour 1997 Nature **386**, 351
[19] S. Nakamura 1997 Solid State Commun. **102**, 237
[20] J. Chen, Z.H. Levine and J.W. Wilkins 1995 Appl. Phys. Lett. **66**, 1129
[21] J. H. Edgard (Ed.), Properties of Group-III Nitrides, EMIS Data reviews Series, IEE, London (1994).

[22] B. Daoudi, M. Sehil, A. Boukraa and H. Abid 2008 Int. J. Nanoelectronics and Materials **1** 65-79
[23] J.P. Perdew and Y. Wang 1992 Phys. Rev. B **46** 12947
[24] P. Hohenberg and W. Kohn 1964 Phys. Rev. B **136** 864
[25] W. Kohn and L.J. Sham 1965 Phys. Rev. A **140** 1133
[26] P. Blochl, O. Jepsen and O.K. Andersen 1994 Phys. Rev. B **49** 16223
[27] F.D. Murnaghan, Proc. Natl. Acad. Sci. USA **30** (1947) 244; J.R. Macdonald and D.R. Powell 1971, J. Res. Nat. Bur. Stand. A **75** 441
[28] V.L. Solozhenho, in: Properties of Group III Nitrides, Electronics Materials Information Service (EMIS) Data Reviews Series, Ed. J.H. Edgar, Institution of Electrical Engineers, London 1994, p. 10.
[29] D. Vogel, P. Kruger and J. Pollmann 1997, Phys. Rev. B **55**, 12836
[30] A. Trampert, O. Brandt, K.H. Ploog, in: Crystal Structure of Group III Nitrides Semiconductors and Semimetals, Eds. J.I. Pankove and T.D. Moustakas, Academic Press, San Diego 1998
[31] M.E. Sherwin and T.J. Drummond 1987 J. Appl. Phys. **69**, 8423
[32] R. Riane, Z. Boussahl, A. Zaoui, L. Hammerelaine and S.F. Matar 2009, Solid State Sci. **11**, 200
[33] C. Stamp and C.G. Van de Walle 1999 Phys. Rev. B **59**, 5521
[34] B. Daoudi, M. Sehil, A. Boukraa and H. Abid 2008 Int. J. Nanoelectron. Mater. **1**, 65
[35] L.E. Ramos, J. Furthmuller, F. Bechstedt, L.M.R. Scolfaro and J.R. Leite 2002, J. Phys. Condens. Matter. **14**, 2577
[36] T. Lei, T.D. Moustakas, R.J. Graham, Y. He and S.J. Berkowitz 1992 J. Appl. Phys. **71** 4933
[37] L.E. Ramos, L.K. Teles, L.M.R. Scolfaro, J.L.P. Castineira, A.L. Rosa and J.R. Leite 2001 Phys. Rev. B **63** 165210

Institute of Physics
Dirac House, Temple Back
Bristol BS1 6BE UK

ISSN: 1742-6588
ISBN 978-1-5108-3179-7

V International Workshop on Mathematical Models and their Applications 2016

IOP Conference Series: Materials Science and Engineering
Volume 173

Krasnoyarsk, Russia
7 - 9 November 2016

IOP
Institute of Physics